Understanding Student Participation and Choice in Science and Technology Education

Ellen Karoline Henriksen • Justin Dillon
Jim Ryder
Editors

Understanding Student Participation and Choice in Science and Technology Education

Springer

Editors
Ellen Karoline Henriksen
Department of Physics
University of Oslo
Oslo, Norway

Jim Ryder
School of Education
University of Leeds
Leeds, UK

Justin Dillon
Department of Education
and Professional Studies
King's College London
London, UK

ISBN 978-94-007-7792-7 ISBN 978-94-007-7793-4 (eBook)
DOI 10.1007/978-94-007-7793-4
Springer Dordrecht Heidelberg New York London

Library of Congress Control Number: 2014948455

Printed on acid-free paper

Springer is part of Springer Science+Business Media (www.springer.com)

Foreword

New technologies are changing the world in profound ways and science-based innovations in health and medicine open possibilities that did not previously exist. People communicate in ways that were once considered science fiction. The world has become a global village and information flows rapidly between people. These new technologies entail new challenges, both ethical and moral. Young people use new technologies for different purposes, sometimes social and personal, but also with political, environmental, religious or even terrorist agendas. And, as we know, the flow of information is also used for the surveillance of citizens, for spying on potential enemies as well as friends. We also live in a world with serious challenges, such as global environmental threats towards climate and biological diversity.

In short, the world is changing, and young people grow up to live in an unknown place and are likely to get jobs that we do not yet have names for. But one thing seems certain: the future will be shaped and influenced by innovations in science and technology. Socio-scientific issues, controversies and argumentation have become key concepts also in school science, and rightfully so. Traditional school science has to adapt to this changing world, where science and technology becomes important in new ways, permeating nearly all realms of private and public life.

In this situation, most predictions of the demands of society and the workplace are likely to fail. Young people are likely to find jobs that do not yet exist, but many of these jobs will be in the fields of science and technology. Other jobs in the future will, to a large extent, be shaped by innovations in science and technology, even more so than in the past decades.

In a changing and unpredictable world, young people need to have the knowledge and skills that can enable them to adapt to changes as well as to understand and influence this development. The science, technology, engineering and mathematics (STEM) subjects are therefore important, not only for those who want a job in this broad and rapidly developing sector, but also for future citizens in all kinds of other jobs.

Young people no longer act as their parents want or expect, neither do they make their choices based on what society might expect of them. Values, personal motives, desires, likes and dislikes all play a part in this process. The choices of young people may even seem irrational, emotional and chaotic. And there is no simple recipe for getting things "right". Young people cannot any longer be forced to go in particular directions, and they do not necessarily accept well-meant advice from adults or from the rest of society. Our democratic societies do (luckily) not have the means to impose particular choices. It is of paramount importance to understand how and why certain choices are made – why certain options are chosen and others are avoided. Only when understanding these processes can one give informed advice on how to influence the choices that are made.

This book provides a broad perspective on issues concerning how young people relate to science and technology and how they make their choices of subjects and studies. In order to understand such issues, one needs a broad and interdisciplinary approach. One also needs to address the different critical periods when choices are made and remade. Young people do not make clear-cut and one-off decisions and plans which they stick to. One needs to see young people's choices as an ongoing and dynamic process.

This book is based on the EU-supported project IRIS (Interests & Recruitment in Science) and provides a multifaceted contribution to understanding the processes of choice related to STEM subjects. The scope of the book is broad and comprehensive. The various chapters present and use theory taken from many relevant fields: sociology, social-psychology and feminism. The book also presents chapters that provide up-to-date reviews of theoretical frameworks as well as empirical studies that shed light on young people's STEM-related educational choices. The final part of the book draws the book together and provides theoretical insights as well as policy-relevant advice on how to achieve improved participation and a better gender balance in STEM disciplines.

The IRIS project grew out of another project – the ROSE project (The Relevance of Science Education) – where the focus was on how 15-year-olds relate to science and technology. The IRIS project is more focused on older students, and on their interests, motivations and educational choices. IRIS therefore adds another flower to the previous ROSE, and the organisers used the network of international ROSE contacts developed in this project in its planning phase. The initial cooperation developed into a successful application to become a FP7 project under the 'Science in Society' programme.

While we have an abundance of studies of young people's conceptual understanding of science as well as more or less global comparative studies of their achievements in science and technology (PISA, TIMSS), much less is known about the factors that actually determine the choices that are made by young people. In modern, free and democratic societies, young people make their own choices. In doing so, their attainment in the subject is only one of many factors that influence their choices.

This is an important book for everyone who is concerned with how young people can be stimulated to choose STEM disciplines at school, in higher education, and as careers. Moreover, the underlying values in the book are to promote gender balance and equity, and how to reach students who are today feeling somewhat alienated by their perceptions (and misperceptions) about what science and technology mean for society.

This book should have a wide audience; most certainly other researchers in the field, but also a variety of stakeholders, such as policy-makers, curriculum planners, career advisors, as well practitioners working in school science and informal science education.

Department of Teacher Education
and School Research
University of Oslo
Oslo, Norway

Svein Sjøberg

Preface

The present book is the result of a fruitful collaboration during a research project – IRIS (Interests and Recruitment in Science) – which brought together researchers from six academic institutions in five European countries. Our aim was to achieve a better understanding of how young people evaluate science, technology, engineering and mathematics (STEM) as an option when making their educational choices. The book aims to provide more developed theoretical frameworks, a more robust empirical knowledge base, and policy-relevant guidelines concerning how both levels of participation and gender balance in the STEM disciplines may be improved. All the chapters are written by IRIS project members with the exception of two (Chapters 6 and 10) which are contributions from related research projects.

Our main target audience is researchers in STEM education in both secondary and higher education. In a sense, we have written the book that we as researchers would have liked to have been available when we entered the field of educational choice research. However, the book also aims to be of relevance to a number of stake-holders: educational policy-makers, faculty and administrative staff at universities, companies and professional societies wanting to improve STEM participation and extend the future workforce, teachers and administrators at schools, textbook writers, curriculum makers, career advisers, the media, and the informal science sector (museums, science centres, science fairs).

IRIS began as a response to a call for applications to the European Commission's *Science in Society* programme within the 7th framework programme. We would like to acknowledge the support from the European Commission without which the project (and this book) would not have been realised. Also, we acknowledge the administrative staff at our home institutions for facilitating our work with IRIS.

Dr. Camilla Schreiner, Prof. Svein Sjøberg and Prof. Jonathan Osborne played important roles in establishing the project but moved on to other duties during early phases of the project. Thank you for your invaluable contribution in this early phase! We also wish to thank colleagues at Springer for their enthusiastic support and helpful guidance.

Finally, we would like to thank our colleagues within education research for providing the stimulating environment, the discussions and perspectives that have helped us move research forward through the IRIS project. Royalties from the sale of this book will be donated to UNICEF's work within Basic Education and Gender Equality.

Oslo, Norway — Ellen Karoline Henriksen
London, UK — Justin Dillon
Leeds, UK — Jim Ryder

Contents

Chapter 1
Introduction: Participation in Science, Technology, Engineering and Mathematics (STEM) Education: Presenting the Challenge and Introducing Project IRIS

Ellen Karoline Henriksen

Improved Participation in Science, Technology, Engineering and Mathematics (STEM): What Does It Mean, and Why Is It Needed?

Science and Technology Shape Our Lives and Provide Career Opportunities

Science and technology[1] matter. They shape our daily lives, as they have done in the past and will continue to do in the future. They influence the way we work, how we keep healthy, how we spend our time, how we communicate – and how we think. Just as science and technology have contributed to some of the great challenges the world faces today, such as climate change, it is also impossible to meet these challenges without employing science and technology in developing solutions. Scientific and technological advances – from the heliocentric worldview, the theory of evolution and the invention of the steam engine to current computer technology and genetics – have profoundly influenced our views of ourselves and the world we live in.

[1] In the present book, the terms "science and technology" and "science, technology, engineering and mathematics (STEM)" are used more or less interchangeably. In most cases, when the focus is on young people's relationship to this broad area of study as opposed to other disciplinary and professional areas like humanities, law, or crafts professions, no strong distinction is made between the different STEM disciplines. In some cases, STEM disciplines like physics, biological sciences or computer science are specified when the difference between STEM disciplines appears relevant or when the results reported concern only a subset of the STEM disciplines.

E.K. Henriksen (✉)
Department of Physics, University of Oslo, P.O. Box 1048, Blindern, N-0316 Oslo, Norway
e-mail: e.k.henriksen@fys.uio.no

E.K. Henriksen et al. (eds.), *Understanding Student Participation and Choice in Science and Technology Education*, DOI 10.1007/978-94-007-7793-4_1

Arguably, young people in developed countries today experience greater freedom than ever before in shaping their life in general and their education and career in particular. In a globalised world, they can take all or parts of their higher education abroad and they can choose between a wide range of different studies and disciplines. For modern youth, choosing an education may be seen as part of their identity project: through this potentially important life choice, they seek to express who they are and which values they wish to pursue. How do the science and technology disciplines present themselves in the marketplace where young people evaluate different educational options? Do young people see the potential of STEM to contribute to their personal development through a challenging education and diverse career opportunities? Do they see how science and technology will contribute to changing the world and their personal lives during their lifespan? This book is about understanding how young people make their educational choices and how they evaluate science, technology, engineering and mathematics in this context.

Science and Technology Contribute to Solving Global Challenges and Promote Economic Growth and Equitable Societies

In year 2000, world leaders adopted the United Nations Millennium Declaration, committing their nations to work towards a series of targets – with a deadline of 2015 – that have become known as the Millennium Development Goals (UN 2012). Among these eight goals are: to end poverty and hunger; secure maternal and child health, combat HIV/AIDS, and promote environmental sustainability. In order to reach these goals, science and technology are central in developing and improving renewable energy technology, communication systems, agricultural technology, medical treatments, systems for transporting and storing food and medical supplies, etc. Thus, a competent and responsible workforce within the science and technology disciplines is essential in order to meet some of the greatest challenges the world faces in the first decades of the twenty-first century.

STEM is also identified as an important sector for economic growth and stability in individual countries and regions. In its strategy "EUROPE 2020", the European Union (EU 2010) put forward three mutually reinforcing priorities:

- Smart growth: developing an economy based on knowledge and innovation.
- Sustainable growth: promoting a more resource efficient, greener and more competitive economy.
- Inclusive growth: fostering a high-employment economy delivering social and territorial cohesion.

The same report identified five measurable targets for 2020: for employment; for research and innovation; for climate change and energy; for education; and for combating poverty. Working towards these targets will involve "investing in research and development as well as innovation, in education and in resource efficient technologies" (EU 2010).

Along similar lines, there are recognised needs for strengthening STEM participation in most European countries (see for instance Norwegian Ministry of Education and Research (2009), Ministry of Higher Education Science and Technology of Republic of Slovenia (2010), Royal Academy of Engineering (2012)).

Improved Participation in STEM: Diversity and Equity Issues

Improved participation in STEM is not only a question of numbers of STEM students and practitioners; it is also a question of diversity. The FP7 Capacities Work Programme for Science in Society (EU 2008) states that "the pursuit of scientific knowledge and its technical application towards society requires the talent, perspectives and insight that an increasing diversity in the research workforce will ensure. Therefore, a balanced representation of women and men at all levels in research projects is encouraged". Similar views are expressed in the report "Land of Plenty: Diversity as America's Competitive Edge in Science, Engineering and Technology" (NSF 2000):

> Our economy will not only be positively affected by bringing more women, underrepresented minorities, and persons with disabilities into the SET workforce, but our high-tech, scientific, and engineering industries will benefit from their diverse viewpoints and approaches, as well as their skills.

UNESCO (2007) stated that "increasing women's involvement, input and access to S&T is essential to reducing poverty, creating job opportunities and increasing agricultural and industrial productivity". Thus, there is a broad recognition of the role of STEM in creating a sustainable and knowledge-based economy in an equitable society.

Increased STEM participation concerns not only the needs of the STEM establishment and society at large; it also concerns the interests of individuals. Bøe et al. (2011), discussing reasons why participation in STEM is an important issue, contended that involvement in STEM gives people literacy, empowerment and economic freedom to shape their world and everyday life and that women and other under-represented groups need to engage in STEM to be empowered to influence their own lives and the development of the world. These groups should be encouraged to participate with their priorities on the arenas where decisions are made regarding research and technology development. A classic example concerns women's health, which has received increasing focus in medical research as a result of women's engagement (NIH 1999). Another aspect of this argument is that the failure of women to pursue STEM careers limits their career opportunities and earning potential.

Furthermore, equitable access to STEM education and career means that everyone should be given the chance to engage in the scientific and technological world, which may enrich their lives and contribute to their individual development, in line with the ideals for a liberal education as described for instance by Carson (2002). Access to STEM is seen as a means of empowering individuals and opening up opportunities for self-development as well as a profession and career.

Finally, in an equitable society, everyone should have a real, not only a formal, free choice of education (Bøe et al. 2011). This requires that youth has access to sufficient and reliable information about the various educational and occupational options available, and that there are no formal or informal obstacles to a free choice. An example of the latter would be norms, stereotypes or expectations which young people are confronted with and which limits their perception of the options available to them and the roles they are expected to take on, as for instance discussed by Mujtaba and Reiss (2013) for the case of physics. Stereotypical views of scientists are still prevalent and fit poorly with the ideals that are held up, particularly for young women, by contemporary culture. Young people will not have a real free choice of education before these cultural barriers are reduced.

In this book, we use the term "improved participation in STEM" to denote a situation where society's needs for scientific and technological expertise are fulfilled and where each individual has a real opportunity to participate in STEM practice and STEM-related decision-making, and to pursue a STEM education and career, regardless of gender, ethnicity, class or other potentially inhibiting factors.

By *improved participation* in *STEM* we mean a situation where:

1. a larger and more diverse group of young people, based on reliable information and realistic impressions of their STEM education and career opportunities, consider STEM a viable possibility when making their educational choice;
2. a larger proportion of students complete a higher education within STEM after having entered.

Current Situation and Projected Needs for STEM-Educated Persons

Young People's Participation in STEM on Various Levels of the Educational System

In project IRIS (and in this book), the main focus is on educational choice in the transition from upper secondary to higher education. However, STEM participation in higher education can only be understood if school and childhood experiences as well as discipline cultures, possibilities and barriers in higher education STEM departments are taken into account. Thus, students' experiences with STEM in school as well as in higher education up to PhD level are all addressed in this volume. Moreover, as discussed in several of the chapters, educational choice can be seen as an on-going and dynamic process extending in both directions from the educational decision-point of entering higher education. Notably, the choice process continues through the first years of higher education, greatly influencing retention rates of STEM students, as discussed in Part III of this book.

STEM participation challenges vary between countries and disciplines and between levels of the educational system, as discussed by Bøe et al. (2011). We do not provide a detailed discussion of these variations here. However, the following paragraphs provide an overview of some tendencies that are found in several countries and several disciplines. These serve to document that despite variations, the challenges concerning STEM participation warrant continued attention from policymakers as well as the research community.

There is substantial evidence that many young people are disengaging from science from the first possible decision point during secondary education. In the UK, for instance, the proportion of students taking A-level physics and chemistry fell markedly between 1990 and 2008 (Joint Council for Qualifications [JCQ] 2009) but increased slightly again from 2008 to 2010 (JCQ 2010). In Norway, one third of the students choosing Level 1 physics in upper secondary school, choose *not* to continue to Level 2, and girls are overrepresented among the "leavers". In contrast, 97 % of biology Level 1 students continue to biology Level 2 (Norwegian Directorate for Education and Training [NDET] 2009). In Australia, proportionally fewer students have been choosing science at the first decision point. According to Ainley et al. (2008), between 1992 and 2007 the proportions of senior high school students taking physics, chemistry and biology courses declined by 26 %, 22 % and 29 % respectively. More recent figures suggest a stabilisation (Lyons and Quinn 2010). Researchers in New Zealand have also reported early student disengagement from science and mathematics (Hipkins and Bolstad 2005).

In many countries, increased student disengagement from STEM has been most apparent in the secondary to tertiary transition. In France, for example, the percentage of high school graduates enrolling in first year university science courses (excluding health and medicine) almost halved from 8.4 % in 1995 to 4.3 % in 2007 (Arnoux et al. 2009).Over the last decade, universities in Japan have been increasingly concerned about the 'flight from science', since the number of students studying science and engineering at university decreased by 10 % between 1999 and 2007 (Fackler 2008). Since 2000, the proportion of tertiary graduates specialising in science, mathematics and computing in Europe has been reduced from around 12 % to 9 % (Eurydice 2012).

In order to increase the number of STEM graduates, retaining the students who have started on the STEM track is as important as increasing the number of entrants. According to the OECD (2008), science and technology are among the disciplines where non-completion rates are highest. Understanding – and responding to – students' reasons for leaving their chosen STEM study thus becomes an important task. Ulriksen et al. (2010) discussed student drop- out/opt-out from STEM higher education programmes. They suggested that instead of describing drop-out as a problem belonging to the student, retention should be addressed as a relation between the student and the institution. In this context, there is a need to study how teaching approaches and department culture influence student achievements, attitudes, self-image, well-being and – ultimately – completion.

Projected Needs for STEM Professionals in the Future Workforce

The STEM participation challenge arises not only from falling enrolments in some disciplines, as described above, but also from projected needs for *increases* in the STEM workforce in the future. The widely cited report "Europe needs more scientists" (EU 2004) called for a substantial increase in the percentage of science and technology researchers in the total workforce. Projections from a number of countries indicate that the demand for STEM-educated labour will increase during the years to come, and there is widespread concern that the supply of people educated within STEM will be too small to meet future demands (see for instance Bjørnstad et al. (2008), Confederation of Danish Industry (2010), ERT (2009), Confederation of British Industry (2010)).

The STEM participation challenge does not apply equally to all STEM disciplines. University enrolments in life and health sciences are considered sufficient to meet projected demands in most developed countries (Organisation for Economic Co-operation and Development [OECD] 2008), while there are predictions of shortages in most engineering disciplines (United Nations Educational, Scientific, and Cultural Organization [UNESCO] 2010). Engineering graduates are in great demand in Australia, Germany, the US, Canada, Norway, the UK and New Zealand (Kaspura 2010; Manpower 2009), while serious shortages of physics and chemistry teachers have been reported in the UK, Norway, Denmark, the Netherlands (Osborne and Dillon 2008) the US (Hodapp et al. 2009) and Australia (Department of Education, Science and Training [DEST] 2006). For biology teachers, on the other hand, the demand in England has been decreasing in relation to that for physics and chemistry teachers, with the result that the amount of funding available to biology student teachers has been reduced in recent years (Department for Education 2013).

However, there are challenges to the suggestion that we need more students to be following post-compulsory science courses in order to address a shortage of scientists. Smith and Gorard (2011) analysed the educational and career trajectories of those students following post-compulsory science courses in England. They showed that for the majority of science graduates the occupations they enter immediately after study are not directly science-related. One possible interpretation of this is that the demand for scientists is indeed met by the supply of science graduates. However, it is also likely that many science graduates *choose* not to follow science-related careers. These findings call into question the assumption that increasing the numbers of science graduates will necessarily lead to an increase in the number of STEM graduates entering science-related careers.

Women's Participation in STEM

As discussed above, there are good reasons to promote increased participation of under-represented groups in STEM. Such groups may be defined through parameters such as gender, ethnicity, class or socioeconomic status. Although several of these parameters may interact in forming young people's relationship to STEM, as discussed by Archer et al. (2012), the main focus in the present book is the under-representation of females. Women continue to be under-represented in a number of STEM disciplines, notably in physics, mathematics and engineering (NSB 2010; Eurydice 2010). In the report "Europe needs more scientists" (EU 2004), it was remarked that increasing the number of women entering science and engineering careers would go a long way towards filling the demands for an increased R&D workforce in terms of numbers.

The publication "She Figures" (EU 2009) documented the under-representation of women in most STEM disciplines. For instance, in the fields of engineering, manufacturing and construction, women accounted for only a quarter of PhDs earned in 2006 in the EU countries (ibid., p. 51). In the USA, men earned four out of five bachelor's degrees awarded in engineering, computer sciences, and physics in 2007 (NSB 2010). Among fields with notable increases in the proportion of bachelor's degrees awarded to women were earth, atmospheric, and ocean sciences, agricultural sciences and chemistry, whereas women's share of bachelor's degrees in computer sciences, mathematics and engineering declined in recent years (ibid). The 2011 "Education at a glance" report (OECD 2011) shows that in 26 of 33 countries, women represented fewer than 30 % of graduates in the fields of engineering, manufacturing and construction.

In the report "Why so few?", the American Association of University Women states:

> By graduation, men outnumber women in nearly every science and engineering field, and in some, such as physics, engineering, and computer science, the difference is dramatic, with women earning only 20 percent of bachelor's degrees. Women's representation in science and engineering declines further at the graduate level and yet again in the transition to the workplace. (AAUW 2010)

Will more women and other under-represented groups come to the STEM field if given sufficient opportunities and encouragement? In the Scandinavian countries (two of which are represented in IRIS), a situation persists which has been termed "the Scandinavian paradox". These societies are among the world's most gender equitable according to the "Gender Inequality Index" in the UN Human Development Report (UNDP 2011), but educational and occupational patterns in these countries are distinctly "gendered", with men clustering in science and technology and women in teaching and health care; see for instance NMCE (2006) and Holt et al. (2006). Thus, there are indications that improving women's STEM participation is not a simple matter of removing inequitable practices.

The project UPGEM (Understanding Puzzles in the Gendered European Map) looked at the differences (in terms of numbers, experiences and outlook) between female researchers in physics in five European countries and described how the career paths of female physicists are conditioned by cultural patterns both within the discipline and in society at large. For instance, the UPGEM researchers compared women in physics in Italy and Denmark and noted that Italy has a higher proportion of female physicists, and also a higher proportion of women who stay on in research after having children. The reasons they suggested for this difference included different paths from school into higher education; differences between societies in how class interacts with gender; different family patterns, and variations in workplace cultures (Hasse and Trentemøller 2008).

The study "Meta-analysis of gender and science research" collected and analysed research on horizontal and vertical gender segregation in research careers, as well as the underlying causes and effects of these two processes. The project's synthesis report (European Commission 2012) stressed "the reality of horizontal and vertical segregation, the existence of pay gaps, stereotypes, and the biased nature of criteria of excellence" and stated that "the key challenge is not to change women but, on the contrary, to change the culture of science and research".

The issue of females in STEM will be discussed in greater depth in a number of chapters in this volume, notably in Chap. 4 and in Part IV.

Project IRIS

Introducing IRIS

The Science in Society programme under EU's 7th framework programme states on its web pages[2]:

> Science is part of almost every aspect of our lives. Although we rarely think about it, science makes extraordinary things possible. At the flick of a switch, we have light and electricity. When we are ill, science helps us get better. It tells us about the past, helps us with the present, and creates ways to improve our future.
>
> (...)
>
> With the pace that the world keeps and the speed with which technology advances, an understanding of science is a crucial part of a rounded education. Moreover, Europe needs more scientists and more people skilled in science and technology in order to compete in the global arena. It is, however, becoming increasingly difficult to attract young people to science careers. There is also a clear gender imbalance in science, engineering and technology: while 59 % of graduates in EU universities are female, only 18 % of professors are women.

In response to this challenge, project IRIS was established in 2009 with support from the European Commission's "Science in Society" programme and with the following overall aim:

> The objective of IRIS is to develop knowledge and recommendations informed by evidence on how the participation of young people, women in particular, in STEM higher education may be improved.

[2] http://ec.europa.eu/research/science-society/, Accessed May 2013.

To approach this overall aim, we have addressed three broad research questions, each with a view to the role of gender:

1. Which priorities, values and experiences are prominent in young people's educational choice processes?
2. How can education, interventions, information and outreach be designed in order to improve young people's participation in STEM higher education?
3. Which factors are important for retaining students in the STEM higher education they have embarked on?

The IRIS consortium consisted of researchers from six academic institutions in five European countries:

- **Norway (co-ordinator):**
  University of Oslo and Norwegian Centre for Science Education
- **United Kingdom:**
  King's College London
  University of Leeds
- **Denmark:**
  University of Copenhagen
- **Italy:**
  Observa – Science in Society
- **Slovenia:**
  Institute for Innovation and Development, University of Ljubljana

IRIS Research Activities

The IRIS project draws on a range of theoretical frameworks in order to address different aspects of young people's educational choice processes and their relationship to STEM. The most important of these are described in Part I of this book (see below).

Data has been collected and analysed within the IRIS project using a variety of quantitative as well as qualitative approaches. A data collection instrument common to all IRIS partners was the questionnaire IRIS Q, which was completed by almost 7,000 first-year STEM students in the five IRIS consortium countries in 2010–2011. The questionnaire was based on the theoretical perspectives adopted and on previous projects, notably the ROSE study (Schreiner and Sjøberg 2007). The questionnaire comprised a total of 65 items (most of them multiple choice; some open-ended) covering school science experiences, sources of inspiration for choice of education, expectations for future job, first experiences as a STEM student, and attitudes to gender equity in STEM. The target population for IRIS Q was first-year STEM students within eight selected disciplines defined through the International Standard Classification of Education (ISCED). IRIS Q may be found in the Appendix, together with details about questionnaire development, target population, data collection etc. Several chapters in this volume use data from IRIS Q.

In addition to the questionnaire, a range of qualitative and quantitative modules contribute to the overall results of project IRIS. These include for instance a combined questionnaire, focus group and individual interview study of the impact of school science curriculum content on students' subject choices in post-compulsory schooling, an interview study of how first-year students make sense of their experiences (with a view to identifying factors of importance for completion), and a study of first-year female STEM students' written narratives ("life stories") of their relationship to STEM and how they came to choose a STEM education. Analyses of students' choice narratives are prominent in several of the qualitative modules.

It should also be mentioned that an international network of *IRIS associated partners* has been established and that several of these have used the IRIS questionnaire to collect data in their respective countries. At the time of writing, only the Australian associated partner has published results (Lyons et al. 2012). This partner is also represented with a chapter (Chap. 10) in the present volume. The IRIS associated partners may in time provide possibilities for wider international comparisons.

The Structure of this Book

The book's **Part I** describes the most central theoretical frameworks that have been employed in IRIS. The Eccles et al. expectancy-value model of achievement-related choices (Eccles and Wigfield 2002; Eccles et al. 1983) offers a comprehensive framework for analysing the various factors influencing young people's considerations concerning educational choice. Another theoretical perspective informing the work in IRIS is sociological theories about youth in late modern societies, particularly the focus on education as a component of young people's identity development. Expectancy-value theory and perspectives on late modern youth are described in Chap. 2. Choice of education is a dynamic process (Cleaves 2005) in which young people constantly negotiate their choice in interaction with their surroundings, developing a narrative describing their choice to themselves and to their "significant others". The narrative approach to studying educational choice, which is detailed in Chap. 3, focuses on identity and emphasises choice as a process rather than as a decision taken at one particular point in time. Finally, project IRIS employs feminist perspectives (Haraway 1991; Harding 1986; Schiebinger 1999) on the processes that contribute to women's continued low participation in STEM. Such perspectives are presented in Chap. 4.

In **Part II** of this book, a range of results concerning students' participation in STEM and factors of importance for their educational choice are presented, based on results produced within IRIS as well as in related projects. Chapter 5 sums up previous research related to STEM attitudes, interests and participation.

Chapter 6 is an invited chapter from researchers in the ASPIRES project, which looks at how gender and identity shape young adolescents' science aspirations. In Chap. 7, the role of the school science curriculum in students' educational choice trajectories is addressed, whereas Chap. 8 examines the role of *place* in students' educational decision-making process. Short narratives where first-year university students describe the background for their educational choice are analysed in Chap. 9 with attention to how they describe their development of interest in STEM. Chapter 10 is again an invited chapter from one of the IRIS associated partners and adopts a systemic perspective on educational choices, examining how educational reforms contribute to shaping STEM participation in Australia. Chapter 11, building on questionnaire data from Slovenia, examines factors influencing STEM students' decision to pursue a PhD, whereas Chap. 12 looks at how three different recruitment and outreach efforts in Norway are received by their target groups and, based on these results, discusses success factors for initiatives aimed at improving STEM participation.

The book's **Part III** treats the issue of students' completion, or non-completion, of the STEM higher education they have embarked on. The first chapter in this part, Chap. 13, presents an overview of central research and theoretical approaches to this issue, with focus on Tinto's model and the work of Seymour and Hewitt. Chapter 14 builds on an extensive quantitative data material to track Danish students' movement from upper secondary school into (STEM) higher education, and specifically addresses background factors relating to completion or non-completion. In Chap. 15, focus is on individual students' meeting with STEM higher education and how the teaching environment influences their social and academic integration. Data collected using the IRIS Q questionnaire in Italy forms the basis of Chap. 16, which treats educational choice motivations as well as first-year experiences of STEM students with a view to factors that promote completion.

Although gender and STEM is a theme running through many of the chapters, **Part IV** of this book specifically shows how feminist perspectives may be used to analyse and understand female students' relationship to, and participation in, STEM. Chapter 17 displays nuances and variations in female STEM students' relationship to their chosen education and warns against communicating broad generalisations about STEM participation patterns based on gender/sex. In Chap. 18, Italian STEM students' short narratives about their educational choice are analysed in a similar manner to Chap. 9, but with particular attention to gendered patterns in the responses. Students' negotiation of their identity as minority-gender participants in first-year STEM educational environments is described from a Danish setting in Chap. 19, whereas Chap. 20, again from an Italian setting, employs a feminist interpretative framework to analyse the views that students express about gender and STEM.

The final part of the book, **Part V**, sums up theoretical, empirical and methodological outcomes and experiences of the IRIS project and presents insights that may be of importance for understanding, and improving, STEM participation. In Chap. 21, focus is on what the studies presented in this book have contributed to our understanding of students' educational choices and of how these choices may be approached through research. Chapter 22 presents insights that may be of use to stakeholders aiming to improve STEM participation through the educational system or through information, outreach, campaigns or other measures.

The various approaches described in the chapters of this book have yielded multifaceted results and interpretations concerning STEM participation. The book constitutes the main outcome of the IRIS project. However, IRIS data, perspectives and analyses have been, and will continue to be, used in a number of research articles, conference presentations, policy documents, and other dissemination arenas.

References

AAUW (2010). *Why so few? Women in science, technology, engineering, and mathematics* (C. Hill, C. Corbett, & A. St. Rose, Eds.). Washington: American Association of University Women.

Ainley, J., Kos, J., & Nicholas, M. (2008). *Participation in science, mathematics and technology in Australian education*. Camberwell: Australian Council of Educational Research (ACER).

Archer, L., DeWitt, J., Osborne, J., Dillon, J., Willis, B., & Wong, B. (2012). Science aspirations, capital, and family habitus: How families shape children's engagement and identification with science. *American Educational Research Journal, 49*(5), 881–908.

Arnoux, P., Duverney, D., & Holton, D. (2009). The rise and fall of mathematical enrolments in the french educational system: A case study. *International Journal of Mathematical Education in Science and Technology, 40*(1), 43–57.

Bjørnstad, R., Fredriksen, D., Gjelsvik, M. L., & Stølen, N.M. (2008). *Tilbud og etterspørsel etter arbeidskraft etter utdanning, 1986–2025* [Supply and demand for labour according to education, 1986–2025] (Vol. 2008/29, SSB-rapport). Statistics Norway

Bøe, M. V., Henriksen, E. K., Lyons, T., & Schreiner, C. (2011). Participation in science and technology: Young people's achievement-related choices in late modern societies. *Studies in Science Education, 47*(1), 37–71.

Carson, R. N. (2002). The epic narrative of intellectual culture as a framework for curricular coherence. *Science & Education, 11*, 231–246.

Cleaves, A. (2005). The formation of science choices in secondary school. *International Journal of Science Education, 27*(4), 471–486.

Confederation of British Industry (CBI). (2010). *SET for growth: Business priorities for science, engineering and technology*. www.cbi.org.uk/media/878512/16B6A4AD035FF3D88025778B0053D94D__20100801-set-for-growth.pdf

Confederation of Danish Industry. (2010). *Fremtiden kalder. Uddanner vi nok?* [The future is calling. Are we educating enough?]. http://di.dk/SiteCollectionDocuments/Shop/Fremtiden%20kalder%20-%20uddanner%20vi%20nok_WEB.pdf?productid=8653&downloadType=Produkt. Accessed 21 May 2012.

Department for Education. (2013). *Funding for postgraduate teacher training*. Accessed May 2013.

DEST. (2006). *Audit of science, engineering & technology skills: A summary report*. Canberra: Department of Education, Science and Training.

Eccles, J., & Wigfield, A. (2002). Motivational beliefs, values, and goals. *Annual Reviews of Psychology, 53*, 109–132.

Eccles, J., Adler, T. F., Futterman, R., Goff, S. B., Kaczala, C. M., Meece, J. L., et al. (1983). Expectancies, values, and academic behaviours. In J. T. Spence (Ed.), *Achievement and*

achievement motives. Psychological and sociological approaches. San Francisco: W. H. Friedman & Co.
ERT. (2009). *Mathematics, science & technology education report.* Brussels: European Round Table of Industrialists.
EU. (2004). *Europe needs more scientists!* Brussels: Directorate-General for Press and Communication.
EU. (2008). *Work programme 2008, capacities, part 5: Science in society.*
EU. (2009). *She figures. Statistics and indicators on gender equality in science on gender equality science.* Brussels: European Commission. Science and Society.
EU. (2010). *Europe 2020. A European strategy for smart, sustainable and inclusive growth.* Brussels: European Commission.
European Commission. (2012). *Meta-analysis of gender and science research: Synthesis report.* Luxembourg: Publications Office of the European Union.
Eurydice. (2010). *Gender differences in educational outcomes: Study on the measures taken and the current situation in Europe.* Brussels: Eurydice.
Eurydice. (2012). *Key data on education in Europe 2012.* Brussels: Eurydice.
Fackler, M. (2008, May 17). High-tech Japan running out of engineers. *New York Times.*
Haraway, D. J. (1991). *Simians, cyborgs, and women: The reinvention of nature.* London: Free Associations Books.
Harding, S. (1986). *The science question in feminism.* Milton Keynes: Open University Press.
Hasse, C., & Trentemøller, S. (2008). *Break the pattern! A critical enquiry into three scientific workplace cultures: Hercules, caretakers and worker bees.* Tartu: Tartu University Press.
Hipkins, R., & Bolstad, R. (2005). *Staying in science. Student participation in secondary education and on transition to tertiary studies.* Wellington: New Zealand Council for Educational Research.
Hodapp, T., Hehn, J., & Hei, W. (2009). Preparing high-school physics teachers. *Physics Today, 62*(2), 40–45.
Holt, H., Geerdsen, L. P., Christensen, G., Klitgaard, C., & Lind, M. L. (2006). *Det kønsopdelte arbejdsmarked* [The gender-segregated labour market]. Copenhagen: Socialforskningsinstituttet [Danish National Centre for Social Research].
JCQ. (2009). *Entriesto A-level examinations in sciences and mathematics 1985–2008.* http://www.iop.org/policy/statistics/education/page_43188.html. Accessed 12 Feb 2010.
JCQ. (2010). *GCE A-level trends: Appendix.* http://www.jcq.org.uk/attachments/published/1303/02.%20Appendix%20-%20GCE%20Tables.pdf. Accessed 12 Feb 2010.
Kaspura, A. (2010). *The engineering profession – a statistical overview 2010.* Barton: Engineers Australia.
Lyons, T., & Quinn, F. (2010). *Choosing science. Understanding the declines in senior high school science enrolments.* Armidale: University of New England.
Lyons, T., Quinn, F., Rizk, N., Anderson, N., Hubber, P., Kenny, J., et al. (2012). *Starting out in STEM. A study of young men and women in first year science, technology, engineering and mathematics courses.* Armidale: University of New England, SiMERR National Research Centre.
Manpower. (2009). *Talent shortage survey results.* Milwaukee: Manpower Inc.
Ministry of Higher Education Science and Technology of Republic of Slovenia. (2010). *Resolution on research and innovation strategy of Slovenia 2011–2020.* http://www.mvzt.gov.si/fileadmin/mvzt.gov.si/pageuploads/pdf/odnosi_z_javnostmi/01.06.2011_dalje/01.06._RISSdz_ENG.pdf. Accessed 21 May 2012.
Mujtaba, T., & Reiss, M. J. (2013). Inequality in experiences of physics education: Secondary school girls' and boys' perceptions of their physics education and intentions to continue with physics after the age of 16. *International Journal of Science Education, 35*(11), 1824–1845. doi:10.1080/09500693.2012.762699.
NDET. (2009). *Elevers fagvalg i videregående opplæring 2008–2009* [Students' upper secondary subject choices 2008–2009]. Oslo: Norwegian Directorate for Education and Training.

NIH. (1999). *Agenda for research on women's health for the 21st century*. Bethesda: National Institutes of Health.
NMCE. (2006). *Implementation of the Convention on the Elimination of all forms of Discrimination against Women (CEDAW) – Norway's seventh periodic report*. Norwegian Ministry of Children and Equality.
Norwegian Ministry of Education and Research. (2009). *Climate for research. Report no. 30 to the Storting (2008–2009)* (Report no. 30 to the Storting (2008–2009)).
NSB. (2010). *Science and engineering indicators 2010* (Vol. NSB 10-01). Arlington: National Science Foundation.
NSF. (2000). *Land of plenty: Diversity as America's competitive edge in science, engineering and technology*. Arlington: National Science Foundation (NSF).
OECD. (2008). *Encouraging student interest in science and technology studies*. Paris: Global Science Forum.
OECD. (2011). *Education at a glance*. Paris: OECD Publishing.
Osborne, J., & Dillon, J. (2008). Science education in Europe: Critical reflections (N. Foundation, Ed.). London: King's College.
Royal Academy of Engineering. (2012). *Jobs and growth: The importance of engineering skills to the UK economy*. London: Royal Academy of Engineering. www.raeng.org.uk/jobsandgrowth
Schiebinger, L. (1999). *Has feminism changed science?* Cambridge: Harvard University Press.
Schreiner, C., & Sjøberg, S. (2007). Science education and youth's identity construction – Two incompatible projects? In D. Corrigan, J. Dillon, & R. Gunstone (Eds.), *The re-emergence of values in science education* (pp. 231–247). Rotterdam: Sense Publishers.
Smith, E., & Gorard, S. (2011). Is there a shortage of scientists? A re-analysis of supply for the UK. *British Journal of Educational Studies, 59*(2), 159–177.
Ulriksen, L., Madsen, L. M., & Holmegaard, H. (2010). What do we know about explanations for drop out/opt out among young people from STM higher education programmes? *Studies in Science Education, 46*(2), 209–244.
UN. (2012). *Millennium development goals*. www.un.org/millenniumgoals/. Accessed Jan 2012.
UNDP. (2011). *Human development report 2011. Sustainability and equity: A better future for all*. New York: United Nations Development Programme.
UNESCO. (2007). *Science, technology and gender. An international report*. Paris: United Nations Educational, Scientific and Cultural Organization.
UNESCO. (2010). *Engineering: Issues, challenges and opportunities for development*. Paris: United Nations Educational, Scientific and Cultural Organization.

Part I
Theoretical Perspectives on Educational Choice

Chapter 2
Expectancy-Value Perspectives on Choice of Science and Technology Education in Late-Modern Societies

Maria Vetleseter Bøe and Ellen Karoline Henriksen

Introduction

"Am I clever enough? Am I really interested? Do I enjoy working with mathematics problems? Is science "me" – am I comfortable with an identity as a science student? Will this study lead to an exciting career and high wages? How much time and effort will it demand?" These are questions that young people considering a higher education in science, technology, engineering or mathematics (STEM) might ask themselves. Their answers to these questions lead them towards or away from STEM studies beyond school. The IRIS project aims to understand young people's choices to participate in – or not participate in – STEM education.

In this chapter, we present the Eccles et al. expectancy-value model of achievement-related choices (Eccles et al. 1983; Eccles and Wigfield 2002) and argue as to why it provides helpful perspectives for understanding young people's choices about participation in STEM. Sociological theories on late-modernity offer additional insights into how STEM-related choices are negotiated in rich, developed societies where participation problems are most pronounced (Bøe et al. 2011). Identity development is a particularly important part of young people's lives, according to such theories, and must be taken into consideration when understanding their educational choices. The Eccles et al. model acknowledges the importance of identity. However, a more thorough discussion of identity development related to educational choice is provided by Holmegaard and colleagues in Chap. 3.

This chapter is partly based on an article by Bøe et al. (2011) which presented the Eccles et al. model in more depth and demonstrated how a large body of research

M.V. Bøe (✉)
Norwegian Centre for Science Education, P.O. Box 1106, Blindern, N-0317 Oslo, Norway
e-mail: g.m.v.boe@naturfagsenteret.no

E.K. Henriksen
Department of Physics, University of Oslo, P.O. Box 1048, Blindern, N-0316 Oslo, Norway
e-mail: e.k.henriksen@fys.uio.no

E.K. Henriksen et al. (eds.), *Understanding Student Participation and Choice in Science and Technology Education*, DOI 10.1007/978-94-007-7793-4_2

literature on young people's relationship to STEM could be interpreted in the light of this model, combined with perspectives on late modernity, to inform the discussion about STEM-related choices.

The Eccles et al. Expectancy-Value Model of Achievement-Related Choices

Young people's educational decision-making is a complex process, and many approaches have been taken to understand it. In psychology, theorists have, for example, linked educational choices to individuals' personality types (Costa et al. 1984; Head and Ramsden 1990). In sociology, educational and vocational behaviour have been understood as products of socio-economic factors such as social class (Ball et al. 2002; Bourdieu and Passeron 1990). Other approaches to academic motivation include self-efficacy theory (Bandura 1997), intrinsic and extrinsic motivation (Ryan and Deci 2000), interest development (Hidi and Renninger 2006; Krapp 2005), attribution theory (Weiner 1985), and expectancy-value theory. The Eccles et al. expectancy-value model of achievement-related choices (Eccles et al. 1983) (Fig. 2.1) is founded in social psychology, and incorporates social, psychological and cultural aspects that affect young people's motivational behaviour.

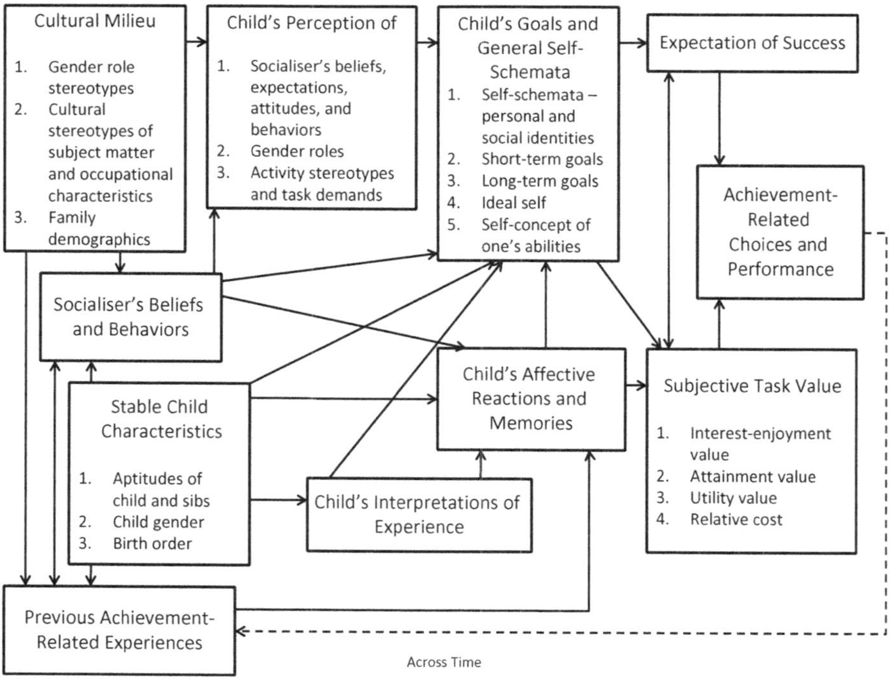

Fig. 2.1 Eccles and Wigfield (2002) expectancy-value model of achievement-related choices

The underlying premise of expectancy-value theory is that choice, persistence and performance can be explained by individuals' beliefs about how well they will perform in a particular activity and the extent to which they value the activity, that is, the subjective value they attach to the activity (Wigfield and Eccles 2000). The Eccles et al. model predicts that students are most likely to choose courses they think they can master, and that have high subjective value for them (Eccles et al. 1999). This is represented in the model through the boxes in the rightmost end of Fig. 2.1, where expectation of success and subjective task value represent the components that directly influence choice (referred to below as the "agency-related" components of the model). Subjective task value is further subdivided into interest-enjoyment value, attainment value, utility value, and relative cost. These will be presented in more detail later in this chapter and related to empirical findings from science education research.

Eccles et al. (1999) previously showed that both expectation of success and subjective value predict career choices. A specific focus on differences between the sexes is often seen, for example in studies of women's educational and occupational choices in relation to physical sciences, engineering and applied mathematics (Eccles 1994; Eccles et al. 1999). According to the model, young people's thoughts about their identity and identity development affect the expectation of success and the subjective values that they attach to different educational options (Eccles 2009).

An example of how the model has been used is the longitudinal study by Simpkins et al. (2006) of the links between mathematics and science choices and expectations and values. They collected data among 227 young people: in fifth grade the students reported on participation in various activities, and in sixth and tenth grade they reported on their expectations of success and subjective values related to mathematics and science. The study found that participation in activities predicted expectations and values, which in turn predicted enrolment in high school mathematics and science courses. Another example of possible uses of the model is provided by Denissen et al. (2007). They studied intraindividual coupling between academic achievement, interest and self-concept of ability in approximately 1,000 children between grades 1 and 12. They found that individuals tended to feel competent and interested in areas where they achieved well, and that the strongest coupling was between interest and self-concept of ability. They also observed an increase in the coupling across time.

Among the strengths of the model are that it is comprehensive, inclusive and based on empirical evidence. It is comprehensive in the sense that it includes different levels of influential factors: for example, young people's surroundings, such as their cultural *milieu* and the beliefs and behaviours of socialisers, young people's perceptions of the beliefs, expectations and stereotypes in their surroundings, and young people's personal goals and identity, and their affective reactions and memories. These factors influence their expectation of success and the subjective values that affect educational choices. The model includes constructs that overlap with concepts from other motivational theories. These include Bandura's (1997) *self-efficacy*, Ryan and Deci's (2000) *intrinsic* and *extrinsic motivation*, and the concept of *interest* (Hidi and Renninger 2006; Krapp 2002, 2005). The link between these concepts and the Eccles et al. model are described by (Eccles and Wigfield 2002). The model has been developed and tested over many years and in many studies (see Eccles et al. 1999, 1983; Nagy et al. 2008; Meece et al. 1990).

The strengths mentioned above made the Eccles et al. model well-suited to guide the development of the IRIS questionnaire (see Appendix), and to strengthen analyses and interpretation of results for some of the IRIS data and research questions. The focus on both expectation of success (or self-efficacy) and subjective task value is particularly pertinent, as both are reported to be very influential in young people's decisions about whether or not to study STEM (see e.g. Simpkins et al. 2006; Bandura et al. 2001). Another favourable trait of the model is its acknowledgement of the many other social and psychological aspects that affect the formation of students' expectations and subjective values. How family and friends talk about science, for example, impact on students' own relationship with science. Moreover, cultural stereotypes of STEM subjects and occupations affect how students relate to these subjects and occupations. It is important to note that the choice process is dynamic, that expectations and subjective values develop and change over time. At specific decision points, for example when applying for higher education, these expectations and values are brought to the fore and influence what courses students choose to apply for. However, the choice process continues beyond the decision point. That is, choices are negotiated and renegotiated as students gain experience with the study they have selected. This on-going nature of the choice process is clearly illustrated in Chap. 3, and is also relevant for understanding why some students choose to leave their STEM study before graduating (Part III).

A few issues require consideration when the Eccles et al. model is used (Bøe et al. 2011). First, expectation of success and subjective values are affected by constantly changing society and cultural *milieu*. As a result, measures of expectation of success and subjective values are sensitive to cultural changes. Second, social background variables such as ethnicity and class may not be sufficiently clearly stated in the model. Third, researchers should note that the expectancy-value structure of the model does not imply that choices are made through a fully informed calculation of all the available options. For example, interest-enjoyment value has major affective components and might be based on a "gut-feeling". Moreover, the utility value that a youngster ascribes to, for instance, an engineering course, may not be based on actual facts about the employability and career prospects of engineers; it may be based on stereotypes or hearsay. Yet it is this subjective perception of the utility value of the study in question which guides the choice.

Sociological Theories on Late-Modernity

We draw on some perspectives from sociology about late-modernity and identity development to understand better how cultural traits of highly developed societies may be recognised in young people's STEM-related choices. These theories provide additional insight into the importance of identity development for educational choices, and into young people's pronounced search for education and careers that appear interesting, self-realising and personally meaningful.

Sociologists such as Giddens (1991), Beck (Beck and Beck-Gernsheim 2002) and Inglehart (1997) described late-modernity in the 1990s, and the period's characteristic traits have also been recognised in more recent work (Bauman 2008; Furlong and Cartmel 2007). Characteristics of late-modern societies include less emphasis on material values and more emphasis on personal ones, such as self-realisation and quality of life (Inglehart 1997). Each individual is more culturally liberated and can to a larger extent than in more traditional societies make choices in relation to, for example, education and job. This liberation is a result of a less tradition-bound society, where identity is no longer inherited or given but must be constructed by the individual (Côté 1996; Giddens 1991). Young people see their interests, their favourite school subjects, their job plans, their activities and their views (on science and technology and everything else) as part of their identity, of who they are (Beck 1999; Goffman 1959). Note that identity is not a fixed entity that has implications for young people's choices. Rather, identity is in constant development and is negotiated against young people's choices and everyday behaviour, as is also described in Chap. 3. Schreiner and Sjøberg (2007) drew on late-modernity theories and claimed that late-modern young people tend to evaluate STEM education in terms of its contribution to their identity and self-development.

It is important to note that late-modernity provides an *idea* of free choice. Class, gender and other constraints of social life continue to limit young people's life chances, but have been obscured (Furlong and Cartmel 2007; Atkinson 2008). Although young people of different sexes or from different family backgrounds may have equal formal access to, for example, higher education, informal constraints in terms of cultural expectations and stereotypes still restrict access for certain groups. These restrictions may very well not be perceived as such by young people themselves, and are thus obscured.

STEM-Related Expectation of Success and Subjective Task Values

We will now briefly present the most central agency-related constructs of the Eccles et al. model: expectation of success and subjective value (the two rightmost boxes in Fig. 2.1). We use the term "agency-related" to express the idea that expectations of success and subjective values are constructs that students consider – more or less consciously – when they make an educational choice. In contrast, cultural *milieu* is a more structural construct that is very likely to influence students' choices, but potentially in ways that are less direct and less recognised by the students themselves. Subjective values and expectation of success are therefore readily recognised in students' self-reports concerning how they came to make an educational choice (see Chaps. 12 and 18).

In the model, subjective task value is split into four components: interest-enjoyment value, attainment value, utility value, and relative cost. Each of these components and expectation of success will be presented separately below and linked to insights from the STEM education research literature to inform our understanding of the values underlying educational choice. Late-modernity perspectives are included to add insight into how these components are formed and negotiated among young people in highly developed societies.

Expectation of Success

Expectation of success concerns how well students believe they will perform in, for example, a school subject they may choose to take. It includes both the students' self-concept of ability and their impression of the difficulty of the subject. As an example, consider the choice between two subjects in upper secondary school: advanced mathematics and English. What students see as success in advanced mathematics and English, depends on their self-concepts of ability in the two areas. Achieving a mark just above average in mathematics may be seen as a success if they see themselves as average mathematics students, but a failure if they consider themselves to be very good mathematics students. Expectation of success also includes the students' estimations of how difficult the subjects are. If they regard advanced mathematics as more difficult than English, they may characterise a slightly above average mark in advanced mathematics as a big success, while an equal level of success in English would require a top mark.

Physical science and mathematics subjects are often regarded as particularly difficult and demanding (Angell et al. 2004; Tytler et al. 2008; Carlone 2003; Osborne and Collins 2001). Due to this reputation, students might have to be particularly confident in their own abilities to expect to succeed. Females are more likely than males to have a low expectation of success in science and mathematics (Cavallo et al. 2004; Lloyd et al. 2005; Lyons 2006; Barnes et al. 2005; Preckel et al. 2008; Simpkins et al. 2006), especially compared to other school subjects (Häussler and Hoffmann 2000). The impact of expectation of success on choices of STEM education and occupations is widely documented (Bandura et al. 2001; Bennett and Hogarth 2009; Eccles et al. 2004; Kjærnsli and Lie 2011; Nagy et al. 2008).

Late-modern students feel responsible for the outcome of their choices (Furlong and Cartmel 1997), and it may, therefore, be difficult to develop an expectation of success that is strong enough to outweigh the potential costs related to a failure and lost opportunities. Late-modern individualisation means that each person has a unique character with special potentials that may or may not be fulfilled (Frønes and Brusdal 2001). Young people's perception of their talents and abilities are reflected in what they see as their potential, and in how well they expect to succeed in various activities. Expectation of success in STEM subjects is challenged by their reputation as particularly difficult and demanding, which causes some students to shy away from them.

Interest-Enjoyment Value

Interest-enjoyment value concerns how interested students are in the subject in question and the enjoyment they expect to experience when engaging with it. Students who are interested in literature, for example, may attach a higher interest-enjoyment value to an English course than to advanced mathematics.

Some claim that interest in school science among young people in developed countries is low (OECD 2008; Osborne 2008; Tytler 2007), and that it tends to decrease as students progress through school (Osborne et al. 2003). Liking of mathematics has also been found to decline throughout school (Fredricks and Eccles 2002; Wigfield et al. 1991). It is worth noting that interest in science topics *per se* tends to be higher than interest in school science and mathematics as experienced in the classroom (Hazari et al. 2008; Lyons and Quinn 2010; Häussler and Hoffmann 2000; OECD 2007; Schreiner 2006). A number of studies show that females and males, on average, express different interests in science topics (Cerini et al. 2003; Osborne and Collins 2001; Scantlebury and Baker 2007). On a general level, females are more likely to report interest in issues to do with human health and well-being, whereas males are more likely to be interested in things to do with, for example, technology and physics. Females also appear to be generally less engaged by science (Tytler et al. 2008). Several studies have found interest to be among the most important factors for choices of education and occupations in STEM subjects (Archer et al. 2010; Bøe 2012; Purcell et al. 2008; Hipkins and Bolstad 2006; Kjærnsli and Lie 2011; Lindahl 2003; Maltese and Tai 2011).

Late-modern societies emphasise individual self-realisation, personal meaning and subjective well-being (Beck and Beck-Gernsheim 2002; Inglehart 1997). Educational institutions are seen as arenas for self-realisation, where talents are developed and interests fostered. As described in Chap. 3, students expect to be passionate about their chosen education; tediousness is perceived as betraying their identity. It is not surprising, therefore, that many young people who choose science often highlight their interests as the main driving force, and, correspondingly, that students who opt out of science refer to a lack of interest.

Attainment Value

Attainment value concerns how well a subject or career choice fits into a person's identity development and the importance the individual attaches to attaining the goals (in this case a STEM education) they have set for themselves. People will attribute higher value to options that are easily negotiated into their identity development (Eccles 2009). For instance, for someone who wants 'very intelligent' to be part of their identity, an advanced mathematics course may have higher attainment value than an English course, because mathematics is generally considered to be more difficult than most other subjects. Similarly, it may be important not to choose subjects that appear to be in conflict with the direction students want their identity development to take. If physics is perceived to be for brainy and unpopular geeks, physics will have low attainment value for someone who rejects such an identity trait.

School science and STEM career identities appear to be less attractive to many young people (Archer et al. 2010; Hazari et al. 2010; Schreiner 2006; Taconis and Kessels 2009), in particular to females (Buck et al. 2008; Lyons and Quinn 2010; Eccles 2009). Identity development lies at the heart of late-modern youth culture (Illeris et al. 2002), and an educational or occupational choice is an identity choice. In recent years, increased attention has been paid to identity in research considering young people's relationship to STEM (Taconis and Kessels 2009; Schreiner and Sjøberg 2007; Hazari et al. 2010; DeWitt et al. 2011).

Utility Value

Utility value concerns how helpful a certain educational option is in reaching external goals, such as admission to higher education or entry to a future career. Physics in upper secondary school may have high utility value for students who hope to gain entry to medical school, even if they have no personal interest in the subject.

Utility for future careers often emerges as an important reason for choosing these subjects in upper secondary school (Angell et al. 2004; Bøe 2012; Miller et al. 2006; Lie et al. 2010; Hutchinson et al. 2009; Lyons 2006; Osborne and Collins 2001). As upper secondary STEM subjects often have a 'gate keeping' function for entry into prestigious higher education programmes such as medicine and engineering science, some students choose them to gain entry whereas others just want to keep their options open. The utility value of higher education in STEM may concern the prospect of a secure and well-paid job, if such an expectation is reasonable within the economic climate. Due to their perceived high costs, however (see below), STEM education programmes are unlikely to be considered as easy ways to economic security or other job benefits in most developed societies.

Late-modern identity development happens reflexively, in constant negotiation with a rapidly changing society, filled to the brim with information, choices and trends (Giddens 1991). To ensure that educational choices are easily included into this on-going identity development, young people are likely to want a lot of helpful information and as many open options as possible. The school sciences tend to open doors towards many different university studies, giving them high utility value. Concerning STEM higher education the potential utility value for future careers may be obscured for many students, since studies indicate that young people's knowledge about what STEM careers may involve is often limited (Cleaves 2005; Bøe and Henriksen 2013).

Relative Cost

Relative cost refers to negative aspects related to one educational choice compared to other options. It could, for example, be the time and effort that is required to do well in advanced mathematics compared to in English. It could be fear of failing advanced mathematics, or fear of disappointing parents.

Physical science and mathematics subjects on all levels are generally perceived to have higher costs in terms of difficulty and workload than most other subjects (Angell et al. 2004; Tytler et al. 2008; Carlone 2003; Osborne and Collins 2001). Females are more likely than males to perceive the costs of pursuing STEM careers to be high (OECD 2008; Carlone 2003; Warrington and Younger 2000; Angell et al. 2004; Frome et al. 2006).

Late-modern young people who choose an education feel that they themselves are responsible for the outcome. Should something go wrong, they have only themselves to blame and must themselves handle the consequences (Furlong and Cartmel 1997). A study may turn out to be too demanding, it may fail to live up to students' expectations, or for other reasons lead to non-completion. Young people tend not to explain this unhappy situation in terms of destiny, limitations of social class or lack of options, but by their personal failure, even if their problems are actually rooted in social constraints (Furlong and Cartmel 2007). Students are, therefore, likely to balance their STEM-related choices against the risks and costs they entail.

Conclusion

This chapter has presented the Eccles et al. expectancy-value model of achievement-related choices, and argued for its relevance when investigating young people's STEM-related choices. The relevance of the model is demonstrated by linking core constructs of the model to science education research literature, increasing our understanding of current participation problems in STEM. Due to this relevance, the Eccles et al. model has provided guidance in the development of the IRIS questionnaire. It has also been used as a tool for analysing data and interpreting results in parts of the IRIS work where the research questions and type of data material fit such perspectives. In addition, this chapter has introduced perspectives from sociology on late-modernity to provide more insight into how cultural traits of rich, developed societies can be recognised in students' expectations of success and subjective values related to STEM, and thus help us understand their participation in STEM in a cultural context. The Eccles et al. model and/or late-modernity perspectives are used in several of the chapters in the present volume, notably Chaps. 9, 12, 16, and 18.

References

Angell, C., Guttersrud, Ø., Henriksen, E. K., & Isnes, A. (2004). Physics: Frightful but fun. Pupils' and teachers' views of physics and physics teaching. *Science Education, 88*(5), 683–706.

Archer, L., DeWitt, J., Osborne, J., Dillon, J., Willis, B., & Wong, B. (2010). "Doing" science versus "Being" a scientist: Examining 10/11-year-old schoolchildren's constructions of science through the lens of identity. *Science Education, 94*(4), 617–639.

Atkinson, W. (2008). Not all that was solid has melted into air (or liquid): A critique of Bauman on individualization and class in liquid modernity. *The Sociological Review, 56*(1).

Ball, S. J., Davies, J., David, M., & Reay, D. (2002). 'Classification' and 'Judgement': Social class and the 'cognitive structures' of choice of higher education. *British Journal of Sociology of Education, 23*(1), 51–72.

Bandura, A. (1997). *Self-efficacy: The exercise of control*. New York: Freeman.

Bandura, A., Barbaranelli, C., Caprara, G. V., & Pastorelli, C. (2001). Self-efficacy beliefs as shapers of children's aspirations and career trajectories. *Child Development, 72*(1), 187–206.

Barnes, G., McInerney, D., & Marsh, H. (2005). Exploring sex differences in science enrolment intentions: An application of the general model of academic choice. *Australian Educational Researcher, 32*(2), 1–24.

Bauman, Z. (2008). *The art of life*. Cambridge: Polity Press.

Beck, U. (1999). *World risk society*. Cambridge: Polity Press.

Beck, U., & Beck-Gernsheim, E. (2002). *Individualization*. London: SAGE Publications Ltd.

Bennett, J., & Hogarth, S. (2009). Would you want to talk to a scientist at a party? High school students' attitudes to school science and to science. *International Journal of Science Education, 31*(14), 1975–1998.

Bøe, M. V. (2012). Science choices in Norwegian upper secondary school: What matters? *Science Education, 96*, 1–20.

Bøe, M. V., & Henriksen, E. K. (2013). Love it or leave it: Norwegian students' motivations and expectations for postcompulsory physics. *Science Education, 97*(4), 550–573.

Bøe, M. V., Henriksen, E. K., Lyons, T., & Schreiner, C. (2011). Participation in science and technology: Young people's achievement-related choices in late modern societies. *Studies in Science Education, 47*(1), 37–72.

Bourdieu, P., & Passeron, J.-C. (1990). *Reproduction in education, society and culture* (R. Nice, Trans., 2nd ed.). London: Sage Publications.

Buck, G., Clark, V. L. P., Leslie-Pelecky, D., Lu, Y., & Cerda-Lizarraga, P. (2008). Examining the cognitive processes used by adolescent girls and women scientists in identifying science role models: A feminist approach. *Science Education, 92*(4), 688–707.

Carlone, H. B. (2003). Innovative science within and against a culture of "achievement". *Science Education, 87*, 307–328.

Cavallo, A. M. L., Rozman, M., & Potter, W. H. (2004). Gender differences in learning constructs, shifts in learning constructs and their relationship to course achievement in structured enquiry, yearlong college physics course for life science majors. *School Science and Mathematics, 104*(6), 288–300.

Cerini, B., Murray, I., & Reiss, M. J. (2003). *Student review of the science curriculum. Major findings*. London: Planet Science, Institute of Education, University of London, Science Museum.

Cleaves, A. (2005). The formation of science choices in secondary school. *International Journal of Science Education, 27*(4), 471–486.
Costa, P. T., McCrae, R. R., & Holland, J. L. (1984). Personality and vocational interests in an adult sample. *Journal of Applied Psychology, 69*(3), 390–400.
Côté, J. E. (1996). Sociological perspectives on identity formation: The culture-identity link and identity capital. *Journal of Adolescence, 19*, 417–428.
Denissen, J. J. A., Zarrett, N. R., & Eccles, J. (2007). I like to do it, I'm able, and I know I am: Longitudinal couplings between domain-specific achievement, self-concept, and interest. *Child Development, 78*(2), 430–447.
DeWitt, J., Osborne, J., Archer, L., Dillon, J., Willis, B., & Wong, B. (2011). Young children's aspirations in science: The unequivocal, the uncertain and the unthinkable. *International Journal of Science Education*. doi:10.1080/09500693.09502011.09608197.
Eccles, J. (1994). Understanding women's educational and occupational choices. *Psychology of Women Quarterly, 18*, 585–609.
Eccles, J. (2009). Who am I and what am I going to do with my life? Personal and collective identities as motivators of action. *Educational Psychologist, 44*(2), 78–89.
Eccles, J., & Wigfield, A. (2002). Motivational beliefs, values, and goals. *Annual Reviews Psychology, 53*, 109–132.
Eccles, J., Adler, T. F., Futterman, R., Goff, S. B., Kaczala, C. M., Meece, J. L., et al. (1983). Expectancies, values, and academic behaviours. In J. T. Spence (Ed.), *Achievement and achievement motives. Psychological and sociological approaches*. San Francisco: W. H. Friedman & Co.
Eccles, J., Barber, B., & Jozefowicz, D. (1999). Linking gender to educational, occupational, and recreational choices: Applying the Eccles et al. model of achievement-related choices. In W. B. Swann Jr., J. H. Langlois, & L. A. Gilbert (Eds.), *Sexism and stereotypes in modern society: The gender science of Janet Taylor Spence* (pp. 153–192). Washington, DC: American Psychological Association.
Eccles, J., Vida, M. N., & Barber, B. (2004). The relation of early adolescents' college plans and both academic ability and task-value beliefs to subsequent college enrollment. *Journal of Early Adolescence, 24*(1), 63–77.
Fredricks, J. A., & Eccles, J. S. (2002). Children's competence and value beliefs from childhood through adolescence: Growth trajectories in two male-sex-typed domains. *Developmental Psychology, 38*(4), 519–533.
Frome, P. M., Alfeld, C. J., Eccles, J. S., & Barber, B. L. (2006). Why don't they want a male-dominated job? An investigation of young women who changed their occupational aspirations. *Educational Research and Evaluation, 12*(4), 359–372.
Frønes, I., & Brusdal, R. (2001). *På sporet av den nye tid. Kulturelle varsler for en nær fremtid* [On the track of the new era. Cultural signs of a near future]. Bergen: Fagbokforlaget Vigmostad & Bjørke AS.
Furlong, A., & Cartmel, F. (1997). *Young people and social change: Individualization and risk in late modernity*. Buckingham: Open University Press.
Furlong, A., & Cartmel, F. (2007). *Young people and social change – New perspectives* (2nd ed.). Berkshire: Open University Press.
Giddens, A. (1991). *Modernity and self-identity. Self and society in the late modern age*. Cambridge: Polity Press.
Goffman, E. (1959). *The presentation of self in everyday life*. New York: Anchor Book.
Häussler, P., & Hoffmann, L. (2000). A curricular frame for physics education: Development, comparison with students' interests, and impact on students' achievement and self-concept. *Science Education, 84*(6), 689–705.
Hazari, Z., Sadler, P. M., & Tai, R. H. (2008). Gender differences in the high school and affective experiences of introductory college physics students. *The Physics Teacher, 46*(7), 423–427.

Hazari, Z., Sonnert, G., Sadler, P. M., & Shanahan, M.-C. (2010). Connecting high school physics experiences, outcome expectations, physics identity, and physics career choice: A gender study. *Journal of Research in Science Teaching, 47*(8), 978–1003.
Head, J., & Ramsden, J. (1990). Gender, psychological type and science. *International Journal of Science Education, 12*(1), 115–121.
Hidi, S., & Renninger, K. A. (2006). The four-phase model of interest development. *Educational Psychologist, 41*(2), 111–127.
Hipkins, R., & Bolstad, R. (2006). *Staying in Science. An investigation of factors that encourage students to choose science as a study and career focus*. Wellington: New Zealand Council for Educational Research.
Hutchinson, J., Stagg, P., & Bentley, K. (2009). *STEM careers awareness timelines: Attitudes and ambitions towards science, technology, engineering and maths*. Derby: International Centre for Guidance Studies, University of Derby.
Illeris, K., Katznelson, N., Simonsen, B., & Ulriksen, L. (2002). *Ungdom, identitet og uddannelse* [Youth, identity and education]. Frederiksberg: Roskilde universitetsforlag.
Inglehart, R. (1997). *Modernization and postmodernization. Cultural, economic, and political change in 43 societies*. Princeton: Princeton University Press.
Kjærnsli, M., & Lie, S. (2011). Students' preference for science careers: International comparisons based on PISA 2006. *International Journal of Science Education, 33*(1), 121–144.
Krapp, A. (2002). Structural and dynamic aspects of interest development: Theoretical considerations from an ontogenetic perspective. *Learning and Instruction, 12*, 383–409.
Krapp, A. (2005). Basic needs and the development of interest and intrinsic motivational orientations. *Learning and Instruction, 15*, 381–395.
Lie, S., Angell, C., & Rohatgi, A. (2010). *Fysikk i fritt fall? TIMSS Advanced 2008 i videregående skole* [Physics in free fall? TIMSS Advanced 2008 in upper secondary school]. Oslo: Unipub.
Lindahl, B. (2003). *Pupils' responses to school science and technology? A longitudinal study of pathways to upper secondary school*. Göteborg: Acta Universitatis Gothoburgensis.
Lloyd, J. E. V., Walsh, J., & Yailagh, M. S. (2005). Sex differences in performance attribution, self-efficacy, and achievement in mathematics: If I'm so smart, why don't I know it? *Canadian Journal of Education, 28*(3), 384–408.
Lyons, T. (2006). Different countries, same science classes: Students' experiences of school science in their own words. *International Journal of Science Education, 28*(6), 591–613.
Lyons, T., & Quinn, F. (2010). *Choosing science. Understanding the declines in senior high school science enrolments*. Armidale: University of New England.
Maltese, A. V., & Tai, R. H. (2011). Pipeline persistence: Examining the association of educational experiences with earned degrees in STEM among U.S. students. *Science Education, 95*(5), 877–907. doi:10.1002/sce.20441.
Meece, J. L., Wigfield, A., & Eccles, J. S. (1990). Predictors of math anxiety and its influence on young adolescents' course enrollment intentions and performance in mathematics. *Journal of Educational Psychology, 82*(1), 60–70.
Miller, P. H., Blessing, J. S., & Schwartz, S. (2006). Gender differences in high-school students' views about science. *International Journal of Science Education, 28*(4), 363–381.
Nagy, G., Garrett, J., Trautwein, U., Cortina, K. S., Baumert, J., & Eccles, J. (2008). Gendered high school course selection as a precursor of gendered careers: The mediating role of self-concept and intrinsic value. In H. M. G. Watt & J. Eccles (Eds.), *Gender and occupational outcomes* (pp. 115–143). Washington, DC: American Psychology Association.
OECD. (2007). *PISA 2006. Science competencies for tomorrow's world* (Vol. 1). Paris: OECD.
OECD. (2008). *Encouraging student interest in science and technology studies*. Paris: Global Science Forum.
Osborne, J. (2008). Engaging young people with science: Does science education need a new vision? *School Science Review, 89*(328), 67–74.
Osborne, J., & Collins, S. (2001). Pupils' views of the role and value of the science curriculum: A focus-group study. *International Journal of Science Education, 23*(5), 441–467.

Osborne, J., Simon, S., & Collins, S. (2003). Attitudes towards science: A review of the literature and its implications. *International Journal of Science Education, 25*(9), 1049–1079.

Preckel, F., Goetz, T., Pekrun, R., & Kleine, M. (2008). Gender differences in gifted and average-ability students: Comparing girls' and boys' achievement, self-concept, interest, and motivation in mathematics. *Gifted Child Quarterly, 52*(2), 146–159.

Purcell, K., Elias, P., Ellison, R., Atfield, G., Adam, D., & Livanos, I. (2008). *Applying for Higher Education – The diversity of career choices, plans and expectations. Findings from the first futuretrack survey of the 'class of 2006' applicants for higher education.* Coventry: IER, University of Warwick.

Ryan, R. M., & Deci, E. L. (2000). Self-determination theory and the facilitation of intrinsic motivation, social development, and well-being. *American Psychologist, 55*(1), 68–78.

Scantlebury, K., & Baker, D. (2007). Gender issues in science education research: Remembering where the difference lies. In S. K. Abell & N. G. Lederman (Eds.), *Handbook of research on science education* (pp. 257–285). Mahwah: Lawrence Erlbaum Associates.

Schreiner, C. (2006). *Exploring a ROSE-garden. Norwegian youth's orientation towards science – Seen as signs of late modern identities.* Doctoral Thesis, University of Oslo, Oslo.

Schreiner, C., & Sjøberg, S. (2007). Science education and youth's identity construction – Two incompatible projects? In D. Corrigan, J. Dillon, & R. Gunstone (Eds.), *The re-emergence of values in science education* (pp. 231–247). Rotterdam: Sense Publishers.

Simpkins, S., Davis-Kean, P., & Eccles, J. (2006). Math and science motivation: A longitudinal examination of the links between choices and beliefs. *Developmental Psychology, 42*(1), 70–83.

Taconis, R., & Kessels, U. (2009). How choosing science depends on students' individual fit to "science culture". *International Journal of Science Education, 31*(8), 1115–1132.

Tytler, R. (2007). *"Re-imagining science education": Engaging students in science for Australia's future.* Camberwell: Australian Council for Educational Research.

Tytler, R., Osborne, J., Williams, G., Tytler, K., & Clark, J. C. (2008). *Opening up pathways: Engagement in STEM across the primary-secondary school transition.* Canberra: Australian Department of Education, Employment and Workplace Relations.

Warrington, M., & Younger, M. (2000). The other side of the gender gap. *Gender and Education, 12*(4), 493–508.

Weiner, B. (1985). An attributional theory of achievement motivation and emotion. *Psychological Review, 92*(4), 548–573.

Wigfield, A., & Eccles, J. (2000). Expectancy-value theory of achievement motivation. *Contemporary Educational Psychology, 25*, 68–81.

Wigfield, A., Eccles, J. S., Iver, D. M., Reuman, D. A., & Midgley, C. (1991). Transition during early adolescence: Changes in children's domain-specific self-perceptions and general self-esteem across the transition to junior high school. *Developmental Psychology, 27*(4), 552–565.

Chapter 3
A Narrative Approach to Understand Students' Identities and Choices

Henriette Tolstrup Holmegaard, Lars Ulriksen, and Lene Møller Madsen

Narrative Research to Understand the Complexity in Students' Choices

Since the 1970s, research in students' educational choices has been carried out to inform policymakers and to help them predict, plan and affect student enrolment, recruitment activities and student marketing (Paulsen 1990). As a consequence, large-scale quantitative studies have been carried out to identify and map the components that affect young people's educational choices, some with comprehensive models of student choices as an outcome. The Eccles model presented in Chap. 2 is such an example (Eccles and Wigfield 2002). However, whilst this extensive research has provided a reasonably clear picture of these components (Bergerson 2010), as higher education institutions are facing an increasingly diverse student body (Reay et al. 2005) there is a call to move towards more qualitative research that explores how the students themselves handle and make meaning of their choices (Archer et al. 2010; Hsu et al. 2009). In the period 2000–2010 research addressing this purpose has been carried out studying how students' identities relate to their choice of education (Archer et al. 2010; Brunila et al. 2011; Illeris et al. 2002; Schreiner 2006; Schreiner and Sjøberg 2007). A key finding in this research is that for the students it is not only a question about what they want to study, but also of who they wish to become, i.e. of constructing an attractive identity (Illeris et al. 2002). However, there is still a call for qualitative studies 'that can delve into the how and why questions that blur our clear understanding of how students experience the process of making post-secondary education decisions' (Bergerson 2010).

H.T. Holmegaard (✉) • L. Ulriksen • L.M. Madsen
Department of Science Education, University of Copenhagen,
Øster Voldgade 3, DK-1350 Copenhagen C, Denmark
e-mail: hh@ind.ku.dk

E.K. Henriksen et al. (eds.), *Understanding Student Participation and Choice in Science and Technology Education*, DOI 10.1007/978-94-007-7793-4_3

As we will show, narrative theories in general and narrative psychology in particular provide new insights in the study of upper secondary school students' choices of, and transition into, higher education. The focus is on how the students in their construction of an identity balance and negotiate the options they recognize as available and suitable, and how these acts of balancing, negotiating, and constructing eventually lead them to decide whether or not to enter a STEM study programme. In this chapter, we will present two core ideas within the theory that influence our comprehension of students' choices; namely the concept of identity as an ongoing process, and the concept of time; in particular how identities move across time and how individuals make meaning of the present by negotiating what was before (in retrospect) and what is expected in the future (prospective).

The examples used in the chapter are drawn from a longitudinal study presented in Chap. 15, where 38 students in Denmark were followed for 3 years from the end of upper-secondary school. All of the students had attended an upper-secondary programme with a particular emphasis on science and (for some of the students) technology. (For a more extensive presentation of the method see Holmegaard 2013).

Narrative Theories and Narrative Psychology

Awareness of narrative structures has a long history in the research literature (Czarniawska 2004). However, in the 1970s 'the narrative turn' and the emphasis on person-centred approaches introduced a variety of narrative theories across the social sciences (Andrews et al. 2008). The research community's interest in narrative research was initiated by the philosopher Paul Ricoeur (1976, 1990) and his ideas on understanding narratives and experiences as storied structures. His thoughts have been influencing the development of narrative theories in various disciplines such as literature, rhetoric, psychology, political science and anthropology. Despite the different nature of the disciplines, narrative theories can be divided into three main categories. One category consists of theories focusing on narrative syntax or structures and can be applied to understand storylines in for example films, literature and personal narratives. A second category consists of approaches with a focus on meaning making in narratives (semantics or content). Finally, a third category is made of theories focusing on the narrative in a particular context (for instance, studying narrative configurations in particular historical periods) (Andrews et al. 2008; Smith and Sparkes 2008). In this book the chapters drawing on narrative research belong to the second category, since they take their point of departure in an analysis of the way students make meaning of their experiences.

Within psychology, the use of narrative theories followed what is known as the 'crisis in psychology' in the 1970s, breaking with the widespread experimental tradition and moving towards new criteria for conducting science (Sarbin 1986). By examining and measuring the self as traits, abilities, and personality, other theoretical positions arose that described identity as something multifaceted and complex and being produced in a social and cultural context. Identity as a research object moved from the lab into social situations now requiring qualitative research methods (Potter and Wetherell 1987).

Narrative psychology is far from a field characterized by consensus, but covers various ideas of what narratives are and how they should be studied. However, there is agreement in terms of the understanding that 'identities and selves are shaped by the larger socio-cultural matrix of our being-in-the-world and, at the least, narrative implies a relational world' (Smith and Sparkes 2008, p. 3). However, the theoretical positions vary in terms of how the relation between identity/selves on the one side and, on the other side, socio-cultural surroundings are balanced in a spectrum ranging from 'thick individual' and 'thin social' to 'thin individual' and 'thick social' (Smith and Sparkes 2008). In the former, theories perceive individuals to possess a self and highlight narratives as the coherent story constructed from the inside and out: 'That the storied accounts we hear reflect an inner sense of narrative identity' (McAdams 2005, p. 129). Some of the theoretical positions on this side of the spectrum are inspired by psychoanalytic ideas. The other side of the spectrum (thin individual and thick social) emphasises the production of narratives within a socio-cultural context. This perspective is inspired by post-structuralist and social-constructivist theories:

> (. . .) identities are viewed as multiple, fragmentary, unfinished, always changing. They are performative, destabilized and deferred, rather than an inherent, unified property of the individual. (Smith and Sparkes 2008, p. 24)

In the middle of the spectrum are theories that take both the social and the individual perspectives into account. These theories perceive narrative identities on the one hand to be constructed inter-subjectively in interaction with others, constituted by political power-laden processes and social relationships, and mediated through institutional structures (Ezzy 1998). On the other hand, they find that each individual has different resources and possibilities available; each subject is involved with specific persons, capacities, and circumstances (Crossley 2000) and carries with them a history. These theories in the middle of the spectrum, therefore, both look into the structures and cultures in the environment where the narratives are produced, but also how narratives are related to the students' own and the students' surroundings sense of self, i.e. the perception of who he/she is, how she/he will become recognised as him/herself (Bruner 1990).

The narrative theories applied in this book position themselves ranging from the middle of the spectrum (Chaps. 7 and 15) to a 'thick socio cultural' and 'thin individual' position (Chap. 17). The chapters vary in how they use narrative theory and narrative psychology, as theory, method or methodology. Chapters 7 and 15 use narrative psychology as a methodology, that is, both as the underlying conceptual framework for understanding the notion of identity, but it also guiding the way interviews are conducted and the tools for analyzing them. Chapter 17 reads narrative psychology into a broader post-structuralist framework.

In the following we wish to unfold the central concepts within narrative psychology as perceived from the middle of the spectrum introduced above, and with concrete examples to illustrate how these concepts can be applied to empirical data.

Identity, Meaning and Choice

When students are about to choose what to study after upper-secondary school they undergo a meaning making process where they struggle to make sense of who they are and who to become (Illeris et al. 2002; Schreiner 2006). Within narrative psychology this meaning making process is a central concept. Not only do narratives say something about events and experiences, but more precisely they offer an approach to study how individuals understand these events and experiences and what meaning they ascribe to them. Meaning making is understood as a way of structuring the world (Ulriksen et al. 2013). Through narratives, the complexity in our experiences of the world is fixed into a sense of coherence. We construct a causality in terms of what caused the events and why and how we responded to them. A central question, then, is how this meaning making takes place. In this chapter we argue that meaning making is both embedded in the cultural context where the narratives take place, and constructed in relation to the individuals' own and their surroundings' sense of the individual's self. Individuals cannot freely invent narratives that aren't recognizable in terms of these two central aspects – the culture and other people.

We understand our lives as a single progressive story, and our identities are cumulative over time (Polkinghorne 1988). This means that we conceive of others, and ourselves, as possessing a coherent self. Consequently, there is a limit to how flexible and fluid our narratives can appear. Therefore, we present ourselves in a way that appears reliable and valid to other people's expectations of meeting a stable self. Therefore, we present ourselves in a way that appears reliable and valid to other people's expectations of meeting a stable self (Bruner 1990). Therefore, the notion of identity on the one hand must be understood as possessing a culturally embedded stability, but on the other hand as constantly changing, flowing backwards and forwards, a continuous process in which we keep on working to retell ourselves: 'We are always lost in transitions' (Quinn 2010).

Individuals are made and at the same time make themselves recognisable through narratives. The self, then, is a meaning rather than a substance or a thing (Polkinghorne 1988, p. 152). Narratives are then both what structures the world, and what relates us to it. Narratives are a way of framing events, beliefs, and desires into a coherent story. Therefore, when applying narrative psychology to studying students' choices we learn about how they ascribe meaning to their choice but also gain access to how they relate themselves to it. We label this process a 'choice narrative'. This covers students' work on their identities to construct a narrative of why they considered choosing a particular study programme and how they relate themselves to it. To illustrate how this identity work takes place we present part of the choice-narrative of Louise. In an interview just before completing upper-secondary school, Louise explained why she considered choosing to study international business at university:

> I really can imagine myself in a business-suit as a leader. I am always like a leader in my class when working in groups but also in general. I am also the one who takes care of

> coordinating when we meet outside class. (. . .) I think the kind of working culture and job will suit me well, getting to travel a lot and live in the city (. . .) I think I will have a lot of opportunities later on if I study business (Louise, upper secondary school)

Louise presents an example of how the students construct a narrative by relating themselves to their expectations of what kind of future they could see becoming available through this particular study programme. During most of the interview, Louise explained how physics was her favourite course, especially the abstract parts, but, contrary to business, Louise found it hard to see any attractive identity if she were choosing a STEM programme at higher education:

> I've always thought I was going to study engineering, physics or nanotechnology or something. But I just think it will become too boring for me. I like being around people. But physics is just so very fixed. Unless you are really clever, and get to do research in the things that are not explored yet – it is fixed (. . .) It is just too superficial, really. There are no perspectives of personal development in it, and I could not see myself not having anything to do with other people at all. (Louise, upper secondary school)

Louise makes meaning of a future study programme and career as contexts where you need to develop yourself, be around people, and relate to the content in a way that does not appear to be fixed and superficial. Louise finds it more likely to have those requirements met in the study of international business than of physics. Through her choice-narrative she works on relating herself to her expectations of what business will be like, both within the study programme at higher education but also the career opportunities she expects will be available.

Through narrative psychology we do not just learn about how students make meaning and how they relate themselves to that meaning. We also learn about how choosing what to study is embedded in culturally shared understandings. This is illustrated in the following quote from the interview with another student, Filip. For Filip, choosing what to study is perceived to be his own personal task:

> Personally, I'm sort of uneasy about being influenced by a career counsellor. He is not neutral. It would be nice if he was, but nobody would be neutral. A counsellor also has an idea about what would be good to study. I would be nervous, then, to be influenced by it (Filip, upper secondary school).

Filip's idea of having to make his choice by himself also suggests his notion of how a proper choice should be made, namely without the influence of anyone else. Furthermore, the idea of making one's choice by oneself is an example of how the students construct their choices in culturally embedded truisms. The notion of the autonomous choice is not challenged by the students, and even though they are nervous or unsure about the choice, it does not fit well with asking an unknown counsellor for help.

In narrative psychology, narratives are understood as culturally embedded: 'We live publicly by public meanings and by shared procedures of interpretation and negotiation' (Bruner 1990, p. 13). If narratives are to be understood as sensible and recognized by the students' surroundings, they need to be embedded in cultural ways of performing a choice-narrative, and further be consistent with what the student and the student's surroundings consider as a proper match for the particular student:

> We begin with the premise that identities are lived in and through activity and must be conceptualized as they develop in social practice. But we are also interested in identities as psychohistorical formations that develop over a person's lifetime, populating intimate terrain and motivating social life. (Holland et al. 1998, p. 5)

We now turn to how the culturally embedded stability in students' identities affects students' choices. When choosing what to study by the end of upper-secondary school, the students construct narratives that align their expectations of a certain study programme with who they perceive themselves to be. An example is Ian. He was encouraged by his parents to choose his future study programme according to his interests, but when he told them that he considered studying law, his parents questioned whether becoming a lawyer was an attractive choice for him after all. His parents found it hard to see how studying law would suit the person they recognised him to be. From their perspective lawyers were not decent people and could not be trusted, which conflicted with their perceptions of Ian. Therefore his choice narrative of studying law was not recognized as proper and suitable for him. Ian eventually chose to study bio-chemistry, a choice which particularly his mother, herself a bachelor of biomedical laboratory science, found sensible and suitable. Another student put it like this:

> To me choosing the right thing is about getting some kind of acceptance from my family. I also consider what my friends can picture me doing. As when I say 'I would like to study medicine' they reply: 'that is a great idea. We also picture you as a medical doctor (Asger, upper secondary school).

Even though the choice of study is being considered a responsibility of the individual (and the responses in the IRIS Q questionnaire suggest that the students themselves describe the influence from families and other persons on their choice as limited, cf. Chap. 9), the students' narratives show that the choice-narratives are constantly tried out and negotiated in the students' social relations where they are informed, adjusted and revised based on how these social relations meet and recognize the choice-narratives and whether they are considered suitable to who they expect the student to be. This requires the students to make their choice-narratives recognizable, even though not all students present their choice-narratives to their friends and families as explicitly as in the examples above. But most students do need to become recognized by their circle of acquaintances as somebody who is about to make a sensible and well-reasoned choice that suits who they are. Hence, they need to construct a choice-narrative that corresponds with who they are and who their surroundings perceive them to be.

Constructing a choice narrative implies identity-work for the student to gain a sense of match between what the student expects a certain programme to be like, the student's expectations of who to become, and a choice-narrative that combines past narratives with future horizons. In the next section we will show how narrative psychology provides a tool to perceive this identity-work as a continuing process rather than a specific decision point.

Choice as a Process; The Concept of Time

In this section we provide empirical examples of what this process of choosing what to study looks like, and how the concept of time can be used as a tool to explore students' choices.

Christine was really interested in architecture and design, and strongly considered studying engineering. She also thought of studying mathematics at university, but she had difficulties with seeing other prospects of studying mathematics than becoming a teacher, which she did not consider to be an attractive choice:

> I'm convinced that I would kill the children before I got to teach them anything (laughs). I don't think I would fit that well as a teacher (. . .) Now that I think about it I don't think I could stand becoming a teacher. (Christine, upper secondary school)

Five months later, Christine sent the researchers a text-message: 'I have started at teacher education [to become a primary and lower-secondary school teacher with mathematics as speciality]. I have always wanted to become a teacher'.

Christine's major revision of her narrative seemed surprising. In an interview at the beginning of her teacher education, she explained how she was not sure whether engineering would be the right fit for her, and whether it at all was worth spending 4 h on a train each day to go to the technical university. Her relationship with her boyfriend and their new apartment prevented her from moving away from the smaller city she lived in, and the teacher education institution was nearby. Christine's narrative is an example of how students' choices are embedded in their surroundings, that they are influenced by different material and relational components as well as study and career-related considerations, and how choice narratives also have to be aligned with life in general. Even when her choice changed over a short period of time, she managed to negotiate her narrative to keep a sense of stability in her understanding of herself. The phrasing 'I always' indicates a choice well-reasoned, motivated, and stable. Generally, we found that the students often used 'always' in their choice-narratives, but that the 'always' in some cases (as for Christine) was constructed retrospectively in the sense that students used it even when their choice-narrative dramatically changed over time (Holmegaard et al. 2014b). To understand what happened in Christine's narrative it is helpful to draw upon narrative psychology.

From the perspectives of narrative psychology we are always situated in the middle of our stories, and since we are not sure how they will turn out, we constantly revise the plot as new events occur and as new perspectives of how these events will turn out become visible (Polkinghorne 1988). Not only do our perspectives of the future constantly change when new meaning occurs, we also change our narratives retrospectively (Bruner 1990, 2004). Imagine a car driving on a winding road and as it turns new landscapes become visible not only through the windscreen but also through the rear-view window. Similarly, as our narratives change new perspectives become visible and new episodes from our lives are highlighted. Disruptive elements are removed from the narrative to maintain a degree of meaning and stability (Crossley 2000), and new elements are added in a process of negotiation and identity work. Another example from the interviews illustrates this perspective. The following are three quotes from Filip's narrative. The first two quotes are from two interviews made during the first semester. The final quote is from an interview during the third semester of his engineering course:

> I am looking forward to working on management. How to manage craftsmen when building something (. . .) People don't think engineers work with humans, but I think they just do it in another way, they work with management. (Filip, studying engineering, October 2009)

> My professor says: 'Don't focus too much on management. It is too arrogant to enter the labour market as a new engineer and say 'I want to become a leader'. Get some more clear-cut engineering skills instead'. My conclusion is to study energy and then combine it with some management later. It is an important challenge for the world to face in the future (. . .) I also began recognising that management is also tough and hard work. (Filip, studying engineering, November 2009)

> I've been interested in energy for many many years. When I was a kid I found motors to be really cool and later nuclear power. It was really many years ago (. . .). And now I have learned about how companies work and where in the world is more exiting to work with energy. (Filip, studying engineering, September 2010)

The examples shows how Filip kept on working on and negotiating his narrative about why he studies engineering and his perspectives on doing it, but also on retelling his past rationales for entering the programme. Management as an interest in engineering is negotiated in the narrative from being his major perspective for studying engineering, through management being too tough and hard, and, finally, to become totally excluded. Conversely, energy engineering becomes included in the narrative as a future challenge for the world, and something he has always been interested in. Filip's perception of the future changed from management to energy engineering as he interacted with a cultural norm at the study programme mediated through his professor: 'engineering is about engineering'. As a consequence he changed both his perception of the future in terms of who to become and why (energy engineering being an important challenge in the future) but also his retrospective narrative of why he wanted to become an energy engineer (because he always had been interested). Filip is one example of how culture and norms interact very explicitly with his meaning making and his way of understanding himself. It shows how choices are not well-defined decisions taking place at a certain point in time, but rather ongoing processes of negotiations over time where new meanings, new identities and choices are produced.

Discussion and Conclusion

In this chapter we have shown how narrative theory in general, and narrative psychology in particular, provide lenses to look further into how students make meaning of their choice of study after upper-secondary school. Contrary to research in students' choices aiming at identifying and mapping the variables that affect students' choices, theories about narratives provide a framework to study the complexity when students make choices. Through the theory we get to approach students' construction of their choice-narrative. This includes first of all a focus on the students' making meaning of what a particular study programme would be like and how they relate themselves to that meaning, but also what kind of future they recognize as becoming available through a particular study programme. Secondly, the students need to make their choice-narratives recognizable to their surroundings but also to their sense of self. Therefore, their narratives are tried out and negotiated with their social acquaintances where they are informed, adjusted, and revised based on how these acquaintances meet and recognize the choice-narratives as suitable to who they expect the student to be. Finally, the students' narratives are embedded in culturally shared understandings of what a good and proper choice consists of, and in their choice narratives the students need to relate to these truisms in order to be recognized. By applying a narrative psychological approach to students' upper-secondary choices, we can understand students' identity work when they are struggling to find a suitable study programme.

In this chapter we have shown how narrative psychology provides an understanding of choice of study as a process taking place over time where individuals work on their identities in terms of constructing a coherent choice narrative. Choice of study is not an isolated event linked to a formal decision point at a particular time, that is, when students send their application forms listing their desired courses of study. This formal decision point highlights the choice of study for the students, but applying a narrative psychological approach we learn how the choice of study is a continuing process also after entering higher education (see also Chap. 15).

Therefore the findings reached by this approach may differ from the ones reached when using, for instance, the Eccles model. It depends, however, on which version of the Eccles model is used, and what is emphasised. In earlier versions of the model (Eccles 1983, 1994) the different components of the model (cultural milieu, students' perceptions of socialisers' attitudes, etc.) are linked with one-way arrows that end in achievement related behaviour and choices. These versions suggest that the choice of study is an activity that unfolds through a number of steps to reach the final decision point. Consequently, the model presumes that there is a place in time where the choice is made, and hence that it is possible to map which information, persons, teaching, subjects etc. contributed to the particular choice. However in the later version of the model (Eccles and Wigfield 2002) which is the one applied in this book (see Chap. 2) this one-way assumption of choice is modified by a dashed arrow leading from the achievement related choices and behaviour back to a box at the left-hand side of the model referring to 'previous achievement-related experiences'. Hence, the latter version suggest that the process of choice is in fact not a one-time decision, but rather a process where experiences continuously feed into further decisions. Also, the concept of identity has a stronger focus in the later versions of the model (through attainment value).

Narrative psychology emphasises this relation between experiences and decisions and that this is a continuous process. It approaches choice as ongoing identity-work, and hence the way information, persons, teaching, subjects affect the choice depends on how the student makes meaning of it – and that this changes over time. The process of choosing involves an ongoing negotiation of who you are (present), who you wish to become (prospective) and how it suits your notion of self (retrospective).

The results that can be reached through narrative psychology contribute with new knowledge on how students' choices are an ongoing process of identity-work being negotiated throughout time. In this book we show how the results reached through this framework supplement findings from previous research, adding a more complex understanding of how students choose to enter a STEM programme or not, and whether to stay or not (see Chap. 21). The findings emphasise a need for caution concerning methodologies in future studies. Firstly, care needs to be taken when interpreting student responses that they always wanted to study a particular subject. Narrative psychology illustrates how we need to contextualize this always in terms of the students current position, context and meaning making. The framework allows for new interpretation of how the students' descriptions of always wanting to study a certain study programme change over time, including after students begin higher education. Secondly, the results reached by applying a narrative psychological framework in this book add to the discussion on previous research which states that students' choices are constructed already in childhood (Archer et al. 2010; Head 1997). While it is by no means unlikely that many students eventually entering a STEM higher education programme have acquired an interest and inclination towards STEM subjects in the early years of schooling, the point that choices are made over time and involve construction of narratives draws attention to the need for students to be able to continue to construct a viable, recognisable, and convincing narrative through upper-secondary school and beyond. Further, it opens an opportunity to offer students who may not have had access to narratives containing STEM as a possible path of study and career an opportunity to construct such a narrative during upper-secondary school. This, however, presupposes that students' experiences with STEM subjects during upper-secondary school in fact provide the student with material for such a narrative – both in terms of present interest, future perspectives, and the reconstruction of past experiences.

In this chapter we have argued that the choice of study is a continuing process occurring across institutional and cultural contexts. More broadly it could prove fruitful to approach the choice of higher education as a transition process. Previously, transition from upper-secondary school to higher education has been understood as a linear progression from one institution to another (Ecclestone 2007). But new research suggests approaching such transitions in terms of 'transitions of identities'; a process through which students' ongoing work on their identities to become 'somebody' fits into what they recognise as institutionally and culturally accepted pathways and results in a sense of belonging (Ecclestone et al. 2010; Holmegaard et al. 2014a).

As we have shown in this chapter, narrative psychology can be used as a theoretical approach to studying young people's choice of higher education. Future studies could benefit from combining narrative psychology with an approach that aims at understanding how the narratives are situated within institutional, political, social and cultural discourses. In such a framework narrative psychology would be a methodology providing the tools for collecting and interpreting data, and the analysis could draw on a wider array of theoretical approaches to situate the students' identity-work into a larger cultural context.

References

Andrews, M., Squire, C., & Tamboukou, M. (2008). *Doing narrative research*. London: Sage Publications.

Archer, L., DeWitt, J., Osborne, J., Dillon, J., Willis, B., & Wong, B. (2010). 'Doing' science versus 'being' a scientist: Examining 10/11 year old schoolchildren's constructions of science through the lens of identity. *Science Education, 94*(4), 617–639.

Bergerson, A. A. (2010). College choice and access to college: Moving policy, research and practice to the 21st century. *ASHE Higher Education Report, 35*, 1–141.

Bruner, J. (1990). *Acts of meaning*. Cambridge, MA: Harvard University Press.

Bruner, J. (2004). Life as narrative. *Social Research, 71*(3), 691–710.

Brunila, K., Kurki, T., Lahelma, E., Lehtonen, J., Mietola, R., & Palmu, T. (2011). Multiple transitions: Educational policies and young people's post-compulsory choices. *Scandinavian Journal of Educational Research, 55*(3), 307–324.

Crossley, M. L. (2000). Narrative psychology, trauma and the study of self/identity. *Theory & Psychology, 10*(4), 527–546.

Czarniawska, B. (2004). *Narratives in social science research*. London: SAGE Publications.

Eccles, J. S. (1983). Expectancies, values, and academic behaviors. In J. T. Spence (Ed.), *Achievement and achievement motives. Psychological and sociological approaches* (pp. 75–146). San Francisco: W.H. Freeman and Company.

Eccles, J. S. (1994). Understanding women's educational and occupational choices. *Psychology of Women Quarterly, 18*(4), 585–609.

Eccles, J. S., & Wigfield, A. (2002). Motivational beliefs, values, and goals. *Annual Review of Psychology, 53*(1), 109–132.

Ecclestone, K. (2007). *Lost and found in transition: the implications of 'identity', 'agency' and 'structure' for educational goals and practices*. Paper presented at the Researching transition in lifelong learning, University of Stirling, Stirling.

Ecclestone, K., Biesta, G., & Hughes, M. (Eds.). (2010). *Transitions and learning through the lifecourse*. London/New York: Routledge.

Ezzy, D. (1998). Theorizing narrative identity. *Sociological Quarterly, 39*(2), 239–252.

Head, J. (1997). *Working with adolescents: Constructing identity*. Hong Kong: The Falmer Press.

Holland, D., Lachicotte, W., Jr., Debra, S., & Carole, C. (1998). *Identity and agency in cultural worlds*. Cambridge, MA/London: Harvard University Press.

Holmegaard, H. T. (2013). *Students' narratives, negotiations, and choices. A longitudinal study of Danish students' transition process into higher education science, engineering and mathematics*. Department of Science Education, IND's skriftserie no 26/2012, Copenhagen, ISNN: 1602-2149.

Holmegaard, H., Madsen, L., & Ulriksen, L. (2014a). A journey of negotiation and belonging: understanding students' transitions to science and engineering in higher education. *Cultural Studies of Science Education, 9*(3), 755–786. doi: 10.1007/s11422-013-9542-3.

Holmegaard, H. T., Ulriksen, L. M., & Madsen, L. M. (2014b). The process of choosing what to study: A longitudinal study of upper secondary students' identity work when choosing higher education. *Scandinavian Journal of Educational Research, 58*(1), 21–40.

Hsu, P.-L., Roth, W.-M., Marshall, A., & Guenette, F. (2009). To be or not to be? Discursive resources for (dis-)identifying with science-related careers. *Journal of Research in Science Teaching, 46*(10), 1114–1136.

Illeris, K., Katznelson, N., Simonsen, B., & Ulriksen, L. (2002). *Ungdom, identitet og uddannelse*. Frederiksberg: Center for Ungdomsforskning, Roskilde Universitetsforlag.

McAdams, D. (2005). A psychologist without a country or living two lives in the same story. In G. Yancy & S. Hadley (Eds.), *Narrative identities*. London: Jessica Kingsley Publishers.

Paulsen, M. B. (1990). In A. H. Education (Ed.), *College choice: Understanding student enrollment behavior* (Vol. 6). Washington, DC: School of Education and Human Development, The George Washington University.

Polkinghorne, D. E. (1988). *Narrative knowing and the human sciences*. Albany: State University of New Yorke Press.

Potter, J., & Wetherell, M. (1987). *Discourse and social psychology: Beyond attitudes and behaviour*. London/Newbury Park: Sage.

Quinn, J. (2010). Rethinking 'failed transitions' to higher education. In K. Ecclestone, G. Biesta, & M. Hughes (Eds.), *Transitions and learning through the lifecourse*. London/New York: Routledge.

Reay, D., David, M. E., & Ball, S. J. (Eds.). (2005). *Degrees of choice: Class, race, gender and higher education*. Sterling: Trentham Books.

Ricoeur, P. (1976). *Interpretation theory: Discourse and the surplus of meaning*. Fort Worth: Texas Christian University Press.

Ricoeur, P. (1990). *Time and narrative*. London: The University of Chicago Press.

Sarbin, T. R. (1986). *Narrative psychology: The storied nature of human conduct*. New York: Praeger Publishers/Greenwood Publishing Group.

Schreiner, C. (2006). *Exploring a ROSE-garden. Norwegian youth's orientations towards science: Seen as signs of late modern identities* (No. 58 Doctoral Thesis), University of Oslo, Oslo.

Schreiner, C., & Sjøberg, S. (2007). Science education and youth's identity construction – Two incompatible projects? In D. J. Corrigan & R. Gunstone (Eds.), *The re-emergence of values in the science*. Rotterdam: Sense Publishers.

Smith, B., & Sparkes, A. C. (2008). Contrasting perspectives on narrating selves and identities: An invitation to dialogue. *Qualitative Research in Psychology, 8*(5).

Ulriksen, L., Holmegaard, H. T., & Madsen, L. M. (2013). Weaving a bridge of sense: Students' narrative constructions as a lens for understanding students' coping with the gap between expectancies and experiences when entering higher education. *European Educational Research Journal, 12*(3), 310–319.

Chapter 4
Gender, STEM Studies and Educational Choices. Insights from Feminist Perspectives

Alessandra Allegrini

The "Gender Gap In" and the "Gender Dimension Of" STEM

Women continue to be the largest under-represented group in STEM, especially in physics, mathematics, computer science and engineering (see Introduction). This gender gap is visible in every European country, from university choices to later career paths, although girls' disengagement from science already starts in earlier school years. From a feminist perspective, a proper account of this issue would require a wider formulation, that is understanding the gender gap in science, or gender-imbalance, as a double gap, not only affecting women in male-dominated sciences, but also regarding men in increasingly feminised sciences, notably life and health sciences. Such a double-edged approach – on both women and men in science – would be required from a feminist perspective, first of all by considering the meaning of 'gender' as a category of thought and knowledge. Over the last decades, and up to the present, feminist scholars have not shared a single definition of 'gender'; they have instead embraced different conceptualizations of this term, or emphasised specific features of it. Nevertheless they would agree on a basic consideration: gender is not synonymous with being a woman.

Gender

This category refers to both men and women, since it is assumed that both take part in human life, although the relationship between them is neither complementary

A. Allegrini (✉)
Observa Science in Society, Viale A. Fusinieri, 65, 36100 Vicenza, Italy
e-mail: allegrini_a@yahoo.it

E.K. Henriksen et al. (eds.), *Understanding Student Participation and Choice in Science and Technology Education*, DOI 10.1007/978-94-007-7793-4_4

nor symmetrical. It underlines an asymmetry that assigns women a subordinate position to men. Feminist scholars identify the classical age in Greece as the inaugural and most crucial historical-philosophical moment in Western culture, in which this fundamental asymmetry was first formulated, within a theoretical framework where man was viewed as the only one subject, defined as neutral and universal, abstract and transcendental, that removed women, as concrete human beings, and femaleness, as a symbolic principle, from thought and knowledge production (Irigaray 1974, 1977; Cavarero 1987a, b, 1990; Fraisse 1991, 1996; Collìn 1992). Starting from Greek philosophy, women have been reduced to mere body, material, nature, because of their biology, and more specifically their reproductive difference. It is worth underlining that in this theoretical discourse there is a male human being behind what is considered 'neutral' and 'abstract'. Still nowadays the term 'mankind' continues to be a universal representation of both men and women, although it is effectively only male-sexed (Cavarero 1987a, b, 2002).

In order to overcome this naturalised order of thought, feminist scholars, originally in English speaking countries, have used 'gender' as a new conceptual tool. This approach is aimed at pointing out that the asymmetric relationship between men and women takes place within the translation process of biological sexual difference into sociocultural gender system, so that 'gender' is not synonymous with 'sex'. Inequalities among men and women are therefore the sociocultural product of the hierarchical symbolic value attributed to biological sexual difference.

For a long time gender has been largely used in order to distinguish the sociocultural construction of the roles played by, or assigned to, men and women (the gender system) from female and male sex as a biological attribute. This distinction, between biological sex and social-cultural gender, was officially introduced in the mid-1970s by the feminist anthropologist Gayle Rubin with her writing *The Traffic in Women* (1975), and was previously conceptualized by the existentialist philosopher Simone De Beauvoir. In *The Second Sex* (1949) De Beauvoir eminently summed up this distinction with her statement "One is not born a woman, but becomes one" (Moi 1990; Gatens 1991). According to Rubin, the "sex-gender system" can be defined as "the set of arrangements by which a society transforms biological sexuality into products of human activity, and in which these transformed sexual needs are satisfied" (Rubin 1975, p. 159). In this framework the term 'gender' is therefore used to underline the sexed processes through which human beings are perceived and represented, and the way society and culture are structured through sexed relationships.

Starting from the mid-1980s, 'gender' has become a more complex category, as a result of two significant processes within feminist theories. First, the appearance of new fields of knowledge, such as post-colonial studies initiated by feminist women of colour in the United States, began to undermine the predominance of Anglo-American and European feminism, which was accused of re-producing a monolithic discourse centred on Western woman as the normative female subject for all other women (Davis 1981; Hooks 1984; Spelman 1988; Collins 2000; Spivak 1990). In this process, the emergence of other different subjects in a globalised world demanded different points of view to be legitimised.

This necessitated a more complex approach to gender, taking into account other categories conceptualised as multiple and intersected axes of difference that construct subjectivity and identity, and producing a more complex set of inequalities besides gender, including ethnicity and class. In order to underline these interconnected aspects, feminist scholars have used the key-concept "intersectionality", firstly introduced in 1989 by Kimberlé Crenshaw, and later, in the 1990s, reinforced by feminist sociologist Patricia Hill Collins. According to Collins, cultural patterns of oppression are not only interrelated but are bound together and influenced by the intersectional systems of society, such as race, gender, class, and ethnicity (Collins 2000, p. 42). Over the last decade, intersectionality has become a research methodology in feminist studies, aimed at inquiring into the relationships among multiple dimensions and modalities of social relationships and subject formations (McCall 2005). In this methodological approach, categories such as gender, race, class, among other axes of social differentiations and identity formation, interact on multiple levels, contributing to systematic social inequity.

A second, but no less important, process in feminist theories which led to a further reformulation of gender is the relevance increasingly ascribed to new insights coming from other disciplines and scholarships, which are not intrinsically linked to feminist studies. Semiotics, linguistic studies, cultural studies, de-constructionist thought, Foucauldian studies on power and sexuality, Lacanian approaches on sexual identity and language, and several other approaches emerging from post-structuralist French philosophy, strongly redefined the analysis and the theorisation of gender. Since the early 1990s, these new insights tend to undermine the sharp distinction between biological sex and sociocultural gender, in turn affecting the idea of gender as a stable, linear and fixed category. According to Judith Butler, whose conceptualisation of gender is an unavoidable reference in contemporary feminist theory, 'sexuality', 'sex' and 'gender' are terms which cannot be sharply distinguished, since sexed bodies cannot signify without gender. The pre-existence of sex within discursive and cultural patterns is a clear phenomenon, largely caused by the idea that bodies (sex) have an essentially irreducible substance, something which is essentially outside and beyond human thought, while minds, human relationships and language (gender) do not (Butler 1993). Both gender and sex are instead constructed, and more precisely they are "performed" within "regulative discourses" producing stable and coherent genders as well as sexed bodies, which also assure stable and coherent gender identities (Butler 1990). In this framework, 'sexed bodies' become an emerging notion having a pivotal implication for gender, whereas the body is not meant as pure nature (sex), but it is most of all culture, or rather "a point of overlapping between the physical, the symbolic, and the sociological" (Braidotti 1994, p. 4). 'Gender' itself might be better regarded as "a corporeal style" (Butler 1990, p. 140). As I will describe in the following pages, Butler's notion of gender performativity and, more broadly speaking, gender meant as a performative practice, has significant value in understanding the gender dimension of choice processes, including educational choices.

This new approach to gender emphasises the contingency of everyone's location in the world, the body being "not an essence, nor a form of anatomical destiny, but rather one's primary location in the world, one's primary situation in reality" (Braidotti 1991, p. 219). In order to further stress the embodied material nature of any human experience, feminist scholars use a number of concepts. Among these, "situatedness" – or "locating", "positioning", "situating" – together with "embodiment", are key terms specifically employed in order to contrast the idea of an abstract, neutral and universal human experience, that is instead meant as always concrete, different and partial. These notions also underline the epistemological and methodological dimensions of a feminist approach to knowledge, and find their most notable synthesis in the concept of "situated knowledges" offered by Donna Haraway who argues for "a politics and epistemologies of locations, positioning and situating, where partiality and not universality is the condition of being heard to make rational knowledge claims [...]" (Haraway 1988, p. 589).

These feminist conceptualisations of 'gender' are only a selection and there are several others which stress different meanings or connotations of the same concept. Building on feminist thinker Joan Scott (1986), four distinguishing aspects of gender are summarised here in order to outline a pragmatic overview of the different ways this concept might be applied in empirical research.

First, gender is a notion that is fundamentally about the power relations articulating the sociocultural order, in that it produces inequalities between women and men. Moreover, the structural interconnection between gender, ethnicity, race and class, understood as multiple axes of social differentiations, are responsible for a more systematic and complex inequality.

Second, gender has essential implications for subjectivity and identity processes: it is indeed a partial and embodied location attached to the different situated and embodied experiences of human life. From this point of view, ethnicity, race, class and all other significant differences are not uniquely social differentiations producing intersected sets of inequalities, but also subjective sources describing different lived experiences.

Third, symbolic and imagery aspects, such as stereotypical ideas and representations about what is considered 'male' and 'female' in the human imagery, also play a central role in articulating a 'gender discourse', that in turn might enforce the sociocultural gender order. Once again, different embodied locations, due to gender and other differences, might also articulate alternative symbolic orders and representations.

Finally, gender has an important epistemological and methodological dimension. It is a rather recent analytical category which offers a different way of looking at and perceiving reality, by making visible in all areas of human and social experience the relationship between masculinity and femininity, historically and culturally situated, as well as issues of difference and diversity.

Gender is therefore a comprehensive approach that might help to reformulate traditional concepts and analytical tools in different research fields, introducing a new perspective in the interpretations of social and cultural phenomena. Paying systematic attention to gender can provide new insights about women and other traditionally marginalised subjects. Furthermore it makes it possible to formulate a more articulated set of questions, issues and methodologies within research fields.

The Gender Gap in STEM and First-Wave Feminism

Going back to the gender and STEM question, from a feminist perspective this issue would require an approach which simultaneously centres upon women and men. This is mainly because the meaning of 'gender' basically concerns both women and men. Furthermore, this dual approach might be evaluated as more consistent with significant macro-changes which have occurred in the relationship between men and women in science over the last decades.

Women officially entered the institutions of knowledge in the 1960s. Since that time their overall presence has progressively increased in every public and productive field, including paid employment and higher education. It has been estimated that in the United States women made up 43 % of the work force by 1980, compared to 29 % in 1950 (Waite 1981). In more recent years, in almost all European countries, more girls and young women than their male peers have enrolled at university, reaching 55 % of enrolments and 59 % of all graduates (European Commission 2009). This macro-process, briefly summarised here, is the result of multiple sociocultural phenomena, alongside the feminist movement. The latter progressively raised an awareness of gender power dynamics in society and largely undermined a rigid distinction among female and male roles that for a long time in Western history informed the structure of society, on the basis of the so called "male bread winner and female care giver" model (Ferber and Nelson 2003; Picchio 2003).

Despite this increasing feminization of the public and productive spheres, at least until the 1970s there was a serious problem of female access to science, so that the numerical presence of women studying science and engaged in scientific careers was a central issue in the political agenda of the feminist movement, both in Europe and in the United States. Epistemologist Sandra Harding described this "quantitative" issue in science as "the woman question in science", which specifically refers to women as being a minority within sciences (Harding 1986). Why aren't there more women scientists? This was the main question in the 1970s, when articles, conferences and discussions about the woman question in science were sparked by the recognition of women's fundamental absence in the sciences, both as science practitioners and as accepted subjects of study (Zita 1988).

In order to enlighten the political dimension of the "woman question in science", feminist scholars describe it as "equality feminism", or "liberal feminism", while the chronological appearance of it is marked as "First-wave Feminism". Actually, the term 'First-wave Feminism' is largely used with reference to the first women's movement for judicial equal rights which occurred between the end of the 1800s and 1920 in Europe and in the United States. Its conceptual background can be traced back to "First feminism" (1791–1834), just after the French Revolution. At this time, women's right to vote was the crucial woman question, visibly pursued by the suffragettes' movement which was fighting for universal suffrage since the second half of the nineteenth century. First-wave Feminism was grounded on an "equality paradigm", where it was assumed that women's emancipation could be gained through the removal of differences, i.e. obstacles that result in a subordinate condition to men.

Although First-wave Feminism properly defines this early historical phase in feminism, feminist literature often refers to "First-wave Feminism", "equality feminism", "liberal feminism" as a longer period extending up to the first half of the 1970s, in order to stress the continuity of the same political paradigm characterising the re-emergence of the women's movement since the end of the 1960s, alongside the civil rights movements. Some scholars underline its relevance also in contemporary science issues, so that, still nowadays, liberal feminist science studies are mainly concerned with conceptual and practical strategies to be applied in order to bring women and minorities into science, focusing on "a series of external factors that make equal opportunity in science unequal across race, class and gender" (Barton 1998, p. 3).

It could be argued that today the gender gap in STEM, meant as a women's gap in science, is no longer the most relevant question. Rather, the main issue is the persistent gender heterogeneity occurring in STEM studies and later careers. Men and women are indeed both quantitatively present in sciences, but differently enrolled in them. About 58 % of all bachelor's, master's and doctorates in biology are awarded to women in the United States (Drew 2011). More than 50 % of the total number of PhD students in medicine and biological sciences are women, while they are in a minority in physics, mathematics and statistics, computer sciences and engineering (European Commission 2009). This gender heterogeneity follows a traditional gender role division, structuring a gendered order among male-dominated sciences and female-dominated sciences, that might be considered the most topical issue distinguishing the gender gap in STEM studies and careers at present time.

As I will discuss in the last part of the chapter, the political dimension of this issue could still be approached as an "equality" question, not only aimed at addressing issues about women in male-dominated scientific fields, but also considering men who study and work in feminised sciences.

The Gender Dimension of STEM and Second and Third-Wave Feminism

Conceptual and political approaches which merely focus on the gender gap in science are only a part of a broader spectrum of feminist perspectives on science. These explore the systematic intersection of gender and science. Here science itself is understood as a culturally and historically situated human practice of knowledge and thought, which is basically informed by a gender dimension, both at a material level – its gendered practices and institutional manners – and a symbolic level – its gendered representations and discursive orders.

As is frequently underlined in European documents: "the gender issue is not only a matter of quantity and the necessity to balance the number of men and women who gain access to science, participate in knowledge production and occupy senior positions. A gender perspective in science implies a critical viewpoint about existing epistemologies, the proposals for innovative cultural dimensions and the enlarging of intellectual fields that can broaden the functioning of capabilities in knowledge societies. Gender is not purely a matter of 'sameness' between men and women but a criticism against a whole system of values and limited views governing and ordering science and/in society" (MASIS Expert Group 2009, pp. 44–45).

Starting from the 1980s the "woman question in science" has been followed by "the science question in feminism", within a new phase in feminism called "Second-wave Feminism". Sandra Harding described the shift from the "woman question in science" to "the science question in feminism" as a fundamental change in feminist science studies from a numerical, quantitative question, mainly concerned with the low presence of women in science in the 1970s, to a qualitative, ontological and epistemological question, which stems from situating women as different subjects of knowing at the centre of scientific discourse (Harding 1986).

Through the denunciation of the historical and conceptual omission of women as real, political and cognitive subjects, from scientific research to epistemology, feminist science scholars of the Second-wave aim to show that science and the philosophy of science are not neutral, impartial and universal, as they are supposed to be in the so-called "malestream" philosophy of science, from positivist to post-positivist traditions. By means of adopting inter-disciplinary and multi-disciplinary methods of analysis and elaboration, moving across disciplinary borders between science, history, epistemology and psychology, they all share the assumption that science is not free from the influence of cultural values, social constructions, economical and political implications, and most of all, from all the features that shape a gender imagery and practice. A great deal of feminist writings, notably *The man of Reason. 'Male' and 'Female' in Western Philosophy* (Lloyd 1984), *Science and Gender: A Critique of Biology and its Theories on Women* (Bleier 1984), *Reflections on Gender and Science* (Keller 1985) and *The Mind Has No Sex? Women at the Origin of Modern Science* (Schiebinger 1989) show that gender has influenced knowledge production at multiple levels throughout the history of science: scientific theories, experimental choices, the language used to communicate science and the institutional forms in which science has been organised.

Being more committed to increasing the number of women in science, the "woman question in science" is basically informed by an idea of equality and it does not discuss science as a gendered practice and way of thinking. On the other hand, the "feminist question in science" is grounded in a political paradigm of radical difference. Its main aim is not to add women to the different fields of science, but rather to unveil the gender partiality of the scientific system, assuming this partiality as the starting point to orient science towards searching for theoretical and practical alternatives.

Since the second-half of the 1970s, and increasingly during the 1980s, a considerable amount of feminist oriented research in science has appeared, in some cases collecting new empirical evidence by adopting a gender perspective which is able to produce significant changes in the epistemological paradigms of some sciences (Schiebinger 1999), as happened in the well-known case of primatology and human evolution theory (Fedigan 1982; Slocum 1975; Tanner and Zhilman 1976; Dahlberg 1981; Small 1984). During the same period, feminist theoreticians have developed epistemological approaches, mainly by analysing the systematic connection between scientific issues and the social, cultural and historical contexts in which they originate, thus rejecting and challenging positivist notions of objectivity in science, and gender neutrality of scientific knowledge. Feminist rethinking of objectivity embedded in personal, social, cultural values are worth restating: objectivity as "empathy and sympathy" (Bordo 1987; Keller 1983), objectivity as "strong objectivity" (Harding 1991, 1993), objectivity as "situated knowledge" (Haraway 1988, 1991), objectivity as "inter-subjectivity" (Longino 1990, 1996).

Later in the 1990s, new insights re-articulating gender as a more complex analytical tool have also emerged in feminist science studies. These new perspectives are usually located within Third-wave Feminism, largely influenced by post-colonial studies, cultural studies, de-constructionist thought, and several approaches which have arisen from post-structuralist French philosophy. In this phase, key-concepts such as "situated knowledges" and "intersectionality", already described above, are seen as most appropriate feminist methodologies that are able to provide a systematic understanding of the gender *and* science interrelation, therefore offering a comprehensive account of the positional or situated nature of scientific knowledge (Mayberry et al. 2001, p. 11).

An example of these recent feminist approaches to science might be detected in Sandra Harding's latest writings, where she explores the intersection of feminist and postcolonial science and technology studies, arguing for a "borderlands epistemology" which takes into account the distinctive understandings of nature generated by different cultures (Harding 1998, 2008). Another recent feminist epistemological approach is Karen Barad's "agential realism" (Barad 2003, 2007). Drawing on an articulated rethinking of the epistemology of quantum physicist Niels Bohr, along with the feminist notions of "politics of location" and "situated knowledges", Foucauldian and constructionist approaches of science as discourse, and Judith Butler's theory of sex/gender performativity, Barad argues that the embodied knower is always and inevitably entangled in the world she/he analyses. She combines constructivist with conventional scientific approaches to science, thus simultaneously insisting on constructedness and objectivity. This is to demonstrate that a scientist can construct a provisional "cut" between knower and known which allows her/him to give a partially objective, reliable, and ethically committed account of the world "out there"/"in here". Her notion of "ethico-onto-epistemology" aims to show how ontology, epistemology and an ethics of knowing cannot be separated.

As "the gender gap in STEM", outlined above, can be approached from a political perspective with reference to the concept of "equality feminism", "the gender dimension of STEM" might also require political measures which belong to Second and Third Waves of Feminisms, and are thus oriented towards a radical idea of difference, multiple diversity issues in feminist accounts of scientific knowledge, and different ways of knowing. These issues will be briefly discussed in the last part of this chapter.

Gender and STEM Educational Choices

Moving to STEM educational choices, two feminist theorisations of gender, and of gender and science, might be further developed in order to approach STEM educational choices, outlining an interpretative framework of the gender gap in STEM choices. These two approaches stem from distinct feminist scholarships. Nevertheless it is worth linking them to each other in order to configure the gender gap as an ongoing "gender polarisation" among "male sciences" and "female sciences", and highlight the different aspects which shape it.

The first theorisation draws on the feminist critique of identity and gender identity, and sees gender as a performative practice structuring the relationship between masculinity and femininity, rather than an interior essence which belongs to, and defines, men and women. The second approach draws from feminist science studies and focuses on the symbolic representation of male-gendered science as an overall factor affecting the perception of STEM studies.

Gender as a Performative Practice

Choice is a complex and dynamic process, influenced by multiple factors and dimensions, which might be analysed from several theoretical perspectives. Drawing from different approaches, Chaps. 2 and 3 in this book highlight a range of elements which are crucial in the choice process. Among these, identity-building dynamics, already beginning in the early years of human life, are pivotal, since they progressively becomes embedded in, and bound up with, choice process. University choices represent a significant step in this ongoing process: at this stage "students struggle to make sense of who they are and who to become", within a "meaning making process" in which both the cultural context and the perception of the self play a central and intertwined role (Chap. 3).

Feminist theories do not offer systematic and comprehensive explanations for choice mechanisms in general, and more specifically choice processes in STEM studies. Neither do they develop interpretative models to understanding the choice process. Nevertheless, identity and gender identity are central issues in feminism. Thus, since identity is a crucial factor in the choice process, gender identity can be considered as a having an impact on it. Among other noteworthy feminist works, Judith Butler's theorisation of gender (1990, 1993) offers significant insights on the way identity, and more specifically gender identity, might be viewed from a feminist perspective. Building on basic references to Foucault's work of the 1970s, namely *The History of Sexuality* (1978), and drawing on key concepts employed in post-structuralist French philosophy, Butler argues that identity is fundamentally a cultural and discursive process, made up by "regulatory ideals" which provide idealised norms people are expected to live up to (Butler 1993, p. 1).

'Gender', and what is considered 'male'/'female', are also part of these idealised norms. They are not biological facts, but rather categories that everyone creates and recites through performance, that is performative practices. As a result, gender is not something "one has", or "one is". It is a reiterated act structuring the relationship between masculinity and femininity as an apparently stable and coherent polarity, that is considered to be the "true gender". It is a narrative that becomes accountable because it is supported by "the tacit collective agreement to perform, produce, and sustain discrete and polar genders" (Butler 1990, p. 179). In Butler's words "intelligible genders are those which in some sense institute and maintain relations of coherence and continuity among sex, gender, sexual practice and desire" (Butler 1990, p. 17).

Concepts of 'male', 'female', and 'gender', are meant as historically and culturally situated: universal categories of gender, man and woman hardly exist. Butler maintains that regulatory schemas are not timeless structures, but historically revisable criteria of intelligibility which produce and vanquish bodies that matter (Butler 1993, p. 14). Other genders might even be possible, as she argues in more recent works, in which she uses gender as a tool, a strategy for de-constructing gender polar identities, and the Self: "undoing gender" would allow people to overcome sharp categorisations of the human (Butler 2004).

Gender meant as a performative practice is a shared perspective across several feminist and gender studies, especially those which are more oriented to sociological accounts of gender and its functioning. Sociologists Candace West and Don Zimmerman define "doing gender" as a routine, a social activity that is always interactive and situated in everyday life contexts: "in one sense, of course, it is individuals who "do" gender. But it is a situated doing, carried out in the virtual or real presence of others who are presumed to be oriented to its production. Rather than a property of individuals, we conceive gender as an emergent feature of social situations" (West and Zimmerman 1987, p. 126). Apart from pointing out that gender is a performative social practice, these approaches underline the relational structure of gender, which is always positioned in between the concrete or virtual presence of different individuals.

Among others feminist scholars, Linda Alcoff (1988, 2006), specifically uses the concept of "positioning", already mentioned above, in order to emphasize gender identity as discursively constructed, not definitive and unitary, but influenced by relational interactions. She also reminds us that, besides gender, class and race among other features constituting identities, are markers of relational positions, instead of essentially given qualities. Gender is not a unique, stable and fixed category; what is male and female, and polar genders, can change over history and culture. However, other axes of social differentiations, such as age, ethnicity and sexuality, can interact with each other and produce identity. From this perspective, gender might be understood differently, depending on these social differentiations (Butler 1993, p. 116).

Butler's notion of gender performativity and, more broadly speaking, the concept of gender as a performative and relational practice, might have relevant implications for understanding the gender dimension of choice processes. At a general level, by means of conceiving gender as performative and relational, it is indeed possible to configure the choice process as a macro-tendency to perform coherent gender polarities, such as "male sciences" and "female sciences". In this framework, such a tendency is not understood as a "natural" inclination of boys and girls due to essential qualities belonging to their gender identities. It is rather the reiteration of idealised norms which defines what is male and what is female and thereby produce gender polarities.

Within a reiterated intention to "do gender" in everyday life contexts, including educational contexts, "intelligible genders" might play a major role in orienting students' educational choices. Students might indeed choose, or not, to go into a certain subject or to attend a university course, which they regard as appropriate to what the present cultural and social discourse supports as "true gender". As a result, this might confirm or neglect what they perceive to be the right gender identity to perform. Choosing a course which is potentially at odds with a "normal" gender identity might be perceived as a threat to that identity. Students' choice can therefore be interpreted as the sign of what culture accepts and enforces as "intelligible gender" in relation to a specific field of study.

On the other hand, it is important to bear in mind that 'male', 'female' and 'gender' are not timeless structures, so that, as society changes, the cultural meaning of what is male and female might also change. The ways in which boys and girls, women and men, "undo gender" might also be explored and documented, therefore highlighting both continuity and discontinuity factors in the reproduction of gender discourse. Furthermore, these categories might have different meanings to be unveiled in relation to different embodied human experiences such as class, ethnicity and age.

At an analytical level, these concepts might be applied in empirical research on choice processes, not only because they may outline a theoretical framework to orient analysis and interpretation of empirical outcomes, both quantitative and qualitative. From a methodological point of view, they are specifically consistent with a qualitative analysis of narratives, since all these notions underline the importance of language in the cultural reproduction of gender. As Butler maintains, gender performativity is a cultural and discursive mode, not only constructed by culture, but also informed and mediated by language. Performative acts are therefore statements which also produce what they say. Her classic example of that is the midwife cry "it's a girl", which is not merely a reflection of a biological given, but a performative act, binding a gender onto the body (Butler 1993).

The Gendered Representation of Science

Doing, practising and performing gender gains a further and specific meaning to be stressed with reference to science course choices. As already mentioned, this concept allows us to understand the present macro-tendency to reproduce "gender polarities" in science choices, among "male sciences" and "female sciences". Science itself is indeed part of a gender discourse. It might be invested with gendered attributes that can enforce or neglect a coherent gender identity and stable gender polarities, in turn affecting the choice process in science that is deeply bound with gender identity dynamics.

A large part of feminist literature on science and epistemology points out that science is largely represented and perceived as an overall male-dominated field. Already in the early 1980s, Evelyn Fox Keller was among the first feminist scientists who described the multiple manners thorough which science might be considered a male-gendered discourse, a site from which women have been excluded, because they were regarded as being unsuitable for science. Moving from three disciplinary perspectives – psychoanalytic, historic and philosophic – in her most well-known book, *Reflections on Gender and Science* (Keller 1985), Keller traced the fundamental gender stereotypes which produced the long lasting myth of a male-gendered science, what she describes as the system of beliefs in which science acquires a gender, or the "genderization of science" (also in Keller 1978, p. 413). According to Keller, two main stereotypes originated and created the representation of a male-gendered science throughout Western history of thought. The first stereotype is that objectivity and rationality coincide with masculinity, while subjectivity, irrationality and emotionality are female attributes. The second stereotype concerns philosophy and science, meant as neutral human activities, free from personal and affective connotations, in opposition to artistic intuitions and creative practices.

Besides Keller's elaborations, other feminist works mentioned above, such as *The man of Reason. 'Male' and 'Female' in Western Philosophy* (Lloyd 1984), *The Mind Has No sex? Women at the Origin of Modern Science* (Schiebinger 1989) have inquired deeply into the historical steps thorough which modern science, already in its earlier conceptual origins in classic philosophy, built its own epistemology on the exclusion and devaluation of women and femininity. This gendered idea about science has been historically and culturally re-produced over time, thus becoming the long-lasting historical and cultural heritage which might still affect the symbolic representation of science, and of different sciences.

A classic example in feminist science literature attesting to this tendency in the making of modern science is offered by the analysis of Francis Bacon's terminology. His seventeenth-century writings strongly support the idea of strength and power associated with masculinity and science, contributing to create the idea of an inherently masculine character of science through the metaphor of the conquest of male mind over female nature (Merchant 1979; Keller 1985; Harding 1986).

Other feminist and gender scholars use the concept of "hegemonic masculinity" in order to understand the way science might be largely perceived and represented as a male-dominated field, as being crucial in the construction of gender and the reproduction of male power in science. According to Connell, "hegemonic masculinity" is "the configuration of gender practice which embodies the currently accepted answer to the problem of the legitimacy of patriarchy, which guarantees (or is taken to guarantee) the dominant position of men and the subordination of women" (Connell 1995, p. 77).

The gendered representation of science might have significant effects on the gendered order in science course choices. On the one hand, one could argue that the association between science and masculinity – also hegemonic masculinity – no longer holds, by considering what has been already underlined in this chapter: compared to before, the proportion of women enrolled in the life and health sciences has substantially increased. On the other hand, some sciences undoubtedly continue to be male-dominated fields, with a higher number of men studying and working in them, so that the persistence of the gender gap in STEM studies and educational choices might be considered the result of this long lasting idea that science – and more specifically some sciences – is male-gendered. Nowadays it is still the case that the idea of a male-gendered science might be transferred through polarised gender associations, which connect initiative, strength, rationality and autonomy with men, and other traditionally female aspects, such as dependence, emotionality, subjectivity with women.

Empirical studies might document this tendency, by exploring the extent to which gendered representations articulate students' imagery of science and of the different sciences, and more specifically by investigating the relationship between gender and science representations. Indeed, significant changes in contemporary society and culture, as for instance those underpinning late modern societies described in Chap. 2, may influence the way younger generations represent and perform gender polarities, thus "undoing gender" in STEM choices. Even the symbolic representation of science might change, especially if one considers the emerging and pervading impact of technology in contemporary science and society. In so far as the most remarkable gender-imbalance is documented in technological fields, such as computer science and engineering, one key issue to be further investigated in empirical research is not only the gendered representation and perception of science, but also the gendered representation of technology and technical aspects within students' imagery.

The "Gender Gap In" and the "Gender Dimension Of" STEM. What Should Be Done from a Feminist Perspective?

The gender gap in STEM studies and careers might be viewed as a gap regarding both women and men. I have argued that this formulation is coherent with the proper meaning of 'gender', that in feminist theories refers to both women and men, especially if one considers the increasing presence of women in health and life sciences as a significant factor which is redefining the relationship between gender and science. I have approached the epistemological dimension of this issue mostly by focusing on understanding the gender gap in STEM choices as a tendency to reproduce – perform and represent – gender polarities in science studies choices. By means of conceiving gender as a performative act, students' choice processes might indeed be understood as an overall intention to perform a stable and coherent gender identity. Within science studies, gender performativity might be enforced by the long-lasting idea that science is male-gendered. I have also underlined that, as long as society and culture change, gender polarities might also be "undone", and new cultural meanings of 'male', 'female' and 'gender' might be further explored and documented, along with new symbolic representations of science.

Moving to the political dimension of this issue, what practical measures should be pursued from a feminist perspective? Until the first-half of the 1970s the "woman question in science" was an eminently woman's issue, exclusively concerned with the representation of women in science, underpinned by an incontestable acknowledgement of the disproportionate female minority in all sciences, thus predominately approached through an "equality feminism" perspective. Now this perspective could be applied to both women and men. Besides the increasing participation of women in male-dominated sciences, one main measure would be therefore to increase the number of men within feminised sciences, in order to gain a broad gender balance in science, and most of all a plural participation in all sciences. In order to remove visible and invisible obstacles and gain free access to different scientific fields, equal opportunities politics are central tools which might be applied in regard to both genders. This political framework could furthermore consider more complex inequalities caused by other interrelated social differentiations: gender, age, class, ethnicity and any other significant social difference contributing to unequal opportunities.

However, it is worth remembering that, nowadays, "equality feminism" holds a specific value when it is discussed with reference to women's career advancement into senior positions: the "glass ceiling" affecting career opportunities for women, the gender pay gap beginning in the early years after graduation, and the "leaky pipeline" phenomenon, that is the gradual loss of potential female presence in the highest levels of education and careers, most of all in scientific and technological professional fields.

Although equity and equal opportunity politics matters to feminism, strategies which are primarily concerned with effecting a higher participation in science, without questioning the structural intersection of gender and science, are not at the core goals of feminist politics, since they put forward an idea of science as gender-neutral in theory and practice. I have underlined that it is on this issue – "the gender dimension of STEM" – that feminist theories are mostly focused, since the historical-conceptual shift from "the woman question in science" to "the science question in feminism" occurred within Second and Third-wave Feminism. In this theoretical framework, other strategies might be pursued, which are mainly grounded on an idea of difference and diversity, both meant as positive sources for addressing plural and relational knowledge politics. The latter are not structured on the systematic de-valuation of what they exclude, but they are enriched by different, partial and situated perspectives. Within feminist science studies, these approaches generally exceed the "equality issue" in science, since their main aim is to systematically rethink the nature of knowledge and science, as well as to take into account different ways of knowing, rather than trying to reach equality in the sciences. Scientific knowledge is fundamentally understood in feminist theory as culturally and socially bound, so that rethinking science would also imply rethinking science curricula in education, mostly by integrating the study of sociocultural contexts in which scientific theories originate and develop, with plural and diverse accounts of scientific knowledge.

References

Alcoff, L. M. (1988). Cultural feminism versus post-structuralism: The identity crisis in feminist theory. *Signs: Journal of Women in Culture and Society, 13*(3), 405–436.

Alcoff, L. M. (2006). *Visible identities: Race, gender and the self*. London: Oxford University Press.

Barad, K. (2003). Posthumanist performativity: Toward an understanding of how matter comes to matter. *Signs: Journal of Women in Culture and Society, 28*(3), 801–831.

Barad, K. (2007). *Meeting the universe halfway: Quantum physics and the entanglement of matter and meaning*. Durham: Duke University Press.

Barton, A. C. (1998). *Feminist science education*. New York/London: Teachers College Press, Columbia University Press.

Bleier, R. (1984). *Science and gender: A critique of biology and its theories on women*. New York: Pergamon Press.

Bordo, S. (1987). *The flight to objectivity. Essays on Cartesianism and culture*. Albany: State University of New York Press.

Braidotti, R. (1991). *Patterns of dissonance. A study of women in contemporary philosophy*. New York: Routledge.

Braidotti, R. (1994). *Nomadic subjects. Embodiment and sexual difference in contemporary feminist theory*. Cambridge: Columbia University Press.

Butler, J. (1990). *Gender trouble: Feminism and the subversion of identity*. New York/London: Routledge.

Butler, J. (1993). *Bodies that matters*. New York: Routledge.

Butler, J. (2004). *Undoing gender*. New York/London: Routledge.

Cavarero, A. (1987a). Per una teoria della differenza sessuale. In Diotima (a cura di), *Il pensiero della differenza sessuale* (pp. 43–79). Milano: La Tartaruga.

Cavarero A. (1987b). L'elaborazione filosofica della differenza sessuale. In M. C. Marcuzzo, & A. Rossi-Doria (a cura di), *La ricerca delle donne. Studi femministi in Italia* (pp. 173–187). Torino: Rosenberg & Sellier. English translation (1991). The need for a sexed thought. In S. Kemp & P. Bono (Eds.), *Italian feminist thought: A reader*. Oxford: Blackwell.

Cavarero, A. (1990). *Nonostante Platone. Figure femminili nella filosofia antica*. Roma: Editori Riuniti. English Translation (1995). *In spite of Plato: A feminist reading of ancient philosophy*. Foreword by Rosi Braidotti. Cambridge: Polity Press.

Cavarero, A., & Restaino, F. (2002). *Le filosofie femministe*. Milano: Bruno Mondadori.

Collìn, F. (1992). La disputa della differenza: la differenza dei sessi e il problema delle donne in filosofia. In G. Fraisse, & M. Perrot (a cura di), *Storia delle donne in occidente. Il Novecento* (pp. 306–343). Roma: Laterza
Collins, H. P. (2000). *Black feminist thought: Knowledge, consciousness, and the politics of empowerment*. New York/London: Routledge. First Ed. (1990).
Connell, R. W. (1995). *Masculinities*. Berkeley/Los Angeles: University of California Press.
Dahlberg, F. (1981). *Woman the gatherer*. New Haven: Yale University Press.
Davis, A. (1981). *Women, race and class*. London: The Women's Press.
De Beauvoir, S. (1949). *Le Deuxième Sexe*. Paris: Gallimard.
Drew, C. (2011, November 4). Where the women are: Biology. *The New York Times*.
European Commission. (2009). *She figures. Statistics and indicators on gender equality in science*. Luxembourg: Publications Office of the European Union.
Fedigan, L. M. (1982). *Primate paradigms: Sex roles and social bonds*. Montreal: Eden Press.
Ferber, M., & Nelson, J. (Eds.). (2003). *Feminist economics today*. Chicago: The University of Chicago Press.
Foucault, M. (1978). *The history of sexuality*. New York: Pantheon Books.
Fraisse, G. (1991). Dalla destinazione al destino. Storia filosofica della differenza dei sessi. In G. Fraisse, & M. Perrot (a cura di), *Storia delle donne in occidente. L'Ottocento* (pp. 89–123). Roma: Laterza.
Fraisse, G. (1996). *La difference des sexes*. Paris: Presses Universitaires de France.
Gatens, M. (1991). *Feminism and philosophy. Perspectives on equality and difference*. London: Polity Press.
Haraway, D. (1988). Situated knowledges: The science question in feminism and the privilege of partial perspectives. *Feminist Studies, 14*(3), 575–599.
Haraway, D. (1991). *Simians, cyborgs and women: The reinvention of nature*. New York/London: Routledge.
Harding, S. (1986). *The science question in feminism*. Milton Keynes: Open University Press.
Harding, S. (1991). *Whose science? Whose knowledge? Thinking from women's lives*. Buckingham: Open University Press.
Harding, S. (1993). Rethinking standpoint epistemology. "What is strong objectivity?". In L. M. Alcoff & E. Potter (Eds.), *Feminist epistemologies* (pp. 49–82). New York: Routledge.
Harding, S. (1998). *Is Science multicultural? Postcolonialisms, feminisms, and epistemologies*. Bloomington: Indiana University Press.
Harding, S. (2008). *Sciences from below: Feminisms, postcolonialities, and modernities*. Durham: Duke University Press.
Hooks, B. (1984). *Feminist theory: From margin to centre*. Boston: South End Press.
Irigaray, L. (1974). *Speculum. De l'autre femme*. Paris: Les Editions de Minuit.
Irigaray, L. (1977). *C'è sexe qui n'est pas un*. Paris: Les Editions de Minuit.
Keller, E. F. (1978). Gender and science. *Psychoanalysis and Contemporary Thought, 1*(3), 409–433.
Keller, E. F. (1983). *A feeling for the organism. The life and work of Barbara McClintock*. New York: Freeman & Co.
Keller, E. F. (1985). *Reflections on gender and science*. New Haven: Yale University Press.
Lloyd, G. (1984). *The man of reason. 'Male' and 'female' in Western philosophy*. Minneapolis: University of Minnesota Press.
Longino, H. (1990). *Science as social knowledge. Values and objectivity in scientific inquiry*. Princeton: Princeton University Press.
Longino, H. (1996). Cognitive and non-cognitive values in science: Re-thinking the dichotomy. In L. H. Nelson & J. Nelson (Eds.), *Feminism, science and the philosophy of science* (pp. 39–58). Boston/London: Kluwer Academic Publishers.
MASIS Expert Group. (2009). *Challenging futures of science in society. Emerging trends and cutting-edge issues*. Luxembourg: Publications Office of the European Union.

Mayberry, M., Subramaniam, B., & Weasel, L. H. (Eds.). (2001). *Feminist science studies. A new generation*. New York: Routledge.
McCall, L. (2005). The complexity of intersectionality. *Journal of Women in Culture and Society, 30*(3), 1771–1800.
Merchant, C. (1979). *The death of nature. Women, ecology and the scientific revolution*. London: Wildwood House. Second Ed. (1980). San Francisco: Harper & Row.
Moi, T. (1990). *Feminist theory and Simone de Beauvoir*. Oxford: Basic Blackwell.
Picchio, A. (Ed.). (2003). *Unpaid work and the economy*. London: Routledge.
Rubin, G. (1975). The traffic in women: Notes on the 'political economy' of sex. In R. R. Reiter (Ed.), *Toward an anthropology of women* (pp. 157–210). New York: Monthly Review Press.
Schiebinger, L. (1989). *The mind has no sex? Women at the origin of modern science*. London: Harvard University Press.
Schiebinger, L. (1999). *Has feminism changed science?* Cambridge: Harvard University Press.
Scott, J. (1986). Gender. A useful category of historical analysis. *American Historical Review, 91*(5), 1053–1075.
Slocum, S. (1975). Woman the gatherer: Male bias in anthropology. In R. R. Reiter (Ed.), *Towards and anthropology of women* (pp. 36–50). New York: Monthly Review Press.
Small, M. F. (Ed.). (1984). *Female primates: Studies by women primatologists*. New York: A. R. Liss.
Spelman, E. (1988). *Inessential woman: Problems of exclusion in feminist thought*. Boston: Beacon.
Spivak, G. C., & Harasym, S. (1990). *The post-colonial critic: Interviews, strategies, dialogues*. New York/London: Routledge.
Tanner, N., & Zhilman, A. (1976). Woman in evolution. Part I: Innovation and selection in human origins. *Signs: Journal of Women in Culture and Society, 1*(2), 585–608.
Waite, L. J. (1981). *U.S. women at work*. Santa Monica: The Rand Corporation.
West, C., & Zimmerman, D. H. (1987). Doing gender. *Gender & Society Official Publication of Sociologists for Women in Society, 1*(2), 125–151.
Zita, J. N. (1988). The feminist question of the science question in feminism. *Hypatia, 3*(1), 157–168.

Part II
Interest and Participation in STEM from Primary School to PhD

Chapter 5
Attitudes, Interest and Factors Influencing STEM Enrolment Behaviour: An Overview of Relevant Literature

Elaine Regan and Jennifer DeWitt

Introduction

There have been many calls internationally for more people studying science, technology, engineering and mathematics (STEM), although the real need for more STEM students has been challenged at least in the UK (Smith 2010a, b; Smith and Gorard 2011). Since the 1970s, researchers have discussed enrolments in the sciences in terms such as the 'flight from science' (Reitz 1973). Concerns about the trend away from studying the sciences have continued since that time (Cleaves 2005; George 2006; Ormerod and Duckworth 1975; Solomon 1997). Several explanations for the 'swing from science' in Britain have been suggested, including a decrease in interest in science and disaffection with science and technology among students. In particular, concerns about declining enrolments have focused on interests and attitudes, noting that pupils' attitudes likely play a role in their choice of subjects (e.g. Atherton et al. 2009; Ormerod and Duckworth 1975; Vidal Rodiero 2007). An additional, related, focus has also been on the quality of science education in schools (Bull et al. 2010; Fensham 2008; Goodrum et al. 2001; Treasury 2006; Millar and Osborne 2001; National Academy of Science: Committee on Science Engineering and Public Policy 2005).

There are many similarities in enrolment patterns among European Union countries and worldwide, where recruitment to STEM subjects is falling (Ainley et al. 2008; European Commission 2010; OECD 1997; Eurydice 2012; Sjoberg 2002). Several cross-cultural international comparative studies have been conducted which address issues of enrolment in the sciences by exploring influencing factors such as achievement, attitudes and relevance with differences most pronounced when comparing developed and developing countries (Barmby et al. 2008; Jenkins 2006).

E. Regan (✉) • J. DeWitt
Department of Education and Professional Studies, King's College London, Waterloo Bridge Wing, Franklin-Wilkins Building, Waterloo Road, London SE1 9NH, UK
e-mail: elaine.regan@kcl.ac.uk

E.K. Henriksen et al. (eds.), *Understanding Student Participation and Choice in Science and Technology Education*, DOI 10.1007/978-94-007-7793-4_5

Examples include Science and Scientists (SAS) which evolved into the Relevance of Science Education (ROSE) project (Jenkins and Nelson 2005); Trends in Mathematics and Science Education (TIMSS) (Kaya and Rice 2010) and the Programme for International Student Assessment (PISA) (Olsen and Lie 2011; Olsen et al. 2011).

The IRIS project aims to understand young people's choices with regard to participation in STEM education, particularly in the transition from upper secondary school to higher education. This chapter provides an overview of some of the extensive range of literature in the area of STEM subject choice, exploring the question: what factors influence STEM enrolments? In this chapter, as in much of the literature, we describe the concern with enrolments (the result of the actual act of choice) using terms such as subject choice, subject uptake, participation in STEM, enrolment patterns/behaviour and recruitment. Although in many publications, the act of choice is described by the term 'interest', we categorise interest, attitude and identity as factors which explain or help us to understand this behaviour.

The issues relating to STEM participation in higher education are best understood when we consider both school and childhood experiences. In this chapter, we explore the factors affecting STEM subject choice and enrolments, interest, and attitudes towards science, as distinct from a treatment on the research on drop out/opt out from STEM programmes (Ulriksen et al. 2010; see also Part III in this book). We begin with a brief overview of the methodological and theoretical approaches that have been taken by many researchers in the field to describe, understand and explain patterns in STEM enrolments, highlighting the common issue across the constructs – inconsistent and varied conceptualisations of theory. This provides a brief follow-on from the chapters in Part I of this book, which focused on the expectancy-value model and narrative and gender perspectives which dominate the work in the IRIS project. Here we briefly explore models of enrolment and some key constructs for understanding choice of STEM. The remainder of the chapter then explores key factors affecting STEM choices for young people derived from empirical studies).

Exploring Subject Choice and Enrolments

Subject choice for A-levels (or upper secondary school) is of critical importance since it has consequences in terms of career paths (Lamb and Ball 1999; Warton 1997; Warton and Cooney 1997).

> Students who study Physics and Chemistry, for example, have a wide range of further education courses available to them, from engineering to the arts. Those who do not do any science or mathematics courses may have more limited choices both in further education and in the types of jobs they want to pursue. (Lamb and Ball 1999, p. 10)

Studies on subject take-up have mostly centred around the idea of individual choice with emphasis on the influence of career value, interest value, performance expectations and perceptions of the subject (Greenfield 1997; Kelly 1988; Malone and Cavanagh 1997; McEwen et al. 1997; Solomon 1997; Stokking 2000; Ventura 1992; Whitehead 1996; Wikeley and Stables 1999; Woolnough 1994). A vast array of factors has been considered to be influential in enrolments and subject choice (Barnes et al. 2005; Lyons 2006), particularly gender and achievement (Colley and Comber 2003; Davies et al. 2008; Francis 2000; Francis et al. 2003; Lamb and Ball 1999; Skelton 2010; Skelton et al. 2010; Smyth and Darmody 2009). This section presents an overview of some of the key theoretical and methodological approaches to the study of STEM choices, notably, attitudes towards science, modelling the choice process and explanatory constructs derived from psychology such as identity and interest.

Attitudes Towards Science

Within the field of science education concern for falling enrolments and attitudes of secondary school pupils in England led to a research focus on attitudes towards science (Reid 2006). The term 'attitude' has become part of our 'common-sense' language and has arisen out of the need to explain and predict behaviour. It can be defined as feelings towards an object or an evaluative judgement formed by a person (Ajzen 2001; Crano and Prislin 2006; Kind et al. 2007), and is seen as a construct which precedes behaviour and guides our decisions and choice, even though it is not directly observable. Attitude has cognitive, affective and behavioural components and there is a strong relationship between all three since the manner in which a person perceives an object should influence the strength of their feeling about it and thus in turn, influence his or her overt behaviour (Lemon 1973). Attitudes can be defined by their content (for example, attitude towards science), their direction (positive, negative, neutral feelings about science, for instance) and their intensity (such as strongly disagree/agree). Within science education, it can be divided into 'attitudes towards science' (for example towards school science or 'real' science) and 'scientific attitudes' (such as mind-sets about thinking in a scientific way) (Gardner 1975). 'Attitudes towards science' are what is frequently studied when exploring subject choice.

Due to the abstract nature of attitude (Ramsden 1998) the task of measurement becomes complex, since attitude cannot be observed directly but must always be inferred from behaviour. Research on attitudes towards science has received much criticism around the validity of the attitude constructs (Fishbein and Ajzen 1975; Gardner 1975; Munby 1982) and lack of standardised definitions and measurement instruments, which can often lead to contradictory results (Barmby et al. 2008).

Despite these challenges, much research highlights that attitudes towards science form a key factor influencing enrolments and subject choice (e.g. Tytler and Osborne 2012), and a number of seminal review papers have been conducted in the area (Gardner 1975; Osborne et al. 2003; Schibeci 1984; Tytler and Osborne 2012). Gardner (1975) declared in his review of attitudes to science (reiterated later by

Schibeci (1984, p. 26)) that '*the volume of research on attitudes in the field of science education has grown so large that it is no longer possible to produce a comprehensive review of the literature within the confines of a journal article*' (p. 2). More recently, Osborne et al. (2003) highlighted a more pressing issue: that while there is a large volume of work exploring student attitudes towards science '*it has little to say definitively about how the problem could be remedied*' (p. 1073). In the same way as the factors affecting student subject choice are complex and multifaceted, so too are the influences on student attitudes. Consequently many of the research studies on attitude measurement concentrate on the influence of a single variable, which may further contribute to the problem of practical implications. (Due to the enormous volume of studies in this area, we do not attempt to summarise them here.)

Models of Enrolment Behaviour

The process leading to enrolment in STEM study is complex, with various individual, psychological, contextual and social influences (Wang 2013). Owing to the complexity of the subject choice process, there have been limited attempts to explore the relationships between these factors in a manner which could productively integrate empirical results into models of enrolment behaviour. Some recent examples of those that have include: Bøe et al. 2011; Bøe and Henriksen 2013; Cerinsek et al. 2013; Jensen and Sjaastad 2013; Skryabina 2000; Smyth and Hannan 2002; Wang 2013 and several chapters in the present volume. The majority of frameworks used in these models originate in the field of motivational psychology. These include:

- Eccles et al.'s expectancy-value model presented in Chap. 2, which describes how young people base their educational choice on their expectation of success and the interest and enjoyment, attainment value, utility value and cost they ascribe to various educational options.
- Fishbein and Ajzen's Theory of Reasoned Action, which explains that a person's voluntary behaviour is predicted by their attitude toward that behaviour and how they think other people would view them if they performed the behaviour.
- Theory of Planned Behaviour, which is an extension of Fishbein and Ajzen's Theory of Reasoned Action and argues that attitude toward behaviour, subjective norms, and perceived behavioural control, together shape an individual's behavioural intentions and behaviours (Dalgety et al. 2003; Khoo and Ainley 2005).
- Social Cognitive Career Theory (SCCT), which is based on Bandura's (1986) general social cognitive theory and posits that the determination to produce a particular choice can be explained as a result of interests and self-reference beliefs. Key factors in SCCT include self-efficacy beliefs, outcome expectations, interests, environmental support and barriers, as well as choice actions (Lent et al. 2010).

In a study drawing on the Eccles et al. expectancy-value theory to explore the choice of Physics in Dutch secondary education, the main predictors of choice were found to be: future relevance, interest, appreciation, physics achievement, self-confidence, difficulty, and clarity (Stokking 2000). Some of these predictors were also found in an analysis of patterns for career choice among university students in Finland (Salmi 2002). These patterns included: interest in the context of study and future work; interest in future work career, salary, and new positions; social pressure from parents, relatives and peer groups; the effect of career campaigns, school career advice, and work experiences; a special course at school and an exceptionally skilful teacher; own hobby, media, and other informal sources.

Social Cognitive Career Theory (SCCT), has been used in a small number of studies on STEM-related academic choice intentions (e.g. Gainor and Lent 1998; Lent et al. 1993; Lent et al. 2008). Wang (2013) found that intent to pursue STEM was significantly and positively influenced by maths self-efficacy, exposure to maths and science and maths achievement (in the latter case, with the exception of Asian students). Self-efficacy in STEM while at university has also been shown to be a predictor of college majors (Heilbronner 2011).

Psychological Constructs: The Act of Choosing

Solomon (1997) asserts that little effort has been made to explain psychological effects involved in the act of choosing. She feels that there is a need to focus on the construction of personal choosing, and cultural persuasion rather than simple explanations such as 'liking' or finding 'interesting' aspects of the subject (Solomon 1997). Historically, work in this vein has included analysis of the personal process of subject choice which led to Kelly's (1988) extensive work on girls in science and Head's 'personality in the pursuit of science' (Head 1979). More recently work focusing on subject choice process has moved from cognitive preferences (Malone and Cavanagh 1997) to the formation of science choices (Cleaves 2005; see also Chap. 7 in the current volume) and constructing desirable identities (Holmegaard et al. 2014; see also Chap. 3 in the present book). In the next section, we articulate what is meant by identity within the context of STEM enrolments.

Identity

It has been claimed that '*post-16 choices are bound up with the expression and suppression of identities. These choices are one aspect, of varying importance, of the sort of person you may become*' (Ball et al. 2000). Identity in science education has been used to address questions such as '*what does it mean to do science?*' (Carlone 2003, p. 21) or who do '*we think we must be to engage in science?*' (Calabrese Barton 1998, p. 379). The construct is also explored in Shanahan (2009) and in Chap. 3 of this volume with regard to narrative explorations. Similarly to the concept of attitude, researchers have employed a multitude of definitions of 'identity' both in sociology (Lawler 2008) and psychology (Côté and Levine 2002). The nature of the problem stems from the fact that there are at least two aspects to identity, core identity and a concept of the self that is frail, brittle and fragmented (Roth and Tobin 2007). A generally accepted definition of identity is encapsulated by Ehle (1989) who sees identity as: '*how one sees oneself (self-concept), how one evaluates himself (sic) (self-esteem), how one desires to be (self-ideal), and how assured one is at meeting life (self-confidence)*' (p. 46). Studies of identity have recognised that identities are not isolated constructs but co-constructions between the individual, their surroundings and their relationships, consequently, focused on individuals, their actions and their agency (Shanahan 2009). Schreiner and Sjoberg (2007) focused on student 'identity construction', a '*who do you want to be*' rather than '*what do you want to be*' approach also thought to play a stronger role in the way young people in western societies relate to science.

In an exploration of 10–11 year olds' attitudes toward and interest in science, identity has been described as an embodied and a performed construction (Archer et al. 2010). Similar to other work (Jenkins and Nelson 2005), this research found that although children can report enjoying science they may still not choose it as they see it as 'not for me'. Analysing decisions regarding participation in STEM using an identity framework involves exploring relationships with family, teachers, peers and others to determine the degree of synergy or disjuncture experienced by young people between their everyday lives and the pursuit of STEM (Archer et al. 2007). Pike and Dunne (2011) report that pedagogies of secondary school science have a major influence on students' learner identities, their identification with science, and their decision about whether to continue to pursue the study of science as part of their future. Wong (2012) also applies an identity lens, using Bourdieu's notion of habitus, to explore the ways in which 'two high achieving working-class' British Asian girls engage with science, while Archer et al. (2012; see also Chap. 6 in this volume) explore how family habitus contributes to the formation of children's aspirations in science. Taconis and Kessels (2009) argue that the unpopularity of science, at least in industrialised countries, is due to the gap between the subculture of science and students' self-image. They found that Dutch students perceived themselves as less similar to science prototypes than to humanities prototypes, viewing peers who favoured science subjects as less attractive, less popular and socially competent, less creative and emotional and more intelligent and motivated than peers who favour humanities. An additional study examined the nature and extent of participation in science-based courses in Canada through cultural reproduction and gender lenses, showing how the intersection of organisational structures and cultural capital shapes STEM opportunities for students (Adamuti-Trache and Andres 2008). Other researchers have discussed the 'process of enculturation' of individuals to a culture, or a science sub-culture, as the acquisition of values, beliefs, expectations, communicative codes, conventional actions and attitudes of the science culture, in addition to the acquisition of knowledge (Aikenhead 2001; Krogh and Thomsen 2005; Lyons 2006).

Holmegaard et al. (2014) report that that the students who did not choose STEM, perceived STEM as stable, rigid and fixed, and, hence, too narrow a platform for developing and constructing desirable identities. Furthermore, the process itself of choosing was a complex ongoing and social process, rather than an isolated individual event (also see Chap. 3). Hernandez-Martinez et al. (2008) identified four identity repertoires of aspiration and choice in higher education mathematics in the UK: 'becoming successful', 'personal satisfaction', 'vocational' and 'idealist', which were strongly related to background factors such as class, gender and ethnicity. Overall, studies reflect that continuities and discontinuities between students' identities and science are likely to play an important role in the choice process, leading to the use of identity as an important theoretical construct in the study of choice. Interest is another key construct in the study of choice and the next section explores the concept of interest in science education.

Interest

In general, the term *interest* describes the mind-set characterised by a need to give selective attention to something that is significant to a person such as an activity, goal or subject. With regard to science it can be used to describe '*tendencies to engage in science-related activities inside and outside of schools*' (Olsen et al. 2011, p. 2). The notion of interest in science can be interpreted in a number of ways, particularly in relation to students' motivation in science, and similarly in relation to attitude and identity. The lack of a consistent theory of interest has persisted since the 1940s (Allport 1946; Renninger and Hidi 2011) and researchers often fail to articulate their conceptualisation of interest and the connections to their measurements. Despite this, research conducted over the years has articulated five characteristics of interest as a motivational variable: interest is content or object specific; it involves a particular relation between a person and the environment, sustained through interaction; it has both a cognitive and an affective component; a person may not be aware of when their interest was triggered or their interest during engagement; and it has a neurological/physiological basis (see Hidi and Renniger 2006; Renninger and Hidi 2011 for a full explanation of each characteristic).

A number of studies (Ainley 2007; Hidi and Renninger 2006; Krapp 2002) discuss interest as a psychological state, while others (Ainley 2007; Silvia 2006) focus on interest as an emotion. The most commonly referred to form of interest is individual interest: a personal orientation, predisposition or tendency to engage with something, which would seem also to be applicable to individual differences in students' general orientation to science (Ainley and Ainley 2011a). Situational interest however, is a temporary concentration of attention and feeling in response to a specific situation (Hidi 1990) that can be 'triggered' and 'maintained' (Hidi and Renninger 2006) or 'stabilised' (Krapp 2003). Renninger et al. (1992) view interest from a person-object theory perspective, as a relationship between a person and an object. Two models of interest development which have been frequently used in research include: Krapp's (2003) stage model of interest development, an extension of the person-object conceptualisation, and that of Hidi and Renninger (2006). Both distinguish between stages of interest development, from an 'emerging' to a more well developed, individual interest. Knowledge, affect and value are key components of a strong individual interest. A full account of the conceptualisations of interest can be found in Renninger and Hidi (2011) and Krapp and Prenzel (2011). The latter is also consistent with sociological theories of interest development as an integral part of identify formation (Olsen and Lie 2011; Osborne et al. 2003; Schreiner and Sjoberg 2007).

Research has shown that individual interest in science is very important for choosing science and often forms at an early age (Ainley and Ainley 2011b; Maltese and Tai 2010; Tai et al. 2006a). Interest was identified as the dominant influence on enrolment behaviour in many studies examining both the direct and indirect effects of interest (Kelly 1988) and strong predictive relations have been found between personal value of science, enjoyment of science, and interest in learning science (Ainley and Ainley 2011b). Consequently, interest is one of the strongest predictors of decisions in relation to choice of subjects and courses (Olsen et al. 2011, Chap. 9 in the present volume). Regan and Childs (2003) found that students' choice of science at the junior level was determined by their expressed interest in the subject and by their future/career plans. Taking career expectations as an indicator of interest, a large-scale study of 3,300 university-level students found that those who had expected to be working in a science career by age 14 were 3.4 times more likely to earn a physical science and engineering degree than those who had not (Tai et al. 2006a). Maltese and Tai (2010) also found that the majority of scientists reported that their interest in science began before their middle school years. Another recent study used PISA 2006 data to explore student interest in science across different countries. This research found that having a general interest in learning science predicts both current and intended future participation in science related activities and concluded that where science education is perceived as personally important to students, and where they are doing well, a stronger interest in learning science will result (Ainley and Ainley 2011a).

In the next section, we explore in more detail empirical studies of some of the key factors influencing STEM choice and enrolment patterns, based on research drawn from the range of perspectives discussed in this section, both as single variable perspectives and models of decision-making processes.

Factors Influencing STEM Choice

The remainder of this chapter concentrates on the identification of various influences on enrolment behaviour and consideration of their importance. The findings from the studies reviewed here tend to produce fairly consistent results in so far as they have identified the same variables, but differences occur in the relative importance attributed to each variable. This seems most likely to be a consequence of the varying contexts of the research, different school systems and techniques of analysis, which provide a range of additional factors which could be influential on findings. Moreover, the same factors cannot be assumed to exert influence in the same way on take-up of the three main science subjects (Smyth and Hannan 2002). Complicating the picture further, a recent review of STEM choices in the UK highlighted that much of the research in the area relied on small sample sizes, short-term 'snapshot' approaches, inconsistent analysis, imprecise terminology and over-reliance on historical (pre-2000) data (Tripney et al. 2010).

In a study investigating information sources utilised by students selecting subjects, Warton (1997) reports that adolescents indicate a lack of knowledge about their options. Similarly, Wikeley and Stables (1999) cite studies that found that students make naïve links between subjects and careers, that the choices are volatile, that parents act as the '*chief advisers*', that boys and girls can receive differing advice and that schools implement implicit policies of selection. Warton (1997) concurs, stating that she found little evidence that students treat subject choice as a deliberate planned activity where information should be sought. Thus when students are making subject or career decisions they generally attempt to match personal needs with the experiences which are likely to result from the choice (Malone and Cavanagh 1997). Students are more likely to choose a subject that they believe to be useful for a job or career or a requirement for a college course, a subject they find interesting, or a subject that they can achieve a good grade in (Regan and Childs 2003). Similar reasons were also found to underpin students' subject choices at ages 14 and 16 in Britain (Blenkinsop et al. 2006), as well as A-level subject choices (Vidal Rodiero 2007). The latter two studies also highlighted that students' choices (or anticipated choices) vary over time, as well as reflecting the role played by families and school experiences on choices.

While the above studies focus on subject choice more broadly, the sections below focus on factors identified by a number of studies as being key influences on choice of and, consequently, enrolments in science. Some of these factors could be considered individual background characteristics such as age, ability and gender, while others involve the way in which students experience science in school (teaching and learning, other school-related factors) and key individuals of influence (teachers and parents). We also discuss possible influences of the images students have of science and scientists.

Age

Research has suggested that attitudes towards science decline with age, particularly in secondary school (Barmby et al. 2008) with positive attitudes towards school science declining significantly from the age of 10 (Bennett and Hogarth 2009; Murphy and Begg 2005). George (2006) explored attitudinal dimensions in American middle and high school students and found that overall students' attitudes about the utility of science were positive but their attitudes towards school science declined over the middle and high school years. However, other research has challenged this perspective, reflecting that attitudes to school science do not necessarily decline as students move into secondary school (DeWitt et al. 2014; Lyons and Quinn 2010; NFER 2011). Despite mixed findings with regard to potential declines in attitudes to science, research more clearly suggests that positive attitudes to science at a young age are related to later participation in science. For instance, attitudes to subjects at the age of 14 were found to be highly predictive of Biology, Physics and Chemistry take-up (Lamb and Ball 1999). Similarly, Lindahl (2007) found that Swedish students had formed their career aspirations and interest in science by the age of 13 years, highlighting the importance of engaging with students long before they make their choices about what subjects to select in school (Tytler and Osborne 2012).

Attainment

Previous science performance is highly predictive of science take-up: young people with higher levels of prior attainment are more likely than those with lower levels of prior attainment to continue their studies in STEM subjects (Smith and Gorard 2011). For instance, Gill and Bell (2013) found that attainment in physics and maths at age 16 (and attending a grammar school) was associated with a greater probability of uptake of A-level physics. However, other work has suggested that interest and enjoyment may be more predictive of a degree in STEM than achievement alone (Maltese and Tai 2011). Moreover, the link between attainment and participation or take-up may also vary by subject within STEM. For instance, in an Australian study, mathematics and the physical sciences were claimed to be the domain of high achievers and ability grouping at junior level was found to have a significant relationship with take-up of Physics and Biology (Lamb and Ball 1999). All students from ability grouped classes were more likely to take Physics than pupils from mixed ability classes. Students in the top and bottom streams were less likely to choose Biology than students in mixed ability or middle stream classes.

Teaching and Learning

The quality of teaching is a major determinant of student engagement with and success in a school subject (Tytler and Osborne 2012) and a number of studies reflect the critical role that school science experiences may play in influencing STEM choice. Chapter 7 in the current volume also discusses how school science curriculum may impact on STEM uptake. A huge variation in quality of teaching and resources for science in Australian schools has been found, with some schools having outstanding programs supported by well qualified and enthusiastic teachers, while others have no programs in place at all (Sadler 2002). A telephone survey indicated some additional causes of students' retreat from the sciences, including poor transition from primary to secondary school, low levels of experimental work in science classes, an increase in available choice of non-science subjects, advice from career guidance teachers and lack of role models in science (Sadler 2002). Tytler and Osborne (2012) have also suggested that school science lacks purpose for students, making it unappealing, which would in turn dissuade students from continuing with post-compulsory science. Relatedly, a study of students in England found key influences on post-16 choice of science to include school pedagogical experiences, the different ways subjects were perceived (e.g. as 'higher status' or not) and students' future aspirations (Pike and Dunne 2011). Similar factors – such as the ability to imagine themselves in science and experience of school science – have also been identified as possible reasons for the decline in science enrolments in Australia (Lyons and Quinn 2010). Moreover, such factors (e.g. achievement in science, science classes and teachers, as well as interest in science) were identified by practicing scientists as contributing to their decision to pursue science (Venville et al. 2013). A retrospective study conducted in the US with 8,178 university students focusing on science attainment at university (which would be connected to retention) found that one of the significant predictors (in addition to achievement) of success in university science courses was the type of instruction students received in high school (Tai et al. 2006b).

The experience of a school/education activity (such as a science competition, camp, teacher demonstrations, project work, or enrichment activity) was the second most common factor attributed as the initial source of interest and was a greater influence on females than males (Maltese and Tai 2010). The influence of the teacher was also prominent for both male and female students who became interested in science after middle school. In addition, PISA 2006 data indicate an association between students' motivation towards science, enjoyment of science and future orientation towards science, and the frequency of various teaching and learning activities in the classroom (Hampden-Thompson and Bennett 2013).

School Type

Type of school seems to be another factor influencing choice, where girls from single-sex schools are more likely to choose physics than those from mixed schools (Byrne 1993; Solomon 1997). A task force in Ireland examined schools that exhibited a high take-up of the physical sciences in order to identify strategies for increasing take-up. Four approaches were illuminated in these schools, which also had a higher level of laboratory resources than the norm (The Task Force on the Physical Sciences 2002): high priority attached to science at management level; good subject level coordination and planning; emphasis on building positive student experiences at the junior level; and emphasis on practical work. Other work comparing high- and low-uptake schools found that students in high-uptake schools appear to make a proactive choice in relation to career aspirations, rather than a reactive choice on the basis of past experience (Bennett et al. 2013).

Persons of Influence

The influence of parents and teachers on choices, particularly during the earlier years of schooling, has been well-documented (Maltese and Tai 2010; Mujtaba and Reiss 2012; Olszewski-Kubilius and Yasumoto 1994; Raved and Assaraf 2011; Salmi 2002; Solomon 1997). Perceived support from both parents and teachers was identified as the strongest predictor of continuing with a subject (Kelly 1988; Maltese and Tai 2010). Research has also shown that the early years of secondary education are crucial in terms of the impact a teacher can have on students' views of science and careers involving science (Cerini et al. 2004; Munro and Elsom 2000; Osborne and Collins 2001; Watt 2005). Students who report negative interaction with teachers are less likely to choose Physics (Smyth and Hannan 2002).

More recent work has highlighted the important role that teachers play in promoting student interest in science through scaffolding and guidance (Xu et al. 2011). Similarly, Sjaastad (2012), in his exploration of people influencing STEM choice, distinguished between persons acting as *models* (e.g. parents, teachers or others displaying a STEM professional identity) and those acting as *definers* (parents or others helping the young person in the process of setting goals, defining values and identifying personal strengths in the educational choice process). For instance, teachers are models by displaying how STEM might bring fulfilment and by giving pupils a positive experience with the subject, and they are definers who help young people discover their STEM abilities. Teachers, however, are generally not well informed about careers in or outside science (Stagg 2007) and often do not perceive themselves to be sources of careers information for their students (Munro and Elsom 2000).

Other factors, such as the education level of parents or social class background, have also been demonstrated as influencing choice (Ayalon 1995; Solomon 1997; Uerz et al. 1999), however, socio-economic status has been shown to have only an indirect effect on young people's perceptions of their capabilities (Blenkinsop et al. 2006). All the same, students who continue within STEM are more likely to have higher socio-economic status (Thomson and De Bortoli 2008). The degree and extent of relatively low attainment amongst students from lower socio-economic groups across all subjects in school has recently been explored in light of recent curricular reforms in England (Homer et al. 2013). This work explored the impact of curricular reform on the overall pattern of participation and whether the existing stratification persisted following the reform. Using data from the National Pupil Database this study shows that curricular reform 'and the offer of entitlements for particular course can and do impact on stratification' (p. 261), at least for some sub-groups. Relatedly, Lyons (2006) found higher levels of cultural and social capital in students choosing physical sciences relative to those who did not. They had supportive family relationships, parents who recognised the value of education and family members supporting an interest in science. Similarly, other work highlights that the transmission of cultural capital can restrict STEM pathways when parents do not encourage academic pursuits in STEM

(Adamuti-Trache and Andres 2008). Focusing on children's aspirations and attitudes towards science Archer and colleagues (Archer et al. 2010, 2012, 2014; see also Chap. 6 in the present volume) explore similar themes.

Conceptions of Science and Scientists

Conceptions of science and scientists held by students have also been postulated as forming an integral part of the framework of the subject choice patterns. That is, the ability of students to imagine themselves in a science profession (which, in turn, impacts on subject choice) is influenced by the images they hold of science and scientists. One recent study found that US students considered scientific professions to be less creative and less people-oriented than other career choices, holding misperceptions that 'science is difficult, uncreative and a socially isolating pursuit' (Masnick et al. 2010, p. 693). Another study (Dalgety and Coll 2004) found that students saw chemistry and media portrayals of the subject in terms of images of laboratory work, experiments and wearing white coats, consistent with early research on conceptions of science (Mead and Metraux 1957; Schibeci 1986). Within the ASPIRES project, Archer et al (2010) found that science was viewed as a difficult subject that requires a natural interest, which sometimes students associated with natural ability. Although students were found to be mostly enthusiastic about science, aspects of a science identity were rejected by many students along gender, ethnic, and class lines. Steinke et al. (2012) explored traits found in media portrayals of scientists with 370 students from three middle schools in the U.S. The results clearly indicate that when identifying with female scientists, students identified most strongly with characters showing *respect*, *caring*, and *dominance*. When identifying with male scientists, however, students selected those with the traits *intelligence*, *dominance*, and *respectedness*. This suggests that students do identify with scientists on television; it also reflects that these identifications are nuanced and depend on the TV scientists' behaviours, gender, and context. Although few students held 'negative' representations of science/ scientists, other recent work reports that students are influenced by popular constructions of science, such as 'specialist' and 'clever', which in turn feed into their feelings that science is not for them (DeWitt et al. 2013a).

Another key influence on STEM subject choice and retention is that of gender, which was also a focus of the IRIS project discussed in this volume. The next section focuses specifically on what the research has to say about gender as an influencing factor on subject enrolment.

Gender as an Influencing Factor

One of the major factors connected with student interest in school science is gender, with boys consistently showing more positive attitudes to school science than girls (e.g. Brotman and Moore 2008; Haste 2004; Jenkins and Nelson 2005; Jones et al. 2000; Murphy and Whitelegg 2006; Scantlebury and Baker 2007; Sjoberg and Schreiner 2005). Given the relationships between interest in science and decisions to pursue post-compulsory science, this gendered difference in interest is likely to contribute to gendered subject choice. Allegrini explores the construct in more depth in Chap. 4 but an up to date review of the literature on gendered study choice can be found in Yazilitas et al. (2013).

The major concern with regard to the low representation of women studying science is the polarization of choices of boys and girls (Bell 2001; Smyth and Hannan 2002), which leads to potentially large numbers of qualified people missing from the work force (Fouad 1995; Malone and Cavanagh 1997). Moreover, gendered choice of science is also an equity concern, given the opportunities that can be open to individuals who pursue post-compulsory science. The metaphor of a 'leaky pipeline' is often used to describe the manner in which individuals, particularly females, drop out of science at various choice points (Blickenstaff 2005). Gender differences in subject choice have been well documented and persist despite minimal – if any – attainment-related differences between genders (e.g. Hyde and Lynn 2006) and many interventions targeting girls and science (Brotman and Moore 2008; Colley and Comber 2003; Davies et al. 2008; Francis 2000; Francis et al. 2003; Malone and Cavanagh 1997; Murphy and Whitelegg 2006; Sadler et al. 2012; Skelton 2010; Skelton et al. 2010; Smyth and Darmody 2009; Whitehead 1996). Indeed, despite these efforts, undergraduates in physical sciences in the UK remain largely high-achieving, White, middle-class young men (Smith 2010a).

Blickenstaff (2005) summarises the explanations put forward in the research for the underrepresentation of women in STEM careers and post-compulsory study as being: biological differences between men and women; girls' lack of academic preparation for a science major/career; girls' poor attitude toward science and the lack of positive experiences with science in childhood; the absence of female scientists/engineers as role models; science curricula being irrelevant to many girls; the pedagogy of science classes favouring male students/a 'chilly climate' for girls in science classes; cultural pressure on girls to conform to traditional gender roles and an inherent masculine worldview in scientific epistemology. Blickenstaff argues that these explanations 'hold very little water' in explaining the underrepresentation of women in science and concludes with seven suggestions to tackle the under-representation. In the UK, the schools inspection agency has reported that although many girls do not consider gender to be a barrier to participation, their actual choices of subjects and careers remain gender-traditional (Ofsted 2011).

A 2008 special issue in Studies in Educational Evaluation entitled *Narrowing the Gap* explores trends in gender and achievement in the TIMSS study (Neuschmidt et al. 2008; Thomson 2008). In addition, studies in the UK, Ireland and United States reflect that girls are now out-performing boys in most school subjects (McEwen et al. 1997; Smyth and Hannan 2002; Solomon 1997). Nevertheless, gender remains a strong predictor of A-level physics uptake in the UK, even controlling for prior attainment (Gill and Bell 2013). Skelton (2010) claims that despite girls being heralded as educational 'success stories', classroom research continues to find they are less confident than boys. The discussions about girls and science suggest that there is a disparity between the 'masculine' image of the physical sciences and girls' identification with the 'feminine' role in adolescence (Archer et al. 2013; Brotman and Moore 2008; McEwen et al. 1997) and feminist theorists have a long established concern about this association between STEM and 'masculinity' (Burton 1990; Haraway 1988; Harding 1998). For instance, Archer et al. (2013) found that the highly gendered nature of girls' aspirations were indicative of underlying masculine constructions of science careers and how imagined science careers were incompatible with girls' performances of popular femininity; see also Part IV of the present volume for discussions of this.

A consistent picture from early research on choosers of science depicts them as male (Baker and Leary 1995; Buck et al. 2008). Many students consider STEM subjects to be 'for boys' (Adamuti-Trache and Andres 2008; Calabrese Barton and Tan 2009; Mendick 2005) – or at least science careers as being done primarily by men – a perception that could contribute to girls not seeing themselves as 'science people' (Carlone 2003). A recent cross-national analysis of young people's preferences, expectations, and perceptions of ability regarding STEM subjects using PISA data from four countries (Switzerland, Finland, Australia and Korea) found that gender plays a crucial role in students' choices regarding STEM study and careers (Buccheri et al. 2011). Moreover, even within STEM, career choices tend to follow a gendered pattern with females choosing medical/health and biology careers and males choosing engineering and computer sciences (Sikora and Pokropek 2012). One potential reason underpinning such discrepancies, beyond associations between science careers and masculinity, concerns the kinds of values that students seek to fulfil in their subject – and career – choices. For instance, students' physics identity, which strongly predicted career intentions in physics, was closely linked to a desire to pursue a career that involved working with knowledge, skills or products and negatively related to more people-related career motivations (Hazari et al. 2010). Similarly, a study of Slovenian university students found that they wanted to do something interesting and fulfilling using their talents and abilities, with female STEM students favouring inter-personal career priorities (i.e. helping other people, contributing to society and protecting the environment) more than males (Cerinsek et al. 2013). Put differently, it is possible that the differential values that male and female students seek in a future occupation at least partly underpin the gender discrepancy in subject choice, given the link between subject choice and career aspirations (Jones et al. 2000).

Other factors that have been explored as related to gendered choice of science concern the curriculum and students' self-concept or self-efficacy in science. It has also been argued that curricular content in science is unbalanced, favouring boys' interests (Brotman and Moore 2008; Haussler and Hoffmann 2002) and would benefit from a more human-related focus (Krogh and Thomsen 2005) that considered ethical factors and how science was relevant to students' lives (Haste 2004). This situation is also likely to be exacerbated by the way in which the nature and culture of science are portrayed in the classroom (Brotman and Moore 2008).

At the same time, attempts to reform the curriculum to better align with girls' interests have met with mixed success. While some studies have found that curricular interventions have improved girls' interest and achievement in science (e.g. Haussler and Hoffman 2002), Carlone's (2004) ethnographic study of girls' cultural production of science in a reform-based physics class found that the girls were most concerned about maintaining their 'good student' identities and resisted promoted science learner identities.

In addition to classroom experiences in science which are more aligned with boys' interests and portray a masculine view of science, gendered subject choice is also likely to be impacted by students' self-concept in science. Students with strong self-concepts of their abilities in science are more likely to pursue science once it is no longer compulsory and to aspire to science-related careers (Halpern et al. 2007). However, girls often have less positive self-concepts in science than boys (DeWitt et al. 2013b), although this can vary within areas of science (Britner 2008). Moreover, girls' self-concepts in science may also be affected by the extent to which they receive recognition of their science achievements or encouragement to pursue science from teachers, family and friends, with research finding that females often receive less recognition and encouragement than males (Carlone and Johnson 2007; Hazari et al. 2010; Mujtaba and Reiss 2012).

In addition to work on self-concept in science, other studies have focused on more psychological factors as potentially underpinning gender discrepancies in subject choice. For instance, a study exploring girls' decisions about whether or not to pursue science via the lens of cognitive preferences found that a large proportion of the girls who chose not to enrol in science, but who were capable – as identified by their teachers – held cognitive preferences for 'feeling' and 'judging' and negative preferences for 'intuition' and 'thinking'. The cognitive preference differences between the girls who accepted teacher recommendations and those who did not is indicative of a perceived mismatch between the curricula of these subjects and the needs of these students, which may lead them to avoid post-compulsory science (Malone and Cavanagh 1997).

A more recent study (Zeyer and Wolf 2010) is based upon a body of work that characterises individuals as being either primarily empathizers, systemizers, or an equal balance of both. *Systemizing* describes the ability to understand the world in terms of a system, whereas *empathizing* is the ability to identify and perceive the mental states of others. In this study, the authors examined whether gender played a role in determining motivation for science learning or whether the personality attributes of either systemizer or empathizer were more significant. Analyses reflected no statistically significant difference in the motivational levels between male and female students to learn science. However, there were highly significant differences between personality attributes, with female students being more likely to be empathizers. This discrepancy between genders has interesting implications, the authors assert, for student engagement with system-rich disciplines such as the physical sciences, with systematizers possibly having a greater *motivation* to learn science than empathizers. Thus, although gender alone was not found to be significant in determining motivation level for science learning, the authors argue that the secondary correlation – that male students are more likely to be systematizers – may help to explain the observed gender differences in the choices of male and female students regarding STEM study and careers.

Conclusions

The goal of this chapter was to provide an overview of what the research literature has to say about attitudes, interest and factors influencing subject choice in STEM. The message from the literature is "that we are not going to find one single factor which is universally influential; different students are persuaded by quite different factors" (Woolnough 1994, p. 672) and "many different factors contribute to children's decisions, and these factors are inter-related among themselves" making it "difficult to disentangle the effect of any one variable" (Kelly 1988, p. 18). Twenty years after the Woolnough article, and even longer since Kelly, the research reviewed in this chapter highlights that this is indeed still very much the case.

In this chapter we have explored a myriad of factors that have been proposed as impacting on STEM subject choice and ensuing enrolment behaviour. STEM choice and enrolment have been examined in a number of ways, drawing on both qualitative and quantitative methods in an attempt to unpick and understand this phenomenon more deeply. Much of this work has used student attitudes to science as a starting point, given the links between attitudes and choice. Other work has focused on psychological constructs underpinning the act of choosing, such as identity (also a construct in sociology) and interest. Taken together, research conducted over the years has consistently identified a number of factors related to school (pedagogy/teaching, curriculum, type of school, teachers), family background (parental support and resources), and individual characteristics (age, attainment and, critically, gender), as well as widely-held images of science and scientists, as influencing STEM choice.

Given the complexity of these factors in and of themselves, as well as their interrelationships, it is perhaps not surprising that attempts to integrate the findings from this enormous body of research have been relatively limited. Of those that have, a number have originated in motivational psychology, rather than science education, suggesting that theorisation in this area has further room for development. Studies such as the IRIS project have a role to play in this theorisation but even then, they face challenges in achieving a truly broad view of the landscape. Nevertheless, doing so is critical if we are going to make significant progress in developing interventions with the potential to address the multifaceted issues around post-compulsory STEM choice, including the seemingly intractable one of gender. Thus, while the response of young people to STEM courses and careers remains a policy priority across many nations the search for insights into the factors affecting participation, attitudes, interest and choice will continue.

References

Adamuti-Trache, M., & Andres, L. (2008). Embarking on and persisting in scientific fields of study: Cultural capital, gender, and curriculum along the science pipeline. *International Journal of Science Education, 30*(12), 1557–1584.

Aikenhead, G. S. (2001). Students' ease in crossing cultural borders into school science. *Science Education, 85*(2), 180–188.

Ainley, M. (2007). Being and feeling interested: Transient state, mood, and disposition. In P. A. Schutz & R. Pekrun (Eds.), *Emotions in education*. New York: Elsevier.

Ainley, M., & Ainley, J. (2011a). A cultural perspective on the structure of student interest in science. *International Journal of Science Education, 33*(1), 51–71.

Ainley, M., & Ainley, J. (2011b). Student engagement with science in early adolescence: The contribution of enjoyment to students' continuing interest in learning about sciences. *Contemporary Educational Psychology, 36*(1), 4–12.

Ainley, J., Kos, J., & Nicolas, M. (2008). *Participation in science, mathematics and technology in Australian education*. Camberwell: Australian Council of Educational Research (ACER).

Ajzen, I. (2001). Nature and operation of attitudes. *Annual Review of Psychology, 52*, 27–58.

Allport, G. W. (1946). Effect: A secondary principle of learning. *Psychological Review, 53*(6), 335–347.

Archer, L., Hollingworth, S., & Halsall, A. (2007). 'University's not for me – I'm a Nike person': Urban, working-class young people's negotiations of 'style', identity and education. *Sociology, 41*, 219–237.

Archer, L., DeWitt, J., Osborne, J., Dillon, J., Willis, B., & Wong, B. (2010). "Doing" science versus "being" a scientist: Examining 10/11-year-old schoolchildren's constructions of science through the lens of identity. *Science Education, 94*(4), 617–639.

Archer, L., DeWitt, J., Osborne, J., Dillon, J., Willis, B., & Wong, B. (2012). Science aspirations, capital, and family habitus: How families shape children's engagement and identification with science. *American Educational Research Journal, 49*(5), 881–908.

Archer, L., DeWitt, J., Osborne, J., Dillon, J., Willis, B., & Wong, B. (2013). 'Not girly, not sexy, not glamorous': Primary school girls' and parents' constructions of science aspirations. *Pedagogy, Culture & Society, 21*(1), 171–194.

Archer, L., DeWitt, J., & Wong, B. (2014). Spheres of influence: What shapes young people's aspirations at age 12/13 and what are the implications for education policy? *Journal of Education Policy, 29*(1), 58–85.

Atherton, G., Cymbir, E., Roberts, K., Page, L., & Remedios, R. (2009). *How young people formulate their views about the future*. London: Department for Children, Schools and Families.

Ayalon, H. (1995). Math as a gatekeeper: Ethnic and gender inequality in course taking of the sciences in Israel. *American Journal of Education, 104*(1), 34–56.

Baker, D., & Leary, R. (1995). Letting girls speak out about science. *Journal of Research in Science Teaching, 32*(1), 3–27.

Ball, S. J., Maguire, M., & Macrae, S. (2000). *Choice, pathways and transitions post-16. New youth, new economies in the global city*. London: Routledge Falmer.

Bandura, A. (1986). *Social foundations of thought and action: A social cognitive theory*. Englewood Cliffs: Prentice Hall.

Barmby, P., Kind, P. M., & Jones, K. (2008). Examining changing attitudes in secondary school science. *International Journal of Science Education, 30*(8), 1075–1093.

Barnes, G., McInerney, D. M., & Marsh, H. W. (2005). Exploring sex differences in science enrolment intentions: An application of the general model of academic choice. *The Australian Educational Researcher, 32*(2), 1–23.

Bell, J.F. (2001). Patterns of subject uptake and examination entry 1984–1997. *Educational Studies, 27*(2), 201–219.

Bennett, J., & Hogarth, S. (2009). Would you want to talk to a scientist at a party? High school students' attitudes to school science and to science. *International Journal of Science Education, 31*(14), 1975–1998.

Bennett, J., Lubben, F., & Hampden-Thompson, G. (2013). Schools that make a difference to post-compulsory uptake of physical science subjects: Some comparative case studies in England. *International Journal of Science Education, 35*(4), 663–689.

Blenkinsop, S., McCrone, T., Wade, P., & Morris, M. (2006). *How do young people make choices at 14 and 16?* Slough: National Foundation for Educational Research.

Blickenstaff, J. C. (2005). Women and science careers: Leaky pipeline or gender filter? *Gender and Education, 17*(4), 369–386.

Bøe, M. V., & Henriksen, E. K. (2013). Love it or Leave it: Norwegian students' motivations and expectations for postcompulsory physics. *Science Education, 97*(4), 550–573.

Bøe, M. V., Henriksen, E. K., Lyons, T., & Schreiner, C. (2011). Participation in science and technology: Young people's achievement-related choices in late-modern societies. *Studies in Science Education, 47*(1), 37–72.

Britner, S. L. (2008). Motivation in high school science students: A comparison of gender differences in life, physical, and earth science classes. *Journal of Research in Science Teaching, 45*(8), 955–970.

Brotman, J. S., & Moore, F. M. (2008). Girls and science: A review of four themes in the science education literature. *Journal of Research in Science Teaching, 45*(9), 971–1002.

Buccheri, G., Gürber, N. A., & Brühwiler, C. (2011). The impact of gender on interest in science topics and the choice of scientific and technical vocations. *International Journal of Science Education, 33*(1), 159–178.

Buck, G. A., Plano Clark, V. L., Leslie-Pelecky, D., Lu, Y., & Cerda-Lizarraga, P. (2008). Examining the cognitive processes used by adolescent girls and women scientists in identifying science role models: A feminist approach. *Science Education, 92*(4), 688–707.

Bull, A., Gilbert, J., Barwick, H., Hipkins, R., & Baker, R. (2010). *Inspired by science: A paper commissioned by the Royal Society and the Prime Minister's Chief Science Advisor.* Wellington: New Zealand Council for Educational Research.

Burton, L. (1990). *Gender and mathematics: An international perspective.* London: Cassell Educational.

Byrne, E. (1993). *Women and science: The snark syndrome.* Bristol: Falmer Press.

Calabrese Barton, A. (1998). Teaching science with homeless children: Pedagogy, representation, and identity. *Journal of Research in Science Teaching, 35*(4), 379–394.

Calabrese Barton, A., & Tan, E. (2009). Funds of knowledge and discourses and hybrid space. *Journal of Research in Science Teaching, 46*(1), 50–73.

Carlone, H. B. (2003). (Re)producing good science students: Girls' participation in high school physics. *Journal of Women and Minorities in Science and Engineering, 9*(1), 17–34.

Carlone, H. B. (2004). The cultural production of science in reform-based physics: Girls' access, participation, and resistance. *Journal of Research in Science Teaching, 41*(4), 392–414.

Carlone, H. B., & Johnson, A. (2007). Understanding the science experiences of successful women of color: Science identity as an analytic lens. *Journal of Research in Science Teaching, 44*(8), 1187–1218.

Cerini, B., Murray, I., & Reiss, M. (2004). *Student review of the science curriculum.* London: Planet Science.

Cerinsek, G., Hribar, T., Glodez, N., & Dolinsek, S. (2013). Which are my future career priorities and what influenced my choice of studying science, technology, engineering or mathematics? Some insights on educational choice – case of Slovenia. *International Journal of Science Education, 35*(17), 2999–3025.

Cleaves, A. (2005). The formation of science choices in secondary school. *International Journal of Science Education, 27*(4), 471–386.

Colley, A., & Comber, C. (2003). School subject preferences: Age and gender differences revisited. *Educational Studies, 29*(1), 59–67.

Côté, J. E., & Levine, C. G. (2002). *Identity formation, agency, and culture: A social psychological synthesis*. New York: Psychology Press.
Crano, W. D., & Prislin, R. (2006). Attitudes and persuasion. *Annual Review of Psychology, 57*, 345–374.
Dalgety, J., & Coll, R. K. (2004). The influence of normative beliefs on students' enrolment choices. *Research in Science & Technological Education, 22*(1), 59–80.
Dalgety, J., Coll, R. K., & Jones, A. (2003). The development of the Chemistry Attitudes and Experiences Questionnaire (CAEQ). *Journal of Research in Science Teaching, 40*(5), 649–668.
Davies, P., Telhaj, S., Hutton, D., Adnett, N., & Coe, R. (2008). Socioeconomic background, gender and subject choice in secondary schooling. *Educational Research, 50*(3), 235–248.
DeWitt, J., Archer, L., & Osborne, J. (2013a). Nerdy, brainy and normal: Children's and parents' constructions of those who are highly engaged with science. *Research in Science Education, 43*(4), 1455–1476.
DeWitt, J., Osborne, J., Archer, L., Dillon, J., Willis, B., & Wong, B. (2013b). Young children's aspirations in science: The unequivocal, the uncertain and the unthinkable. *International Journal of Science Education, 35*(6), 1037–1063.
DeWitt, J., Archer, L., & Osborne, J. (2014). Science-related aspirations across the primary–secondary divide: Evidence from two surveys in England. *International Journal of Science Education*. doi:10.1080/09500693.2013.871659.
Ehle, M. J. (1989). Self-perception and learning. *Education and Society, 7*(1), 46–51.
European Commission. (2010). *Europe 2020. A European strategy for smart, sustainable and inclusive growth*. Brussels: European Commission.
Eurydice. (2012). *Key data on education in Europe 2012*. Brussels: Eurydice.
Fensham, P. J. (2008). *Science education policy-making: Eleven emerging issues*. Paris: UNESCO.
Fishbein, M., & Ajzen, I. (1975). *Belief, attitude, intention and behaviour: An introduction to theory and research*. Reading: Addison-Wesley Publishing Company.
Fouad, N. A. (1995). Career linking: An intervention to promote math and science career awareness. *Journal of Counseling & Development, 73*(5), 527–534.
Francis, B. (2000). The gendered subject: Students' subject preferences and discussions of gender and subject ability. *Oxford Review of Education, 26*(1), 35–48.
Francis, B., Hutchings, M., Archer, L., & Melling, L. (2003). Subject choice and occupational aspirations among pupils at girls' schools. *Pedagogy, Culture & Society, 11*(3), 425–442.
Gainor, K. A., & Lent, R. W. (1998). Social cognitive expectations and racial identity attitudes in predicting the math choice intentions of Black college students. *Journal of Counseling Psychology, 45*, 403–413.
Gardner, P. L. (1975). Attitudes to science: A review. *Studies in Science Education, 2*(1), 1–41.
George, R. (2006). A cross-domain analysis of change in students' attitudes toward science and attitudes about the utility of science. *International Journal of Science Education, 28*(6), 571–589.
Gill, T., & Bell, J. F. (2013). What factors determine the uptake of A-level physics? *International Journal of Science Education, 35*(5), 753–772.
Goodrum, D., Hackling, M., & Rennie, L. (2001). *The status and quality of teaching and learning of science in Australian schools: A research report*. Canberra: Department of Education, Training and Youth Affairs.
Greenfield, T. A. (1997). Gender- and grade-level differences in science interest and participation. *Science Education, 81*(3), 259–276.
Halpern, D. F., Benbow, C. P., Geary, D. C., Gur, R. C., Hyde, J. S., & Gernsbacher, M. A. (2007). The science of sex differences in science and mathematics. *Psychological Science in the Public Interest, 8*(1), 1–51.
Hampden-Thompson, G., & Bennett, J. (2013). Science teaching and learning activities and students' engagement in science. *International Journal of Science Education, 35*(8), 1325–1343.

Haraway, D. (1988). Situated knowledges: The science question in feminism and the privilege of partial perspective. *Feminist Studies, 14*(3), 575–599.

Harding, S. (1998). Women, science, and society. *Science, 281*(5383), 1599–1600.

Haste, H. (2004). *Science in my future: A study of the values and beliefs in relation to science and technology amongst 11–21 year olds*. London: Nestle Social Research Programme.

Haussler, P., & Hoffman, L. (2002). An intervention study to enhance girls' interest, self-concept, and achievement in physics classes. *Journal of Research in Science Teaching, 39*(9), 870–888.

Hazari, Z., Sonnert, G., Sadler, P. M., & Shanahan, M.-C. (2010). Connecting high school physics experiences, outcome expectations, physics identity, and physics career choice: A gender study. *Journal of Research in Science Teaching, 47*(8), 978–1003.

Head, J. (1979). Personality and the pursuit of science. *Studies in Science Education, 6*(1), 23–44.

Heilbronner, N. N. (2011). Stepping onto the STEM pathway: Factors affecting talented students' declaration of STEM majors in college. *Journal for the Education of the Gifted, 34*, 876–899.

Hernandez-Martinez, P., Black, L., Williams, J., Davis, P., Pampaka, M., & Wake, G. (2008). Mathematics students' aspirations for higher education: Class, ethnicity, gender and interpretative repertoire styles. *Research Papers in Education, 23*(2), 153–165.

Hidi, S. (1990). Interest and its contribution as a mental resource for learning. *Review of Educational Research, 60*(4), 549–571.

Hidi, S., & Renninger, K. A. (2006). The four-phase model of interest development. *Educational Psychologist, 41*(2), 111–127.

Holmegaard, H. T., Madsen, L. M., & Ulriksen, L. (2014). To choose or not to choose Science: Constructions of desirable identities among young people considering a STEM higher education programme. *International Journal of Science Education, 36*(2), 186–215.

Homer, M., Ryder, J., & Donnelly, J. (2013). Sources of differential participation rates in school science: The impact of curriculum reform. *British Educational Research Journal, 39*(2), 248–265.

Hyde, J. S., & Linn, M. C. (2006). Gender similarities in mathematics and science. *Science, 314*(5799), 599–600.

Jenkins, E. W. (2006). Student opinion in England about science and technology. *Research in Science & Technological Education, 24*(1), 59–68.

Jenkins, E. W., & Nelson, N. W. (2005). Important but not for me: Students' attitudes towards secondary school science in England. *Research in Science & Technological Education, 23*(1), 41–57.

Jensen, F., & Sjaastad, J. (2013). A Norwegian out-of-school mathematics project's influence on secondary students' STEM motivation. *International Journal of Science and Mathematics Education*. Published online March 2013 *11*(6):1437–1461.

Jones, G., Howe, A., & Rua, M. (2000). Gender differences in students' experiences, interests, and attitudes towards science and scientists. *Science Education, 84*(2), 180–192.

Kaya, S., & Rice, D. C. (2010). Multilevel effects of student and classroom factors on elementary science achievement in five countries. *International Journal of Science Education, 32*(10), 1337–1363.

Kelly, A. (1988). *Getting the GIST: A quantitative study of the effects of the girls into science and technology project*. Manchester: University of Manchester.

Khoo, S. T., & Ainley, J. (2005). *Attitudes, intentions and participation* (Research Report 41). Camberwell: Australian Council for Educational Research.

Kind, P. M., Jones, K., & Barmby, P. (2007). Developing attitudes towards science measures. *International Journal of Science Education, 27*(7), 871–893.

Krapp, A. (2002). An educational-psychological theory of interest and its relation to self-determination theory. In E. Deci & R. Ryan (Eds.), *The handbook of self-determination research* (pp. 405–427). Rochester: University of Rochester Press.

Krapp, A. (2003). Interest and human development: An educational-psychological perspective. *British Journal of Educational Psychology, Monograph Series II (2) Development and Motivation: Joint Perspectives*, 57–84.

Krapp, A., & Prenzel, M. (2011). Research on interest in science: Theories, methods, and findings. *International Journal of Science Education, 33*(1), 27–50.

Krogh, L. B., & Thomsen, P. V. (2005). Studying students' attitudes towards science from a cultural perspective but with a quantitative methodology: Border crossing into the physics classroom. *International Journal of Science Education, 27*(3), 281–302.

Lamb, S., & Ball, K. (1999). *Curriculum and careers: the education and labour market consequences of year 12 subject choice*. LSAY Research Reports. Longitudinal Surveys of Australian Youth Research Report; n. 12.

Lawler, S. (2008). *Identity: Sociological perspectives*. Cambridge: Polity Press.

Lemon, N. (1973). *Attitudes and their measurement*. London: B.T. Batsford Ltd.

Lent, R. W., Lopez, F. G., & Bieschke, K. J. (1993). Predicting mathematics-related choice and success behaviors: Test of an expanded social cognitive model. *Journal of Vocational Behavior, 42*, 223–236.

Lent, R. W., Lopez, A. M., Lopez, F. G., & Sheu, H. (2008). Social cognitive career theory and the prediction of interests and choice goals in the computing disciplines. *Journal of Vocational Behavior, 73*, 52–62.

Lent, R. W., Sheu, H., Gloster, C. S., & Wilkins, G. (2010). Longitudinal test of the social cognitive model of choice in engineering students at historically Black universities. *Journal of Vocational Behavior, 76*, 387–394.

Lindahl, B. (2007, April). *A longitudinal study of students' attitudes towards science and choice of career*. Paper presented at annual meeting of the National Association for Research in Science Teaching, New Orleans, LA.

Lyons, T. (2006). The puzzle of falling enrolments in physics and chemistry courses: Putting some pieces together. *Research in Science Education, 36*(3), 285–311.

Lyons, T., & Quinn, F. (2010, November 26–27). *Looking back: Students' perceptions of the relative enjoyment of primary and secondary school science*. Paper presented at the STEM in education conference, Queensland University of Technology, Brisbane, Australia.

Malone, J. A., & Cavanagh, R. F. (1997). The influence of students' cognitive preferences on the selection of science and mathematics subjects. *International Journal of Science Education, 19*(4), 481–490.

Maltese, A. V., & Tai, R. H. (2010). Eyeballs in the fridge: Sources of early interest in science. *International Journal of Science Education, 32*(5), 669–685.

Maltese, A. V., & Tai, R. H. (2011). Pipeline persistence: Examining the association of educational experiences with earned degrees in STEM among U.S. students. *Science Education, 95*(5), 877–907.

Masnick, A. M., Valenti, S. S., Cox, B. D., & Osman, C. J. (2010). A multidimensional scaling analysis of students' attitudes about science careers. *International Journal of Science Education, 32*(5), 653–667.

McEwen, A., Knipe, D., & Gallagher, T. (1997). Science and arts choices at A-level in Northern Ireland: A ten-year perspective. *International Journal of Science Education, 19*, 761–771.

Mead, M., & Metraux, R. (1957). Image of the scientist among high-school students. *Science, 126*(3270), 384–390.

Mendick, H. (2005). Mathematical stories: Why do more boys than girls choose to study mathematics at AS-level in England? *British Journal of Sociology of Education, 26*(2), 235–251.

Millar, R., & Osborne, J. (2001). *Beyond 2000: Science education for the future*. London: King's College London.

Mujtaba, T., & Reiss, M. J. (2012). What sort of girl wants to study physics after the age of 16? Findings from a large-scale UK survey. *International Journal of Science Education, 35*(7), 2979–2998.

Munby, H. (1982). The impropriety of "panel of judges" validation in science attitude scales: A research comment. *Journal of Research in Science Teaching, 19*(7), 617–619.

Munro, M., & Elsom, D. (2000). *Choosing science at 16: The influence of science teachers and careers advisers on students' decisions about science subjects and science and technology careers*. Cambridge: National Institute for Careers Education and Counseling (NICEC).

Murphy, C., & Begg, J. (2005). *Primary science in the UK: A scoping study: Final report to the Wellcome Trust*. London: Wellcome Trust.

Murphy, P., & Whitelegg, E. (2006). *Girls in the physics classroom: A review of the research on the participation of girls in physics*. London: Institute of Physics.

National Academy of Sciences. (2005). *Rising above the gathering storm: Energizing and employing America for a brighter economic future*. Washington, DC: National Academy of Sciences: Committee on Science Engineering and Public Policy.

National Foundation for Educational Research (NFER). (2011). *Exploring young people's views on science education*. Slough: NFER.

Neuschmidt, O., Barth, J., & Hastedt, D. (2008). Trends in gender differences in mathematics and science (TIMSS 1995–2003). *Studies in Educational Evaluation, 34*(2), 56–72.

OECD. (1997). *Promoting public understanding of science and technology*. Paris: OECD.

Ofsted. (2011). *The annual report of Her Majesty's Chief Inspector of education, children's services and skills*. Norwich: Ofsted.

Olsen, R. V., & Lie, S. (2011). Profiles of students' interest in science issues around the world: Analysis of data from PISA 2006. *International Journal of Science Education, 33*(1), 97–120.

Olsen, R. V., Prenzel, M., & Martin, R. (2011). Interest in science: A many-faceted picture painted by data from the OECD PISA study. *International Journal of Science Education, 33*(1), 1–6.

Olszewski-Kubilius, P., & Yasumoto, J. (1994). Factors affecting the academic choices of academically talented adolescents. *Talent Development, 2*, 393–398.

Ormerod, B., & Duckworth, D. (1975). *Pupils' attitudes to science: A review of research*. Berkshire: NFER Publishing Company.

Osborne, J., & Collins, S. (2001). Pupils' views of the role and value of the science curriculum. *International Journal of Science Education, 23*(5), 441–467.

Osborne, J., Simon, S., & Collins, S. (2003). Attitudes towards science: A review of the literature and its implications. *International Journal of Science Education, 25*(9), 1049–1079.

Pike, A. G., & Dunne, M. (2011). Student reflections on choosing to study science post-16. *Cultural Studies of Science Education, 6*(2), 485–500.

Ramsden, J. M. (1998). Mission impossible? Can anything be done about attitudes to science? *International Journal of Science Education, 20*(2), 125–137.

Raved, L., & Assaraf, O. B. Z. (2011). Attitudes towards science learning among 10th-grade students: A qualitative look. *International Journal of Science Education, 33*(9), 1219–1243.

Regan, E. M., & Childs, P. E. (2003). An investigation of Irish students' attitudes to chemistry: The promotion of chemistry in school project. *Chemistry Education: Research and Practice, 4*(1), 43–51.

Reid, N. (2006). Thoughts on attitude measurement. *Research in Science & Technological Education, 24*(1), 3–27.

Reitz, J. G. (1973). The flight from science reconsidered: Career choice of science and engineering in the 1950s and 1960s. *Science Education, 57*(2), 121–134.

Renninger, K. A., & Hidi, S. (2011). Revisiting the conceptualisation, measurement, and generation of interest. *Educational Psychologist, 46*(3), 168–184.

Renninger, K. A., Hidi, S., & Krapp, A. (Eds.). (1992). *The role of interest in learning and development*. Hillsdale: Lawrence Erlbaum Associates.

Roth, W.-M., & Tobin, K. (2007). Aporias of identity in science: An introduction. In W.-M. Roth & K. Tobin (Eds.), *Science, learning, identity: Sociocultural and cultural-historical perspectives*. Rotterdam: Sense Publishers.

Sadler, D. R. (2002). *Interdisciplinarity in university teaching and research*. Brisbane: Griffith Institute for Higher Education.

Sadler, P. M., Sonnert, G., Hazari, Z., & Tai, R. (2012). Stability and volatility of STEM career interest in high school: A gender study. *Science Education, 96*(3), 411–427.

Salmi, H. (2002). *Factors affecting students' choice of academic studies: The motivation created by informal learning* (Publication no. http://www.heureka.fi/portal/englanti/about_heureka/

research/factors_affecting_students%60_choice_of_academic_studies_the_motivation_created_by_informal_learning._/).
Scantlebury, K., & Baker, D. (2007). Gender issues in science education research: Remembering where the difference lies. In S. Abell & N. N. Lederman (Eds.), *Handbook of research on science education* (pp. 257–286). Mahwah: Lawrence Erlbaum.
Schibeci, R. A. (1984). Attitudes to science: An update. *Studies in Science Education, 11*, 26–59.
Schibeci, R. A. (1986). Images of science and scientists and science education. *Science Education, 70*(2), 139–149.
Schreiner, C., & Sjoberg, S. (2007). Science education and youth's identity construction – Two incompatible projects? In C. Corrigan, J. Dillon, & R. Gunstone (Eds.), *The re-emergence of values in the science curriculum*. Rotterdam: Sense Publishers.
Shanahan, M. C. (2009). Identity in science learning: Exploring the attention given to agency and structure in studies of identity. *Studies in Science Education, 45*(1), 43–64.
Sikora, J., & Pokropek, A. (2012). Gender segregation of adolescent science career plans in 50 countries. *Science Education, 96*(2), 234–264.
Silvia, P. J. (2006). *Exploring the psychology of interest*. New York: Oxford University Press.
Sjaastad, J. (2012). Sources of inspiration: The role of significant persons in young people's choice of science in higher education. *International Journal of Science Education, 34*(10), 1615–1636.
Sjoberg, S. (2002). Science and technology education in Europe: Current challenges and possible solutions. *Connect : UNESCO international science, technology and environmental education newsletter, 27*, 1–5.
Sjoberg, S., & Schreiner, C. (2005). How do learners in different cultures relate to science and technology? Results and perspectives from the project ROSE (the Relevance of Science Education). *Asia-Pacific Forum on Science Learning and Teaching, 6*(2), 1–17.
Skelton, C. (2010). Gender and achievement: Are girls the "success stories" of restructured education systems? *Educational Review, 62*(2), 131–142.
Skelton, C., Francis, B., & Read, B. (2010). "Brains before 'beauty'?" High achieving girls, school and gender identities. *Educational Studies, 36*(2), 185–194.
Skryabina, E. (2000). *Student attitudes to learning physics at school and university levels in Scotland*. Ph.D. thesis. Glasgow: University of Glasgow.
Smith, E. (2010a). Do we need more scientists? A long-term view of patterns of participation in UK undergraduate science programmes. *Cambridge Journal of Education, 40*(3), 281–298.
Smith, E. (2010b). Is there a crisis in school science education in the UK? *Educational Review, 62*(2), 189–202.
Smith, E., & Gorard, S. (2011). Is there a shortage of scientists? A re-analysis of supply for the UK. *British Journal of Educational Studies, 59*(2), 159–177.
Smyth, E., & Darmody, M. (2009). 'Man enough to do it'? Girls and non-traditional subjects in lower secondary education. *Gender and Education, 21*(3), 273–292.
Smyth, E., & Hannan, C. (2002). *Who chooses science? Subject choice in second-level schools*. Dublin: Liffey Press/ESRI.
Solomon, J. (1997). Girls' science education: Choice, solidarity and culture. *International Journal of Science Education, 19*(4), 407–417.
Stagg, P. (2007). *Careers from science: An investigation for the Science Education Forum*. Warwick: Centre for Education and Industry (CEI).
Steinke, J., Applegate, B., Lapinski, M., Ryan, L., & Long, M. (2012). Gender differences in adolescents' wishful identification with scientist characters on television. *Science Communication, 34*(2), 163–199.
Stokking, K. M. (2000). Predicting the choice of physics in secondary education. *International Journal of Science Education, 22*(12), 1261–1283.
Taconis, R., & Kessels, U. (2009). How choosing science depends on students' individual fit to 'science culture'. *International Journal of Science Education, 31*(8), 1115–1132.

Tai, R. H., Liu, C. Q., Maltese, A. V., & Fan, X. (2006a). Planning early for careers in science. *Science, 312*, 1143–1144.
Tai, R. H., Sadler, P. M., & Mintzes, J. J. (2006b). Factors influencing college science success. *Journal of College Science Teaching, 36*(1), 52–66.
The Task Force on the Physical Sciences. (2002). *Report and recommendations*. Dublin: The Task Force on the Physical Sciences.
Thomson, S. (2008). Examining the evidence from TIMSS: Gender differences in year 8 science achievement in Australia. *Studies in Educational Evaluation, 34*(2), 73–81.
Thomson, S., & De Bortoli, L. (2008). *Exploring scientific literacy: How Australia measures up. The PISA 2006 survey of students' scientific, reading and mathematical literacy skills*. Australian Council for Educational Research: Melbourne.
Treasury, H. M. (2006). *Science and innovation investment framework: Next steps*. London: HMSO.
Tripney, J., Newman, M., Bangpan, M., Niza, C., Mackintosh, M., & Sinclair, J. (2010). *Subject choice in STEM: Factors influencing young people (aged 14–19) in education about STEM subject choices: A systematic review of the UK literature*. London: EPPI.
Tytler, R., & Osborne, J. (2012). Student attitudes and aspirations towards science. In B. J. Fraser, K. G. Tobin, & C. J. McRobbie (Eds.), *Second international handbook of science education* (pp. 597–625). Dordrecht: Springer.
Uerz, D., Dekkers, H., & Dronkers, J. (1999). Wiskunde en Taalvaardigheid als Voorspeller van B-Keuzen in her Voortgezet Onderwijs [Mathematics and language ability as predictors of science choices in secondary education]. *Pedagogische Studieën, 76*, 170–182.
Ulriksen, L., Madsen, L. M., & Holmegaard, H. T. (2010). What do we know about explanations for drop out/opt out among young people from STM higher education programmes? *Studies in Science Education, 46*(2), 209–244.
Ventura, F. (1992). Gender, science choice and achievement: A Maltese perspective. *International Journal of Science Education, 14*(4), 445–461.
Venville, G., Rennie, L., Hanbury, C., & Longnecker, N. (2013). Scientists reflect on why they chose to study science. *Research in Science Education, 43*(6), 2207–2233.
Vidal Rodeiro, C. L. (2007). *A level subject choice in England: Patterns of uptake and factors affecting subject preferences*. Cambridge: University of Cambridge: Local Examinations Syndicate.
Wang, X. (2013). Why students choose STEM majors: Motivation, high school learning, and postsecondary context of support. *American Educational Research Journal, 50*(5), 1081–1121.
Warton, P. M. (1997, April). *Motivational goals, information sources and subject choice in adolescence*. Paper presented at the Biennial Meeting of the Society for Research in Child Development, Washington, DC.
Warton, P. M., & Cooney, G. H. (1997). Information and choice of subjects in the senior school. *British Journal of Guidance & Counselling, 25*(3), 389–397.
Watt, H. (2005). Exploring adolescent motivations for pursuing maths-related careers. *Australian Journal of Educational and Developmental Psychology, 5*, 107–116.
Whitehead, J. M. (1996). Sex stereotypes, gender identity and subject choice at A-level. *Educational Research, 38*(2), 147–160.
Wikeley, F., & Stables, A. (1999). Changes in school students' approaches to subject option choices: A study of pupils in the West of England in 1984 and 1996. *Educational Research, 41*(3), 287–299.
Wong, B. (2012). Identifying with science: A case study of two 13-year-old 'high achieving working class' British Asian girls. *International Journal of Science Education, 34*(1), 43–65.
Woolnough, B. E. (1994). Factors affecting students' choice of science and engineering. *International Journal of Science Education, 16*(6), 659–676.
Xu, J., Coats, L. T., & Davidson, M. L. (2011). Promoting student interest in science: The perspectives of exemplary African American teachers. *American Educational Research Journal, 49*(1), 124–154.

Yazilitas, D., Svensson, J., de Vries, G., & Saharso, S. (2013). Gendered study choice: A literature review. A review of theory and research into the unequal representation of male and female students in mathematics, science, and technology. *Educational Research and Evaluation, 19*(6), 525–545.

Zeyer, A., & Wolf, S. (2010). Is there a relationship between brain type, sex and motivation to learn science? *International Journal of Science Education, 32*(16), 2217–2233.

Chapter 6
Science Aspirations and Gender Identity: Lessons from the ASPIRES Project

Louise Archer and Jennifer DeWitt

Theoretical Approach

Our approach draws on feminist poststructuralist theorizations of identity (e.g. Archer and Francis 2007) and Judith Butler's (1990, 1993) theorizations of gender as 'performance'. Our conceptual framework understands identity as non-essentialised – identity is not 'fixed' or ever 'achieved' (Anthias 2001), rather it is fluid, contested and produced through discourse (Burman and Parker 1993; Gee 1996). That is, identities are constantly developing – they are always' in process' (Hall 1990: 222). We see gender as intersecting with, and mediated by, other aspects of identity, such as 'race'/ethnicity and social class (Archer and Francis 2007; Calabrese Barton and Brickhouse 2006). In this way, identities can be understood as social products, produced within and through discourse and social relations: they are 'real fictions' that are constructed through social life and relations of power (Foucault 1978; Weeks 1981).

Butler's work (e.g. 1990, 1993) has been particularly influential within gender theory, especially her conceptualisation of gender as performative. From this perspective, gender is not the inevitable 'result' or product of a person's sex – it does not emanate 'naturally' from particular (sexed, racialised, classed) bodies. Rather, gender is socially constructed and produced through discursive and bodily 'acts'. Gender is, therefore, not something you 'are' or 'have' – it is something that you 'do' (perform) and continually re-do. Gender identities are powerful illusions (Butler 1990: 185/6) in that they appear 'real' and enduring but the patterned and

L. Archer (✉) • J. DeWitt
Department of Education and Professional Studies, King's College London, Waterloo Bridge Wing, Franklin-Wilkins Building, Waterloo Road, London SE1 9NH, UK
e-mail: louise.archer@kcl.ac.uk

E.K. Henriksen et al. (eds.), *Understanding Student Participation and Choice in Science and Technology Education*, DOI 10.1007/978-94-007-7793-4_6

predictable nature of gender identities is achieved through the repetitive nature of their continual enactment (Renold 2005). In other words, gender is created through a myriad of verbal and bodily performances in which children 'do girl' (or 'do boy') (Butler 1990: 185–6).

We also use Butler's concept of 'intelligibility' to understand the context within which children and adults produce gender identities and the social pressures that they experience to perform particular (normative, socially sanctioned) identities:

> "Intelligible" genders are those which in some sense institute and maintain relations of coherence and continuity among sex, gender, sexual practice, and desire. (Butler 1990: 23)

Consequently, Butler argues, some gender performances are rendered 'unintelligible' (i.e. those which are more subversive or counter-hegemonic). That is, 'the cultural matrix through which gender identity has become intelligible requires that certain kinds of "identities" cannot "exist"' (Butler 1990: 24). For instance, sociological research has shown how children still in primary school can experience considerable pressures to perform particular heterosexualised versions of masculinity and femininity (Renold 2005).

The ASPIRES Study

The ASPIRES project is funded by the UK's Economic and Social Research Council as part of its Targeted Initiative on Science and Mathematics Education (TISME). The study is a 5-year, longitudinal exploration of science aspirations and career choice among 10–14 year olds in England. It comprises a quantitative online survey that was administered to a sample of over 9,000, 10/11 year-old students in the first phase (students will be tracked and surveyed again in subsequent phases at ages 12 and 14) and in-depth, repeat interviews with pupils (at age 10/11; age 12/13 and age 13/14) and their parents (who are interviewed twice, once when their children are age 10/11 and again at age 13/14). This chapter is primarily based on analysis of the Phase 1 qualitative dataset, which comprises 170 interviews with 78 parents and 92 children age 10/11 (Year 6), drawn from 11 schools in England. At points throughout the paper contextual information is provided from the survey as a means for framing the qualitative data analysis, although full details of the survey and its methods, analyses and findings are discussed in separate publications (DeWitt et al. 2010, 2011). In this chapter we focus particularly on data from the girls (see Archer et al. 2010a, b, 2014 for discussion of masculinity and boys).

The students and parents who were interviewed were recruited from 11 primary schools in England (one in the Midlands, two in the Eastern region, two in the South East, four in London and one in the South), which were sampled from the 279 schools that responded to the Phase 1 survey as part of the wider study.[1] A sampling frame was constructed to represent six target categories of school (e.g. 'multiethnic urban/inner city schools'; 'working-class suburban'; 'predominantly white, middle-class suburban schools'; 'independent single sex') to ensure a range of school contexts and populations and prospective schools were purposively sampled from within these target categories. Nine of the schools were state funded primaries and two were private/independent schools (ie. Fee-paying schools, typically attended by high SES students). Students came from a broad range of socioeconomic classes and ethnic backgrounds.[2]

Following extensive reviews of literature from relevant work within the fields of science education and sociology of education, two topic guides (for use with children and parents) were developed and piloted. The children's interviews covered areas such as: aspirations (and sources of these aspirations); interests in school and out; what they like/dislike about school; attitudes towards and engagement in school science; broader perceptions of science. Parental interviews focused on: family context; perceptions and experience of the child's schooling; involvement in education; child's personality and interests; their child's aspirations, their own perceptions of and relationship with science and engineering, including their thoughts about why so few children pursue science post-16.

Interviews were conducted by four of the research team (Louise Archer, Jennifer DeWitt, Beatrice Willis and Billy Wong), with the majority of the interviews being conducted by the second author. Of the interviewers, three [LA, JD and BW] are White middle-class women (with English, American and French national backgrounds, respectively) and one [Billy] is a British-Chinese male PhD student. Interviewees were invited to choose their own pseudonyms, hence the majority of pseudonyms cited in this paper reflect the personal choices of interviewees.

All interviews were digitally audio-recorded and transcribed. In line with the study's conceptual approach outlined earlier, data were analysed using an analysis of discourse approach (Burman and Parker 1993). Initial coding and sorting of the

[1] 9319 Year 6 students from 279 schools (248 state schools; 31 independent schools) completed the Phase 1 questionnaire between October and December 2009. (The Phase 2 survey took place in autumn 2011 and phase 3 will occur in winter/spring 2013.) The sample represented all regions of the country and was roughly proportional to the overall national distribution of schools in England by attainment and proportion of students eligible for free school meals. Of the students who completed the survey there were: 51 % boys, 49 % girls; 846 (9 %) in private schools, 8,473 (91 %) in state schools; 75 % White, 9 % Asian (Indian, Pakistani, Bangladeshi heritage), 8 % Black (Black African, Black Caribbean), 1 % Far Eastern, 8 % mixed or other. The survey itself covered topics such as: aspirations in science; attitudes towards school science; self-concept in science; images of scientists; participation in science-related activities outside of school; parental expectations; parental school involvement; parental attitudes towards science; and peer attitudes towards school and towards school science.

[2] Social class categorisations were assigned by the lead author and second author using the NS-SEC (an official UK government classification system for socio-economic status) as a guide to categorise parental occupations. Ethnicity was assigned based on self/parental reported ethnic background.

data (on key topic areas, themes and by responses to particular questions) was undertaken by two researchers (LA and BW) using the NVivo software package, with the lead author providing a check on reliability of coded extracts for the specified codes. The lead author then searched coded extracts to identify discursive gender repertoires and patterns of aspirations/relationships with science, which were then tested and refined through successive phases of coding and analysis, iteratively testing emergent themes across the data set to establish "strength" and prevalence (Miles and Huberman 1994). In line with the stated conceptual framework, the lead author then developed and tested theoretically informed hypotheses to see if they were supported or challenged by the data, for instance to identify interplays of power and practices of power and gendered discourses within respondents' talk. Draft analyses were then fed back to other team members (especially those who conducted fieldwork) for checking against their readings of the data.

Masculinity, Femininity and Science Aspirations

Our survey of over 9,000 10 and 11 year-olds indicated that the majority (over 70 %) of children reported enjoying science, held positive views of scientists, took part in science-related activities in their spare time and felt that their parents valued science. However, a much smaller proportion (under 17 %) aspired to careers in science, suggesting a disconnect between children's interest in 'doing' science (at school and in their spare time) and 'being' a scientist (Archer et al. 2010a). We found no notable gender difference among the 648 children in the survey sample who were classified as 'uninterested in science' (i.e. there were roughly equal proportions of boys and girls who recorded the lowest scores on all the five science aspirations items on our questionnaire), but notably fewer girls (n = 92, 37 %) than boys (n = 159, 63 %) were classified as being 'science keen' (n = 251[3]) (i.e. those scoring very highly on all five science aspirations items). We were interested, therefore, to explore why girls seem less likely than boys to aspire to careers in/from science at age 10/11, even though both genders generally enjoy science at school at this age – and what makes some girls develop science aspirations but not others?

Which Girls Have Science Aspirations?

Our data suggests that children from 'middle-class' backgrounds are more likely to develop and sustain science aspirations which, as we discuss elsewhere, reflects differences and interactions between family practices, values and science

[3] i.e. 3 % of boys and 2 % of the girls are 'science keen'.

capital[4] (Archer et al. 2012a). For instance, of the 92 'science-keen' girls who completed the survey, only 11 % (n = 10) were classified as having very low/low cultural capital (cf. 25 % of the total sample with very/low cultural capital) whereas 60 % (n = 55) of science keen girls had high or very high cultural capital (cf. 41 % of the total sample with very/high cultural capital).[5] This suggests a link between family cultural capital and the likelihood of a child developing science aspirations. Likewise, in the interview sample, the majority of girls expressing science aspirations were from (upper and lower) middle-class backgrounds,[6] suggesting a class bias in terms of who tends to see science careers as potentially 'for me', irrespective of the majority of children's reported general interest and enjoyment of science both in school and out. Although the 'science aspirant' boys we interviewed came from a range of ethnic and social class backgrounds, it was striking that there was only one clearly working-class girl (Georgia) among our science-aspirant girl interviewees (the others were from upper and lower middle-class backgrounds). Given that science-aspirant girls are proportionally 'over-represented' within the interview sample (as compared to the survey), we found this class imbalance particularly striking.

Of the 55 girls in the interview sample, 17 expressed science aspirations; 13 identified science-related aspirations and 25 expressed aspirations unrelated to science. As we discuss below, the development of girls' science aspirations (or not) seems to be influenced by two dominant popular discourses, namely public perceptions of science as 'clever' and as 'masculine'. These associations were evident within the rationales given by those girls who did not see science careers as 'for me' *and* were prominent in the identity work undertaken by those girls who did aspire to careers in science, with the latter going to considerable lengths to navigate a way through dominant associations so as to present themselves as 'normal' girls.

[4] Science capital is defined as the material and cultural science-related resources that a family may be able to draw on, such as science-related qualifications, knowledge, understanding ('scientific literacy') and social contacts (see Archer et al. 2013a, b, 2014).

[5] Due to the problems in getting children age 10/11 to self-report their parental occupations in order to enable a more accurate assignment of social class, we also used items designed to ascertain measures of 'cultural capital', to provide a rough and ready indication of social class (see DeWitt et al. 2012). In the whole sample, 25 % of children were classified as having low or very low cultural capital and 41 % as having high or very high cultural capital.

[6] In the interview sample there were proportionally more students from upper and lower middle-class backgrounds than from working-class backgrounds, so to an extent this is a reflection of the sample – yet the imbalance is clearly reflected in that just one of the science aspirant girls was from a working-class background as compared to the over-representation of working-class girls among those classified as having no science aspirations (see Archer et al. (2013a, b) for discussion of data from girls with no science aspirations).

The Popular Discourse of Science as 'Clever'

The first, and most prevalent, theme noted within both interview and survey samples was the association of science with '***cleverness***'. Over 81 % of the 9,000+ survey sample agreed or strongly agreed that 'scientists are brainy' and an association of science with 'cleverness' was evident among both parent and child interviewees – including those who personally identified and those who disidentified with science. Although a principal component analysis[7] of the survey items suggests that perceptions of cleverness/braininess were viewed largely as positive attributes (being related to positive views of scientists rather than pejorative views), as we argue elsewhere (DeWitt et al. 2012), these close associations between science/scientists and cleverness/braininess may also contribute to many children's views of science and science careers as unusual, exceptional and 'not for me'.

The girls in our interview sample who held science aspirations all self-identified as, and were identified by their parents, as 'clever', 'bright' high academic achievers. For instance, PJ said 'I like studying' and Preeti explained how 'I'm very interested in science and science lessons in school and er I get some high grades in my science test'. Likewise, Thalia's father explained how his daughter is a 'high achiever' across all subject areas, a view similarly expressed by other parents. These girls tended to be part of academically achieving, like-minded friendship groups. As Luna's mother (Stella) similarly explained "there's a group of them that are all quite clever [. . .] they're all quite similar actually".

In contrast, those girls who did not hold science aspirations were more likely to describe themselves (and be described by their parents) as either 'middling' or 'not clever' pupils. For instance, when asked by the interviewer "who is into science?", Louise (a white, working-class girl who expressed some of the most resistant views of science in our interview sample) replied "Well the clever ones are. Like the ones that are going to the grammar school are into like every subject. . .. They don't mind having lessons'. She continued 'its just strange how all the clever ones are into science'. Likewise, Victoria2 (white Eastern European, working class, Metropolitan School) gave her reasons for not wanting to become a scientist as "cos most scientists are brainy and I don't want to be brainy". Interestingly Victoria2 did like some areas of science (notably animals and biology) but did not enjoy what she called 'the normal subject' of science. Despite her resistance to being 'brainy' she also held some more general, positive views of science, describing it as 'awesome' – suggesting a disconnect between her interest and respect for (some areas of) science and her view of herself as a learner and the capacity to see herself as a 'science person'. Flower (White, Eastern European, working class, Metropolitan School) also agreed that you have to be clever to be

---

[7] Principal component analysis is a way of measuring which items in a survey group together (are responded to in similar ways) and therefore suggest factors or components that underlie responses to survey items.

into science and was adamant that personally she would not want to follow a science career "because I'm not that smart". Likewise, Celina (white, working class, Metropolitan School) described those who are 'really into science' as "brainiacs, because they just want to do Science, they don't want to do anything else in their life".

We therefore suggest that the popular societal association of science with 'cleverness' means that science aspirations are not experienced as viable or appropriate for all students. That is, children who do not self-identify (and who are not recognised by others) as 'clever' and academically successful, may be less likely to see science aspirations as being 'for me'. We found that even where parents attempted to encourage their daughters' science interests and challenged stereotypes of science as being 'geeky' or 'for boys', they still unwittingly associated science with 'cleverness'. A particularly clear example of this is provided by the case of Danielle and her mother, Sandra. Danielle describes herself as a 'middle' student, a view that her mother, Sandra, concurs with ("Um, I think she's more of a middle of the range child. There's nothing really that she excels in"). Danielle has various interests, one of which is science, which she claims is one of her favourites lessons ("I'm not being a kiss-up[8] but my favourite lesson is actually science"). Sandra strongly supports her daughter's science interests and Danielle's father works as a mechanical engineer. Yet, science aspirations are unthinkable for Danielle, who feels "I'm not clever enough to be good at science". As Sandra explained:

Sandra: Yeah, that's what she said to me. I said why? She said oh, you have to be really clever, you have to be a geek.

Int: Mmm, how did you respond?

Sandra: [I said] 'What do you mean, what do you mean you have to be really clever and be a geek?' She said 'well, you do don't you? Everybody sees it. You have to ... you see it on TV and [scientist character], she's a geek, no friends, got glasses'. [...] She said 'well, you have to be really clever and I'm not'. I said you are clever. You could do anything you want.

We suggest that the above example illustrates a disconnect between Danielle's construction of science (as 'clever') and her own self-concept as a 'middling' pupil – and that this contributes to preventing Danielle from perceiving science aspirations as 'for me'. This is likely to be the case for many 'average' pupils. In other words, the powerful popular association of science with 'cleverness' (and its perception as being a highly academic subject) means that identifying with science (seeing oneself as a potential 'science person') requires that students take up (and are recognized by others as embodying) a 'good student' identity

---

[8] 'kiss up' means to falsely flatter or in this case, to express a false opinion in order to gain favour with the interviewer.

(Carlone and Johnson 2007). Existing research suggests that this can be more difficult for working-class learners, girls and those from some minority ethnic backgrounds, due to dominant educational discourses that construct the 'ideal learner' as white, male and middle-class (Archer 2008). Moreover, the popular association of science with cleverness constructs science as an exclusive, distinct and exceptional field – something that is for the 'clever' few, and is not seen as 'for me' by the majority of students.

Science as a 'Boy Thing'/Science as 'Not Girly'

A second key theme to emerge from the wider interviews and survey responses was the discourse of science as masculine ('for boys'). This finding chimes with existing literature (e.g. Boaler 1997; Hughes 2001; Lightbody and Durndell 1996; Mendick 2006; Walkerdine 1990), which discusses how science (especially the physical sciences) and mathematics tend to be associated with masculinity. Although this view was not expressed unanimously (e.g. eight girls and six boys in the interview sample suggested that girls might be more interested in science and many children felt there were no gender differences in terms of interest in science), its purchase as a popular discourse was noted by children and parents alike. Over half of parents described science careers as predominantly masculine (although views differed as to whether this is due to 'biological' differences or social inequalities). We also found that those girls who were very keen on science and who held science aspirations tended to recognize that their interests were not shared by most of their female peers. That is, whether or not they personally subscribed to the view, there was a prevalent recognition among parents and children that science is popularly seen as not 'girly'.

As discussed in Archer et al. (2012b), the 17 girls in our interview sample who expressed science-related aspirations engaged in considerable identity work to 'balance' or accommodate their so-called 'masculine' science aspirations with wider popular discourses of science as 'clever' and 'masculine'. We found that they achieved this in one of two main ways – drawing on identity discourses of either the 'feminine scientist' or the 'bluestocking[9] scientist'. The 'feminine scientist' girls tended to balance their 'masculine' science engagement and aspirations

[9] The term 'Bluestocking' was originally a derisory term applied in eighteenth century England to denote women with scholarly and intellectual interests, but is currently popularly used to denote academic women. The term is used here as a (non-derisory)shorthand to capture and foreground the academic and 'non-girly' nature of these girls' identity performances and their lack of interest in performing more 'popular' hetero-normative femininities. Like Renold's (2005) 'square-girls' who are 'high-achieving, hard-working, rule-following and lacked any interest in popular fashion or 'boys' either as friends or boyfriends' (p. 64), the 'Bluestocking' girls in our study constructed themselves (and were described by their parents) as 'non-girly' and preoccupied with academic success.

with performances of popular femininity (as displayed, for instance, through their interests in fashion, clothes, popular music, sports and so on). Their parents placed particular emphasis on, and were evidently proud of, their daughters' ability to perform these 'balanced' identities that enabled their daughters to be recognised as both clever/liking science *and* popularly accepted as 'normal' girls. Indeed, these girls appeared to be popular and sociable class members. In contrast, the 'blue-stocking' girls foregrounded their academic, 'clever' identities and tended to explicitly define themselves as 'not girly'. While also enjoying a range of interests and by no means being unpopular, these girls were more likely to risk being seen as 'geeks' at school and were more likely to define themselves as quiet and report experiences of being bullied.

As discussed in Archer et al. (2013b), the 25 girls who did not aspire to science careers expressed a range of aspirations (often holding more than one aspiration at a time), but their ambitions tended to coalesce around traditionally gendered careers in the fields of (i) nurturing/caring professions, (ii) expressive/artistic/glamorous jobs and (iii) sports/active jobs (although other areas such as law, business and the police were also mentioned). In line with the findings from wider research on children's aspirations (e.g. Francis 2000), these girls were primarily motivated by vocational concerns (e.g. 'to help others'). As Francis & Skelton (2005) discusses, notions of care (of others and of the self) are integral to 'traditional' (dominant) constructions of femininity and tend not to be voiced by boys to the same extent. The girls' aspirations also revealed high levels of interest in the body and appearance (e.g. aspirations for 'glamorous' and/or jobs in the beauty industry), which similarly resonate with dominant discourses of hetero-femininity (Renold 2005), and intersect with classed discourses (e.g. see Skeggs 1997, 2003).

We suggest that the disconnect between these girls' investments in 'gender traditional' performances of femininity and dominant discursive associations between science and masculinity meant that even those girls who are interested in science could find it difficult to occupy science spaces comfortably–both symbolically and in terms of actual experiences and that this inevitably led them to seeing science as 'not for me'. For example, Sandra described how her daughter, Danielle, had stopped attending an after-school science club because "it was all boys" and how this had impacted on Danielle's perception that science is "a boy thing":

Sandra: I said why can't you do science? She [Danielle] said well, 'oh no it's a boy thing'. And I said 'it's not'. They had [science club name] at school. It's an after school club on Monday and she said 'I'm not going because it's all boys'. You can see what I mean when you're fighting against it aren't you? I said 'well you should at least go along and see if you enjoy it. It's all these experiments' and she said 'oh, it's fun, we did all this' ...

Int: Sorry, is she going to this science after school club?

Sandra: She went twice [Int: She went twice] and then she stopped going because it was all boys and she had no girls to talk to.

We suggest that the above extract illustrates how 'liking science' is not enough to enable many girls to see science aspirations as 'for me' since the popular discursive alignment of science with masculinity mitigates in numerous ways against the development of an understanding or experience of how science aspirations might 'fit' with girls' everyday performances of femininity.

Conclusions

We suggest that the development and cultivation of science aspirations requires girls to engage in considerable identity work, not least to navigate dominant associations of science with 'cleverness' and masculinity, which construct science as an elite field which is only open to women within certain narrow parameters. That is, girls have to identify with, and be able to occupy, a 'clever' learner identity *and* negotiate a socially acceptable performance of femininity that can balance their engagement with the aspects of science that are perceived to be 'masculine' (and masculine notions of 'achievement') if science is to be a 'thinkable' aspiration. We suggest that for the 25 girls in our interview sample who did *not* aspire to science, science aspirations are largely unthinkable because they do not see science as fitting with either (i) their constructions of desirable/intelligible femininity or (ii) their learner identities and student self-concept. Moreover, this lack of fit may be further exacerbated by social inequalities, which render science aspirations less thinkable for working-class girls in particular (not least due to dominant classed associations of 'cleverness' with middle-classness, Archer 2008).

The girls in our sample who did hold science aspirations appeared to achieve this through just one of two options, either through restrained (heterosexualised) 'science femininity' or through an asexualized 'bluestocking' femininity, both of which tend to be associated more often with middle-class femininity. These narrow discursive spaces leave little possibility for other girls (e.g. working-class and/or other minority ethnic girls who may engage in more 'glamorous' performances of working class femininity, for example – e.g. see Archer et al. 2010b; Skeggs 1997) to imagine future science careers as 'for me'. The popular association of science with 'cleverness' (and its perception as being a highly academic subject) means that identifying with science (seeing oneself as a potential 'science person') requires taking up (and being recognized by others for occupying) a 'good student' identity (Carlone and Johnson 2007). This adoption can be more difficult for working-class learners, girls and those from minority ethnic backgrounds because dominant educational discourses construct the 'ideal learner' in particular gendered, classed and racialised ways that normalize the white, middle-class male pupil (Archer 2008; Renold and Allan 2006; Francis 2009; Skelton et al. 2010).

We thus conclude that science aspirations sit in an uneasy tension with femininity and must be continually carefully negotiated and defended against challenges from wider popular discourses which align science with masculinity. The root of continued gender inequalities in girls'/women's participation in, and experiences of, science is, therefore, complex, multiple and highly resistant to change – and is especially problematic for girls who are not middle-class and who do not occupy 'clever' learner identities. Against this, it would seem that those (predominantly middle-class) girls who *do* hold science aspirations need to engage in sustained identity work if they are to be successful both as scientists and as girls (i.e. to perform socially valued forms of femininity) – *see also Part Four, this volume*. This requires careful navigation of dominant cultural associations of science with masculinity and curricula and cultures that are orientated more toward males. Moreover, as our research indicates, these dilemmas are already in place within the elementary school.

Implications

Based on our analyses, we suggest that work might usefully be undertaken to open up popular perceptions of the sciences – that is, to help a wider (and more diverse) range of children and parents to experience and see science-related qualifications and careers as not 'clever' and 'masculine' but 'for me'. There are two key issues associated with this aim: first, to ensure that the cultures operating within post-16 science (in colleges, universities and workplaces) are indeed equitable and do not alienate or disadvantage 'non-traditional' participants. Existing evidence suggests that there are still a number of challenges on this front (e.g. Carlone 2003; 2004; Ong 2005). This will require scrutinizing the cultures that currently operate within the sciences, to make sure that they are fair and inclusive. The second issue concerns how post-16 science qualifications and careers are popularly perceived.

One approach to the latter could be to increase the potential for (and/or families' awareness of) more diverse forms of participation in post-compulsory science. The children and parents in our study largely saw science jobs only in terms of becoming a scientist (or doctor or science teacher), suggesting little public awareness of either the diversity contained within 'being a scientist', nor of the immense diversity of science-related and/or science-informed jobs that exist. If we are to broaden and increase future participation in careers in and from science, it would seem fruitful to broaden teachers', families' and children's awareness of the instrumental benefits and 'transferable' nature of science qualifications (i.e. raising awareness of not only the benefits of careers in science but the diversity of careers in/from science. Indeed, it is particularly ironic that the KS4 programme of science study in England contains not a single reference to the need to educate students about possible future careers in/from science, even though one of the main rationales given for the importance of science to the UK curriculum is the preparation of the next generation of future scientists. Yet changing perceptions of the value of science for future careers is not only a matter of increasing public awareness of diverse routes – there is also a case for increasing the actual diversity of available routes in/from science that go beyond the 'gold standard' of A level[10] and university degree routes in order to broaden participation in the sciences. This is not only a STEM 'pipeline' issue but, in our view, is an important social equity issue. Currently the material and cultural benefits that can derive from post-16 science qualifications and/or careers (not only job opportunities but also the value derived from being a scientifically informed citizen) are largely restricted to particular, privileged social groups (notably white, middle-class men).

Finally, we feel there is a strong case to be made for the implementation of strategies designed to increase science capital (Archer et al. 2012a, 2013a) within UK families, to help make science (and hence science aspirations) more 'known' and familiar within families' everyday lives. In other words, there is still a considerable challenge facing the science education community to enable and encourage more girls to see science aspirations as desirable and 'thinkable' for them (see also Jenkins and Nelson 2005). As Pamela (Black Caribbean girl at Chestnut Junior School, who aspires to be actress, dance teacher or sports teacher) explained, although she enjoys science and does well in it, a science-related future career would be "good for some people but not for me".

References

Anthias, F. (2001). New hybridities, old concepts: The limits of 'culture'. *Ethnic and Racial Studies, 24*(4), 619–641.

Archer, L. (2008). The impossibility of minority ethnic educational 'success'? An examination of the discourses of teachers and pupils in British secondary schools. *European Educational Research Journal, 7*(1), 89–107.

Archer, L., & Francis, B. (2007). *Understanding minority ethnic achievement*. London: Routledge.

Archer, L., DeWitt, J., Osborne, J., Dillon, J., Willis, B., & Wong, B. (2010a). "Doing" science versus "being" a scientist: Examining 10/11-year-old schoolchildren's constructions of science through the lens of identity. *Science Education, 94*, 617–639.

Archer, L., Hollingworth, S., & Mendick, H. (2010b). *Urban youth and schooling*. Maidenhead: Open University Press.

Archer, L., DeWitt, J., Osborne, J., Dillon, J., Willis, B., & Wong, B. (2012a). Science aspirations, capital, and family habitus: How families shape children's engagement and identification with science. *American Educational Research Journal, 49*(5), 881–908.

Archer, L., DeWitt, J., Osborne, J., Dillon, J., Willis, B., & Wong, B. (2012b). "Balancing acts": Elementary school girls' negotiations of femininity, achievement, and science. *Science Education, 96*(6), 967–989.

Archer, L., DeWitt, J., & Wong, W. (2013a). Spheres of influence: What shapes young people's aspirations at age 12/13 and what are the implications for education policy? *Journal of Education Policy, 35*(6), 1037–1063.

Archer, L., DeWitt, J., Osborne, J., Dillon, J., Willis, B., & Wong, B. (2013b). 'Not girly, not sexy, not glamorous': Primary school girls' and parents' constructions of science aspirations. Pedagogy. *Culture and Society, 21*(1), 171–194.

[10] Post-compulsory examinations in the final year of UK secondary education, typically when students are aged 17/18. Used typically as entry requirements to university.

Archer, L., DeWitt, J., & Willis, B. (2014). Adolescent boys' science aspirations: Masculinity, capital, and power. *Journal of Research in Science Teaching, 51*(1), 1–30. doi:10.1002/tea.21122.

Boaler, J. (1997). Reclaiming school mathematics: The girls fight back. *Gender and Education, 9*(3), 285–305.

Burman, E., & Parker, I. (Eds.). (1993). *Discourse analytic research: Repertoires and readings of texts in action*. London: Routledge.

Butler, J. P. (1990). *Gender trouble: Feminism and the subversion of identity*. London: Routledge.

Butler, J. P. (1993). *Bodies that matter: On the discursive limits of sex*. London: Routledge.

Calabrese Barton, A., & Brickhouse, N. (2006). Engaging girls in science. In C. Skelton, B. Francis, & L. Smulyan (Eds.), *Handbook of gender and education* (pp. 221–235). Thousand Oaks: Sage.

Carlone, H. B. (2003). (Re)producing good science students: Girls' participation in high school physics. *Journal of Women and Minorities in Science and Engineering, 9*(1), 17–34.

Carlone, H. B. (2004). The cultural production of science in reform-based physics: Girls' access, participation, and resistance. *Journal of Research in Science Teaching, 41*(4), 392–414.

Carlone, H. B., & Johnson, A. (2007). Understanding the science experiences of successful women of color: Science identity as an analytic lens. *Journal of Research in Science Teaching, 44*(8), 1187–1218.

DeWitt, J., Archer, L., Osborne, J., Dillon, J., Willis, B., & Wong, B. (2010). 'High aspirations but low progression: The science aspirations – Careers paradox amongst minority ethnic students. *International Journal of Science and Mathematics Education, 9*(2), 243–271.

DeWitt, J., Archer, L., Osborne, J., Dillon, J., Willis, B., & Wong, B. (2011). Young children's aspiration in science: The unequivocal, the uncertain and the unthinkable. *International Journal of Science Education, 34*(1), 43–65.

DeWitt, J., Archer, L., & Osborne, J. (2012). Nerdy, brainy and normal: Children's and parents' constructions of those who are highly engaged with science. *Research in Science Education*, 1–22. doi: 10.1007/s11165-012-9315-0.

Foucault, M. (1978). *The history of sexuality: An introduction*. New York: Pantheon.

Francis, B. (2000). The gendered subject: Students' subject preferences and discussions of gender and subject ability. *Oxford Review of Education, 26*(1), 35–48.

Francis, B. (2009). The role of the Boffin as abject other in gendered performances of school achievement. *Sociological Review, 57*(4), 645–669.

Francis, B., & Skelton, C. (2005). *Reassessing gender and achievement: Questioning contemporary key debates*. London: Routledge.

Gee, J. P. (1996). *Social linguistics and literacies: Ideology in discourses* (Taylor & Francis series in pharmaceutical sciences). London: Routledge.

Hall, S. (1990). Cultural identity and diaspora. In J. Rutherford (Ed.), *Identity: Community, culture, difference* (pp. 392–403). London: Lawrence & Wishart Ltd.

Hughes, G. (2001). Exploring the availability of student scientist identities within curriculum discourse: An anti-essentialist approach to gender-inclusive science. *Gender and Education, 13*(3), 275–290.

Jenkins, E. W., & Nelson, N. W. (2005). Important but not for me: Students' attitudes towards secondary school science in England. *Research in Science and Technological Education, 23*(1), 41–57.

Lightbody, P., & Durndell, A. (1996). Gendered career choice: Is sex-stereotyping the cause or the consequence? *Educational Studies, 22*(2), 133–146.

Miles, M. B., & Huberman, A. M. (1994). *Qualitative data analysis: An expanded sourcebook*. Newbury Park: Sage Inc.

Ong, M. (2005). Body projects of young women of color in physics: Interparts of gender, race, and science. *Social Problems, 52*(4), 593–617.

Renold, E. (2005). *Girls, boys and junior sexualities: Exploring childrens' gender and sexual relations in the primary school*. London: Routledge.

Skeggs, B. (1997). *Formations of class & gender: Becoming respectable*. Thousand Oaks: Sage Publications Ltd.
Skeggs, B. (2003). *Class, self, culture*. London: Routledge.
Skelton, C., Francis, B., & Read, B. (2010). Brains before 'beauty'? High achieving girls, school and gender identities. *Educational Studies, 36*(2), 185–194.
Walkerdine, V. (1990). *Schoolgirl fictions*. London: Verso Books.
Weeks, J. (1981). *Sex, politics and society: The regulation of sexuality since 1800*. New York: Longman.

Chapter 7
The Impact of Science Curriculum Content on Students' Subject Choices in Post-compulsory Schooling

Jaume Ametller and Jim Ryder

Introduction

As elaborated in the Introduction (Chap. 1) to this book, the enrolment of appropriate numbers of students onto post-compulsory science courses is an issue of international concern, particularly in the physical sciences (European Commission 2004; NSB 2010). Additionally, attaining an equitable gender balance amongst those students choosing specific post-compulsory science courses has also been identified as a significant challenge. Previous studies have identified a wide range of factors influencing student choice. These include school-related factors such as teacher quality, attainment and enjoyment of the subject (Cleaves 2005). In addition, broader social and cultural factors have also been shown to have a significant impact (Ball et al. 2000; Eccles 2009; Foskett et al. 2008; Foskett and Hemsley-Brown 2001). A recent study conducted in England suggests that a science course providing a strong emphasis on teaching about socio-scientific issues and the nature of science has resulted in increased uptake of science courses within post-compulsory schooling (Millar 2010). Such courses might be affecting uptake by addressing issues known to turn students away from science such as the lack of relevance for their everyday lives (Murphy and Whitelegg 2006). Therefore, a particular interest of the study reported here is on the impact of teaching about socio-scientific issues and the nature of science, alongside other factors, on young people's subject choices within the years of compulsory schooling. Our focus on school students, and the choices that are formed within compulsory schooling, complements other contributions in this book that consider processes of choice beyond compulsory schooling and into higher education.

We do not see students' subject choice as a purely rational decision made at a particular point in time. Rather, we view 'choice' as a dynamic process, influenced

J. Ametller • J. Ryder (✉)
School of Education, University of Leeds, Leeds LS2 9JT, UK
e-mail: J.Ryder@education.leeds.ac.uk

E.K. Henriksen et al. (eds.), *Understanding Student Participation and Choice in Science and Technology Education*, DOI 10.1007/978-94-007-7793-4_7

by a wide range of socio-cultural factors (Foskett and Hemsley-Brown 2001), and a student's developing sense of agency and identity (Archer et al. 2010); a view consistent with perspectives elaborated elsewhere in this book, and particularly in Chap. 3. Hence we focus on two research questions:

1. Through what processes do students come to be following (or not following) a science course within post-compulsory schooling?
2. How do school science experiences feature within students' accounts of these processes?

Viewing subject choice as a dynamic process has influenced the design of our study. We follow Hollway and Jefferson (2000) in using narrative techniques in discussing choices with students. This involves asking students to reflect on *how* they came to be following specific post-compulsory courses, thereby encouraging students to provide stories, or narrative accounts, of the process of choice. This approach can be contrasted with the common approach asking students *why* they made particular choices. This latter approach is likely to lead students to a clipped and overly rationalistic account of choice, with students likely to provide short, standard 'rehearsed answers' such as 'it's my best subject' (Rodd et al. 2010).

Study Design

Overview

We have collected data from two schools known to have a strong focus on the teaching of socio-scientific issues and the nature of science within compulsory schooling. This reflects our interest in the potential impact of such teaching on post-compulsory science choices. We have worked with these two schools as part of a 3 year longitudinal study of school experiences of curriculum reform.[1] Through interviews with teachers in these schools we knew that they had a strong commitment to the teaching of socio-scientific issues and the nature of science.

All students in the first year of post-compulsory[2] schooling (aged 16–17 years) were asked to complete an individual questionnaire probing how they came to be following their particular courses. Based on these responses a purposive sample of 22 students was invited to take part in an individual interview. This sample included those with a science course in their subject choices, and also those with no science

[1] The Enactment and Impact of Science Education Reform (EISER) Project, http://www.education.leeds.ac.uk/research/projects/enactment-and-impact-of-science-education-reform-eiser

[2] Schooling is compulsory up to the age of 16 years in England. All students within compulsory schooling must study science. Thereafter students typically choose to either leave school or choose 4–5 subjects for further study. These may, or may not, include science subjects.

course choices. In interviews students were encouraged to provide a narrative account of their experiences leading to specific subject choices.

Initial Questionnaire

The main purposes of the initial questionnaire were: to identify suitable candidates for the interview sample; to characterise the population of Y12[3] students in these two schools in terms of gender and course choice. In open responses students stated how they had decided which courses to choose at AS-level,[4] why they had chosen/ not chosen science courses, and their career intentions. Students also indicated in closed response questions (using a five point Likert scale, from strongly disagree to strongly agree) their experiences of science lessons (e.g. interest, enjoyment, usefulness), influences on their choices of subject (e.g. attainment, curriculum content, teaching activities, teachers), and when they had first considered following science courses.

In each school the Head of Science was asked to distribute questionnaires to Y12 form tutors. These forms groups comprise 20–30 students, mixed in terms of gender, student attainment and course choices. Form groups are used to address administrative and pastoral issues within schools in England, typically in the first session of the day. Questionnaires were administered by tutors during this form tutor time for completion and return. The response rate is shown in Table 7.1. The gender balance within the questionnaire sample is roughly equal (49 % female). Non-responses were the result of student absences on the day of completion and tutors not conducting or returning questionnaires to the Head of Science. Since the bulk of missing responses are from missing form groups, and form groups are mixed sets of students, we have no reason to expect our sample to be significantly unrepresentative of the student population in these schools.

Table 7.1 Response rate for the student questionnaire

	Questionnaire sample	Y12 student population[a]	Response rate (%)
School A	35	53	66
School B	76	135	56
Total	111	188	59

[a]Based on school reports, cross-referenced with government and Ofsted data. Ofsted is the official organisation in charge of schools' inspections in England. It publishes periodic inspection reports for each school in England

[3] The years of compulsory schooling run from Y7–Y11, followed by 2 years of post-compulsory schooling from Y12–Y13.

[4] Typically, students complete 3–5 courses in Y12. These courses are called 'AS-levels'. Students then continue with 3–4 of these courses to full 'A-levels' in Y13.

Interviews

Based on questionnaire responses we identified 22 students to invite to take part in a 20–30 min interview. Within this sample we ensured a roughly equal proportion of science choosers (10) and non-science choosers (12). We oversampled for female students (15) since the focus of the IRIS study as a whole is on female students' experiences of the sciences. Each interview was conducted by one of the chapter authors or a third researcher.[5]

As discussed earlier we aimed to encourage students to provide narrative accounts of how they had come to make their subject choices. To do so we divided the interview into three sections. The first part of the interview focused on the student's main narrative through the open question: "How is that you came to be following these courses in Y12?" This was followed by a series of probing questions covering topics already included in the questionnaires: school experience over time; school science curriculum, and discussions about socio-scientific issues in particular; and factors that might have influenced the student's decision, such as friends or school grades. In the final section of the interview the student was offered the chance to add new information and to comment on a brief summary of the interview provided by the interviewer.

Questionnaire Responses

Figure 7.1 shows the number of male and female students that included science courses in their AS course choices. Our focus here is on traditional, high status science courses within England. Thus, for this analysis 'science' courses are either AS Physics, AS Chemistry or AS Biology. Students who include science-related courses such as AS Psychology or BTEC[6] Forensic Sciences in their course choices are categorised here as 'non-science' students. Most of these science-choosing students also included non-science courses within their choices.

In this sample female students are underrepresented within science course choosers, as compared with male students Overall, 35 % of female students are science choosers, compared to 47 % of male students. Of those students that choose science courses, 15/19 (79 %) female students include physics and/or chemistry courses, compared to 20/25 (80 %) male students.

The questionnaire asked students to indicate, on a 5-point scale, the extent to which they agree that specific features of school science lesson activities had encouraged them to consider choosing science courses at AS-level. Below we present data for two of these features of science lessons: 'having discussions about socio-scientific issues' and 'learning facts'.

---

[5] We thank our colleague Keith Bradley for his support in conducting these interviews.

[6] BTECs are vocationally-oriented qualifications.

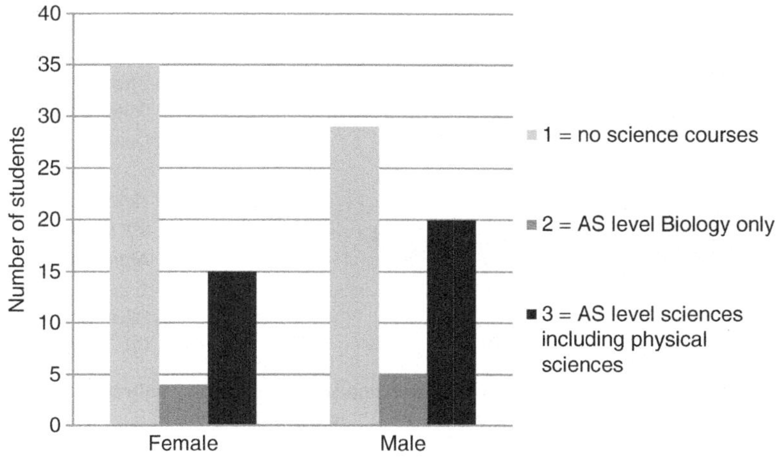

Fig. 7.1 Number of male/female students choosing science courses (Female n = 54; Male n = 54. Data on course selection was not available for three male students.)

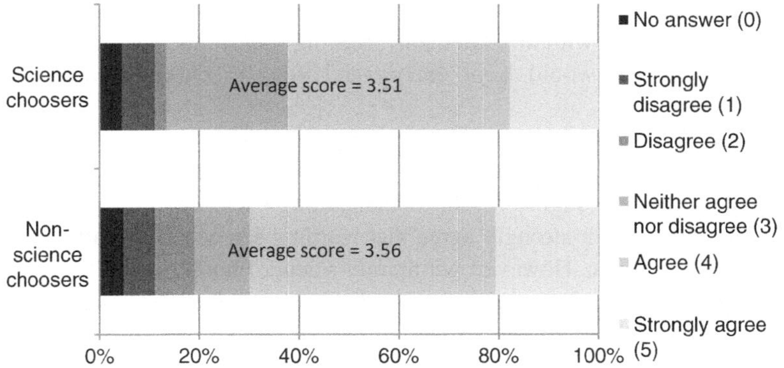

Fig. 7.2 Percentage of science choosers, and non-science choosers, indicating that having discussions about social and/or ethical issues encouraged, or would have encouraged them to choose AS-level science courses (N = 111 students)

Figure 7.2 shows the percentage of science choosers, and non-science choosers, who agreed/disagreed with the statement 'having discussions in science lessons about ethical issues like genetic testing, abortion, mobile phone masts, energy resources, pollution of climate change, encouraged you (or would have encouraged you) to choose science courses at AS-level'.

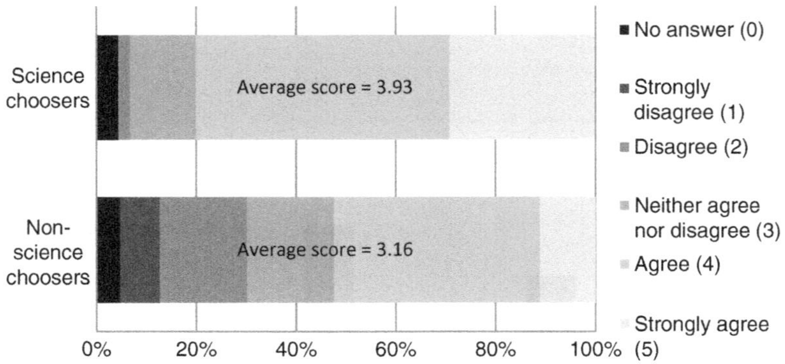

Fig. 7.3 Percentage of science choosers, and non-science choosers, indicating that learning science facts encouraged them to choose AS-level science courses. N = 111

Overall, there is little difference between these two groups. For example, within science-choosers 65 % of students providing a response agree or strongly agree that such activities encouraged them to choose science; within non-science choosers 73 % of students agree or strongly agree that such activities had, or would had, encouraged them to consider choosing a science at AS-level.

Figure 7.3 shows the percentage of science choosers, and non-science choosers, who agreed/disagreed with the statement 'learning science facts in science lessons, encouraged you (or would have encouraged you) to choose science courses at AS-level'.

It might be expected that science-choosers are more encouraged to follow science courses as a result of learning science facts than non-science choosers. This is indeed the case. Figure 7.3 shows that amongst science-choosers providing a response 84 % agree or strongly agree that learning science facts had encouraged them to choose science. However, within non-science choosers only 55 % agree or strongly agree that learning science facts had encouraged them to consider choosing a science at AS-level.

Overall, it appears that, on the basis of this self-reporting of impact, the inclusion of teaching and learning about socio-scientific issues such as the dangers of mobile phone masts, ethical issues related to genetic testing, and climate change within the school science curriculum has had a positive impact on encouraging students to choose, or consider choosing, science courses beyond post-compulsory education. However, science choosers indicate that learning science facts had a more positive impact on their choice (84 % agree or strongly agree), compared to having discussions about socio-scientific issues (65 %). By contrast, non-science choosers indicate that learning science facts had a *less* positive impact on their choice (55 % agree or strongly agree), compared to having discussions about socio-scientific issues (73 %).

Choice Processes Across the Interview Sample

In an initial analysis we attempted to use the categories reported by Cleaves (2005) to characterise the choice trajectories of the 22 students in our interview sample. However, these categories were not directly applicable to our data for two reasons. Firstly, Cleaves employed a longitudinal methodology to capture the changing nature of students' perspectives on choice over time. However, in our interviews students provide a single, retrospective account of the choice process. Hence it is possible that the students offered overly rationalised explanations of their trajectories to account for their present choice of subjects (see Chap. 3). Secondly, we found that two of the Cleaves trajectories (directed and multiple projections) could be clearly attributed to several of the student interview accounts. However, the other three categories were often difficult to assign given our interview data. As a result, we reconfigured the Cleaves trajectories to focus more on the interplay between career and subject/topics as the main drive for the students' choices – issues well represented in our interview data. Table 7.2 summarises these reconfigured categories, and the outcome of coding for the 22 students in our interview sample. The range of trajectories shown in Table 7.2 is comparable

Table 7.2 Categorisation of interview sample using reconfigured Cleaves trajectories

	Reconfigured definition	Main drivers for choice	Male	Female	Science	Non-science
Directed	A clear commitment to a specific career choice, usually over several years. Choice of subjects is determined by this career orientation. Students typically show high attainment in their subjects of choice	Mostly career oriented	2	2	1	3
Multiple projection	These students change future career plans several times, often quite radically. Many are high attaining in most school subjects, and have broad interests	Mostly career oriented	1	2	1	2
Partially resolved	Similar to directed students but with a particular topic/subject as the stable theme over several years, rather than a particular career. They often choose specific AS-level subjects because they have enjoyed them in the past	Topic/subject oriented	2	4	4	2

(continued)

Table 7.2 (continued)

	Reconfigured definition	Main drivers for choice	Male	Female	Science	Non-science
Funnelling identifier	Starting with a broad area of interest, choices are narrowed down over time (as opposed to the more dramatic changes encountered in multiple projection). These students usually provide a detailed account of the process of selection which might include external advice, exam results and growing knowledge about potential future jobs	Career oriented	1	3	1	3
Precipitating	These students usually talk about making a choice as a result of a particular incident, rather than as a process (as in funnelling identifier). They may have given little thought to future career/study plans, or have considered different options, and remain uncommitted to a specific path	Both topic/subject and career oriented	1	4	3	2
Total			7	15	10	12

with that reported for the 69 students in Cleaves's original study. In particular, we find, consistent with Cleaves, that many students do not have a clear, early focus on a specific subject pathway.

Choice Narratives: Two Cases

Here we provide details of how two students talked about the process of course choice and the impact of school science activities on this process. These two students provide contrasting cases in terms of: the student's response to the teaching of socio-scientific issues; the development of the choice process over time. We have selected two science-choosers to focus on the differences that exist, in terms of trajectories, among students who choose to take science subjects in post-compulsory education.

Claire[7]: "Well, It Was Always Picking Between Art or Science"

Claire attained A-grades[8] on her science courses at the end of compulsory schooling. She chose four AS-levels: Biology, Physics, Psychology and History. Claire's trajectory of choice is most closely categorised as 'precipitating' according to Cleaves's categories (Table 7.2). More specifically, in Claire's case, the trajectory of choice shows early subject enjoyment with on-going career uncertainty.

Claire refers to her experiences of school subjects and how these have influenced her AS-level course choices:

> I've always enjoyed the science subjects

> I like biology anyway, and I find that an interesting subject anyway, and I like the stuff about life on earth and animals and stuff. And I think that's partly why I wanted to do physics as well because it was like the solar system. We're not actually doing much about that but I found that sort of thing interesting.

Her references to the influence of school science experiences on course choices reflect a trajectory of early and ongoing enjoyment of school science, and an early inclination to choose sciences at AS-level. However, later in the interview, this choice trajectory on science is allied with a similar choice trajectory for non-science subjects:

> Well it was always picking between art or science. I could have done the art/English side of it, or it was the science, more science and the mathematical side of it.

Overall, Claire has been considering both routes for post-compulsory choices – she has enjoyed a broad range of school subjects. Her resolution of these considerations is related to two key issues: career intentions and school attainment.

> I always knew that I didn't want a career in the art sort of thing, so that's how it finally got to the decision. And when I got my grades as well at [age 16] that's how I made up my mind to specifically not pick art or something, but I did enjoy art, so there was always a different... I could choose either one.

Her reflections on career intentions show a clear and early commitment to not follow a career in the arts. Her school attainment in these subjects confirmed this decision for her. However, beyond that, she talks of an ongoing uncertainty about what career to follow. In several places during the interview she talks about this uncertainty and her changing ideas about career:

> Well I wanted to do forensic psychology, and so then it was criminal law because there is sort of a link between that. And I enjoy history and that fits with the law aspect of it, and I think I would find it interesting. And so, I've had trouble knowing what I want to do, so, I keep changing my mind about it.

---

[7] All student names are pseudonyms.

[8] The grading system runs from A to G, with grade A awarded to the highest attaining students.

Even after making her AS-level course choices she is unclear of her career intentions. Her mixed subject choices at AS-level may be a reflection of this, and a desire to keep her future options open.

Claire describes herself as 'quite a methodical sort of person'. This is reflected in the activities she reports engaging in as she attempts to clarify her career intentions. She has talked to several people about potential careers: a careers advisor in her final year of compulsory schooling, her parents (and particularly her mother) and teachers. At the time of the interview she was arranging a work placement:

> I've definitely got to research it more, and I think I'm trying to get a work experience placement in a law firm.

Her approach here reflects that of a 'rational actor'; someone who is attempting to make an informed choice about a future career, and then basing her school course choices around this. In Claire's reflections we do not see serendipitous events impacting strongly at a particular point in time on her choice trajectory (Ball et al. 2000; Foskett et al. 2001). Claire is someone who is likely to benefit from detailed guidance about careers in the latter years of compulsory schooling.

School Science Experiences: The Attraction of Learning New Explanations

Early in her interview, when responding to open questions such as 'How is it that you came to be following these courses?' Claire described what attracted her to specific subjects in school:

> I think being able to explain like life, and also being able to explain your mind and how it works, or how your heart works or something, and I find that interesting because it's like your body but you don't know about it, and I think it's interesting being able to explain that, being able to explain how different... like the ecosystem, I think that's interesting.

Here we see a clear intrinsic interest in explaining natural phenomena and human behaviour. She does not refer here to relevance or usefulness of school science subjects.

Elsewhere Claire describes her attraction for subjects with clarity, 'scientificness' and structure:

> I was looking at forensic psychology before I started the AS-level course, and then psychological ideas is just... I found Freud's ideas, like they're just too based on nothing, I like the scientificness of... it's like there's no scientific study type things really backing them up (...) there's too many different ways to deal with one problem, and like if you did have a patient or something there's so many different theories, and I think I'd just be stressed out as to which therapy you're supposed to treat the patient with or something, so yeah. So I think I've switched to... I've not really decided but I want to do law now, because, yeah. And that's got a bit more structure to it.

This perspective on school subjects is consistent with her earlier self-identity as 'a methodical sort of person'. Elsewhere Claire describes how her mother dissuaded her from following psychology because 'it got on her [Claire's] nerves'. Again, this reflects an affective, personal, response to school subjects.

In the final part of the interview Claire is asked specifically about her experiences of science lessons relating science to everyday social or ethical issues. She refers to a range of such experiences including radiation in physics, mobile phones, ethical issues in biology, abortion, ecosystems and food chains. Her immediate reflections on these curriculum elements return to her attraction to explanations:

> I find the ones that you can apply to everyday life more interesting than the theory side of it, I think that's why I like physics because you can (...) come away with an explanation of why something happens, and I do find the application of it more like understanding the world. If you come out of the lesson with a better understanding of the world then you feel like you've properly learnt something, as opposed to if it's just some theory that you can't see working really.

She then reflects on what features of a subject attract her to continued study:

> I don't know, I think the ones that definitely lead on to something you can see that there's more to learn about it like if you do say the heart, in AS-level they always, like, they haven't told you everything, like with cells in something there's like fifty different things working inside a cell that you didn't know about before, and you can sort of see that it's going to progress on, and I think if you know that there's more to it, or I always find that if I know there's more to it then it makes me interested to carry on with the lesson, to learn more about it.

Again, the significance of progression in explanations of phenomena is prevalent here. When pushed to reflect on socio-scientific issues her desire for learning, clarity and explanations means that she is less attracted to a current issue such as climate change:

> I think especially with climate change it's kind of a relatively new issue isn't it, and still theories are being thought up, so that's interesting that it's developing. And I think as it develops further the subject of it will probably become more interesting, like as there's more discovered about it and as... But I think in some sort of way the fact that it's happening now and it's such a current topic means that everybody has a basic understanding of it, and so I don't find it as interesting as something that I've never learnt about before.

From these reflections it appears that Claire is a student who is attracted to the explanatory power and clarity of science, as embodied in canonical science knowledge (such as the structure of the heart or the functions within cells). Inclusion of socio-scientific issues in the school science curriculum is unlikely to encourage Claire to pursue post-compulsory science courses.

Anya: "I Have Always Been a Maths or Science Person [Since Y10]"

Anya attained an A and a B grade on her two science courses at the end of compulsory schooling. She chose four AS-levels: Biology, Chemistry, Psychology and Sociology. Anya's trajectory of choice is most closely categorised as 'funnelling identifier' according to Cleaves's categories, reflecting a career-oriented driver of choice (Table 7.2). More specifically, in Anya's case, the trajectory of choice shows strong career-related influence from age 14–15 years.

Anya's course choices are strongly underpinned by her career aspiration. She chose biology and chemistry at AS-level because these courses are needed to become a pharmacist. She identifies this career intention as emerging, and becoming established, in Y9/Y10, i.e. when she was 13–15 years old. Before then, in her early secondary school years, she identified a range of career aspirations including teacher and lawyer.

The aspiration to become a pharmacist has strong family-related ties:

> Well my uncle's a pharmacist and he owns a few chemist [shops] and I've been in there a few times and recently he's more like training me up, telling me like – so I'm working there regularly, and it's interesting.

Elsewhere, Anya shows that her career interest went beyond this particular out of school, family-related experience. She describes how, like Claire, she spent time researching different career options, in Anya's case through internet research. She states that she had very little career guidance from staff in her school.

At the end of the interview Anya is asked whether there is anything that she would like to add to the discussion:

> I want a stable job. Obviously a job with quite good money. Just, you know, that's probably another reason as well. And like pharmacy seemed stable. You can even be full time or you can be like a locum in case a pharmacist is not there, and I could probably work for my uncle if – so there's like a bit of a cushion like in case everything goes wrong.

These comments reflect the strong career oriented features of Anya's choice trajectory.

School Science Experiences: Enjoying Science that Relates to Me

Anya described her school science experiences in Y7–Y9 as 'quite boring'. However, her science lessons became much more interesting in Y10–11. When asked to elaborate on the differences she reflects:

> Oh, well in [Y7-Y9] it was just – I can't really explain it – there was like no pushing you and it wasn't so focussed and it was a bit more like experimenting (. . .) but obviously [in Y10-Y11] they know they've got to make sure we know the stuff otherwise we're not going to pass (. . .) and getting more help was essential.

Here Anya is not referring explicitly to a shift in curriculum content from Y9 to Y10. Rather the focus is on the pressure she perceives from teachers to work hard and progress in Y10–Y11.

In the latter part of the interview Anya states that she has 'always been either a maths or science kind of person'. However, when asked to elaborate, she qualifies this, saying that this only applies to her from Y9/10, consistent with her reflections earlier in the interview (as described above).

Anya reflected, unprompted, on her experiences of different science subjects:

> Physics I just do not like. I just feel like it's the most boring science subject. The reason I like more chemistry is because from the start I've always liked maths, and from young

> I had tuition in maths, and biology is like quite interesting – I like stuff with the body, stuff like that. Physics, like gravity and all that – I just – doesn't relate with me at all.

She goes on to give an example of a science topic she has enjoyed:

> Well, first of all we learnt about different organelles in the body, and now when we talk about – we're doing like health and disease – and like it talks about mitochondria and cilia and that's what we've – so we're putting the before knowledge there now and it just combines nicely like – I know what that is, because I've already been put through it.

It appears that linking the facts of science (organelles, mitochondria) to everyday issues (health and disease) makes Anya feel that the subject relates to her. She doesn't identify these links in her experience of school physics.

When prompted, Anya identifies a wide range of topics relating science to every day issues. Again, she is positive about these lessons because they relate to her:

> We did about genetics, even stuff like cloning, and the mobile phone thing, cancer, yeah. Pollution as well, that's always been in science (...) I think because I could relate to them like everyday things, it was more on my level, so I kind of knew a few things and it was like interesting to know more. I learnt a lot, like a lot more.

She goes on to describe how she enjoys talking to her mother about such issues at home. When asked, Anya says that these lessons did encourage her to choose science at AS-level, because she enjoyed them, but also because she found them 'easy'.

Anya's choice of sciences at AS-level appears to be underpinned by two issues: her enjoyment of science within Y10–Y11, and her desire to become a pharmacist. Both of these align with a choice to follow sciences at AS-level.

School, Curriculum and Choice Across the Interview Sample

Here we consider how the themes identified in the two case studies are reflected in discussions with the whole interview sample of 22 students.

Influence of Teachers and Careers Advisers

The role of science teachers and school-based careers advisers featured in the interviews with all the students and, in almost all cases they appeared to have influenced the students' choices. The style and activities of the teacher are mentioned by several students as a determining factor for their decision to pursue or not pursue science courses. This reinforces findings from previous studies, and highlights the importance of teacher professional development for improving science uptake in post-compulsory education. Furthermore, the interviews show that science teachers and careers advisers provide vital information to students on the relation between science courses and professional careers.

Curriculum Elements

One of our main objectives was to explore the influence of socio-scientific issues on students' experiences of school science and future course choices. Most of the students in our interview sample reported enjoying debates about socio-scientific issues. However, less than half of the students (9/22) stated that these debates had influenced their choice of science subjects in post-compulsory education. Furthermore, only four of these nine students actually chose post-compulsory science courses. This suggests that whilst many students are attracted to socio-scientific debates, this will not necessarily translate into choosing post-compulsory science courses. Our interview sample also included students who reacted more negatively towards teaching/learning about socio-scientific issues. These students were interested in science, obtained good grades, and chose post-compulsory science courses, but did not value the inclusion of socio-scientific issues in the curriculum. This was mostly because they did not see socio-scientific issues as being "real" science. Overall, we do not identify a direct connection between a positive perception of socio-scientific issues by the students and a choice of science subjects and science-related career.

Strategic and Contingent Choices

Analysis of our interviews identifies two kinds of choice, distinguished in terms of time-frame and influencing factors. On the one hand there were *strategic* (long-term) choices based on post-school plans. However, we also identify *contingent* (short-term) choices based on the immediate past and future within the school, and influenced by factors often unrelated to future career choices. In both cases student attainment plays an important role, both because schools often use grades to guide students towards, or away from, science courses, and also because students take their grades into consideration when judging their chances of obtaining good results in future courses. Our interviews show that most students make both contingent *and* strategic choices, heavily influenced by school experiences, as illustrated by the following student statement:

> So all I wanted to do is to keep my options open. So I wanted to have one science at least, maths because it is maths, everyone likes maths, from an employer's point of view. (...) and English literature because I love to read and I found the discussions in our year 12 have been really easy to be quite honest. Like coursework for English literature was the easiest thing to do throughout year 11 and I still don't know how I got an A* in it.

We can see in this quote that the choice of mathematics is strategic, based on the future value of the subject when applying to university and employment. However, the choice of English is contingent, based on the student's experience of studying English. The interplay of these strategic and contingent choices is a further characteristic of choice *as a process* that unfolds over time.

Conclusions

One aim of this study was to examine the extent to which an emphasis on teaching about socio-scientific issues and the nature of science had encouraged students to choose science courses within post-compulsory schooling. Millar (2010) reports that a science course providing a strong emphasis on teaching about socio-scientific issues and the nature of science resulted in increased uptake of science courses within post-compulsory schooling. In our study responses to the questionnaire do indeed suggest that for many students, both science and non-science choosers, teaching about such issues within compulsory science schooling had encouraged them to consider choosing post-compulsory science courses. However, analysis of student interviews suggests that such teaching impacts *differentially* on students. For Claire (a female science chooser) teaching about social and ethical issues did *not* appear to provide encouragement to pursue post-compulsory science courses. Claire was more influenced by her interest in learning scientific explanations. She was more driven by an *intrinsic* interest in the subject, rather than a need to make the science 'relevant' to her everyday life outside of school (Roberts 1988). By contrast, for Anya (another female science chooser), relating science to everyday life did appear to provide encouragement to pursue post-compulsory science courses. However, the impact of this feature of the taught curriculum on Anya's choice of science subjects at post-compulsory level did not appear decisive. Rather, she was most strongly guided by her career aspirations. Her attraction to linking science to everyday life aligned with her strong career aspirations, and hence features in her narrative account of the process of her subject choice.

A further aim of this study was to examine the processes through which students come to be following specific courses within post-compulsory schooling. Our use of Cleaves's categories of choice trajectory supports earlier work showing that students exhibit a broad range of trajectories (Cleaves 2005). For a minority of students in our sample, this trajectory is one of early commitment to a science route through schooling. However, for many students their choice trajectory is characterized by uncertainty and indecision, and includes both strategic and contingent choices. Claire's case study shows a student who has always liked school science, but has been uncertain about whether or not to choose post-compulsory science courses throughout much of her compulsory schooling. We have found it helpful to draw a distinction between an early and ongoing enjoyment of school science ('I've always liked school science') and an early commitment to pursuing a science route through post-compulsory schooling ('I was always going to choose sciences in the future'). Our analysis of a larger number of our case studies has identified many students who enjoy several school subjects (including sciences), and for whom the process of choice is ongoing through compulsory schooling. Thus, our analysis challenges the claim that the majority of students who pursue science courses within post-compulsory schooling develop this commitment early in their school experience (Maltese and Tai 2010).

References

Archer, L., DeWitt, J., Osborne, J., Dillon, J., Willis, B., & Wong, B. (2010). 'Doing' science versus 'being' a scientist: Examining 10/11-year-old schoolchildren's constructions of science through the lens of identity. *Science Education, 94*(4), 617–639.

Ball, S. J., Maguire, M., & Macrae, S. (2000). *Choice, pathways and transitions post-16: New youth, new economies in the global city*. London/New York: Routledge-Falmer.

Cleaves, A. (2005). The formation of science choices in secondary school. *International Journal of Science Education, 27*(4), 471–486.

Eccles, J. (2009). Who am I and what am I going to do with my life? Personal and collective identities as motivators of action. *Educational Psychologist, 44*(2), 78–89.

European Commission (2004). Europe needs more scientists. Report of the high level group on human resources for science and technology in Europe. Retrieved April, 2010, from http://ec.europa.eu/research/conferences/2004/sciprof/pdf/final_en.pdf

Foskett, N., & Hemsley-Brown, J. (2001). *Choosing futures: Young people's decision-making in education, training and careers markets*. London/New York: Routledge-Falmer.

Foskett, N., Dyke, M., & Maringe, F. (2008). The influence of the school in the decision to participate in learning post-16. *British Educational Research Journal, 34*(1), 37–61.

Hollway, W., & Jefferson, T. (2000). *Doing qualitative research differently: Free association, narrative and the interview method*. London: Sage.

Maltese, A. V., & Tai, R. H. (2010). Eyeballs in the fridge: Sources of early interest in science. *International Journal of Science Education, 32*(5), 669–685.

Millar, R. (2010). Increasing participation in science beyond GCSE: The impact of twenty first century science. *School Science Review, 91*(337), 67–73.

Murphy, P., & Whitelegg, E. (2006). Girls and physics: Continuing barriers to 'belonging'. *The Curriculum Journal, 17*(3), 281–305.

NSB. (2010). *Science and engineering indicators 2010*. Arlington: National Science Board.

Roberts, D. A. (1988). What counts as science education? In P. Fensham (Ed.), *Developments and dilemmas in science education* (pp. 27–54). London: Falmer Press.

Rodd, M., Mujtaba, T., & Reiss, M. (2010). Participation in mathematics post-18: Undergraduates stories. *British Society for Research into Learning Mathematics, 30*, 175–182.

Chapter 8
A Place for STEM: Probing the Reasons for Undergraduate Course Choices

Elaine Regan and Justin Dillon

Introduction: Place and Belonging

According to Pretty et al. (2003), the place where young people live contributes to their identity. It is particularly significant for prospective university students who are in a transitional period of their lives. At this time many are suspending ties and attachments to family and friends and moving away from home to attend university. In doing so, they will develop new social and cultural environments and will experience increasing levels of personal autonomy. Moving far away from home to attend university also facilitates independence (Fisher and Hood 1987) and impacts on the student's self-image, identity development and furthermore, 'place attachment' as students consider who they are, how they see themselves, and how they would like others to see them (Cassidy and Trew 2001). Chow and Healey (2008) found that first-year university students in England were concerned with the loss of sense of place, the sense of belonging and the detachment from their home-town. The process of moving from home to university was, in effect, undermining the home-town's capacity to symbolise the student's sense of self.

An individual's evolving self influences decisions being considered such as what course to study at university or even what university to attend. The place where a student chooses to live and the people they choose to spend time with, are likely to either be compatible or incompatible with their self or evolving self. In this chapter we consider the importance and significance of place in the decision-making process. In this way, we propose place as more than an independent variable but as a means of mediating social life (as suggested by Gieryn 2000). Throughout the chapter we draw on literature from a variety of fields, including geography, environmental psychology and sociology to position our argument. For a complete

E. Regan (✉) • J. Dillon
Department of Education and Professional Studies, King's College London, Waterloo Bridge Wing, Franklin-Wilkins Building, Waterloo Road, London SE1 9NH, UK
e-mail: elaine.regan@kcl.ac.uk

E.K. Henriksen et al. (eds.), *Understanding Student Participation and Choice in Science and Technology Education*, DOI 10.1007/978-94-007-7793-4_8

overview of place literature, see Trentelman (2009), Lewicka (2011), Gustafson (2001) and Scannell and Gifford (2010).

Why Is Place Important? And What Is It?

Place plays a key role in the social world and the symbolic meanings and emotional attachments that individuals hold about particular settings. Consequently, places become a key component of identity as individuals 'draw on a range of social processes, symbols and values to describe themselves' and become 'important conduits in the cultural production of the self' (Sampson and Goodrich 2009, p. 902). The concept of place, or sense of place, is found in a wide range of literature within, for instance, social science, geography, architecture, urban planning, sociology, leisure and tourism and environmental psychology (Kudryavtsev et al. 2012; Lee 2011; Eijck and Roth (2010). A settled community, or a locality with a distinct character, is often synonymous with place (Massey 1995). However, place literature contains many terms that are inconsistent (Kudryavtsev et al. 2012), with much of the research concentrating on the relationship between people and places or the connection or disconnection between people and their environment. Examples of the variety of terms in use include sense of place (or Malpas (2008), 'sense of belonging to places and the character of places'), place attachment (or Burdge and Ludtke (1972) 'identification with place'), place meaning and place identity.

Place is defined by Low and Altman (1992) as a physical space imbued with meaning, a meaningful location (Lewicka 2011). Place has three features, a geographic location (the distinction between here and there), a material form, and investment with meaning and value (Gieryn 2000). A geographic location can take a variety of forms from a chair to a room to a district to a country or a destination to go to for a particular purpose or function. It has, therefore, physicality: place is stuff and it is identified or named. It is also the space filled by people, practices, objects and representations and, as such, the meaning or value of the same place is flexible and malleable depending on the person or culture. For a review of the literature on scale of place such as home, neighbourhood, city, region, country, and continent, see Lewicka (2011).

Gieryn (2000) makes a distinction between place and space, the latter being abstract and detached from material form and cultural interpretation. He claims that place is not just a setting or context but a force that exerts effects on social life. Other writers (Canter 1977; Sack 1997; Stedman 2002) have identified the physical setting, the person (an individual's internal psychological and social processes and attributes) and the activities or rituals done at the place (Smaldone et al. 2005). Gustafson (2001) proposes a tri-polar analytical framework of Self—Others—Setting. However, Gustafson (2001) noted that, within his proposed framework, the 'meanings of place are not forced into three discrete categories but mapped around and between the three poles of self, other and environment' (p. 12).

Places bring people together resulting in either engagement or estrangement, which will also depend on whether practices that an individual prefers (identifies with) are appropriate or inappropriate in those locations.

Sense of place is the ability to locate and attribute meaning to a form or spot (Gieryn 2000). It is considered to comprise place attachment (how strongly an individual is attracted to a place) and place meaning (how they describe the reasons they are attracted to the place). Kudryavtsev et al.'s (2012) review of sense of place in environmental education outlines a number of key definitions within the sense of place literature. Place attachment is thought to be the bond between people and places, or the degree to which a place is important to people (Jorgensen and Stedman 2001) and as such also involves social relations. This bond can invoke 'the extent to which an individual values or identifies with a particular environmental setting' (Kyle et al. (2003) and is usually a positive association, linked both by the place and the people attached to it. Place attachment is usually measured using quantitative measures from Likert scale surveys (see Marcouyeux and Fleury-Bahi (2011) for an example). The environment where an individual lives has a strong role in an individual's life, creating a sense of meaning, and is often related to personal identity. A sense of place impacts on identity through the experiences an individual has where they live (Proshansky et al. 1983). As such, places are endlessly made through material form and interpretative understandings or experiences (Gieryn 2000).

Place dependence is the potential a place has to satisfy an individual's needs and it can emerge if the place provides the setting for preferred activities (Halpenny 2010; Vaske and Kobrin 2001). ***Place identity*** is another term that describes the relationship between people and place (Proshansky et al. 1983). It is the extent to which a place becomes a part of an individual's definition of self or their personal identity (Hauge 2007; Korpela 1989; Proshansky et al. 1983), how it influences the individual's sense of self-identity and how it can result in a sense of purpose and belonging. It can be considered a cognitive structure that contributes to the social identity process and self-categorization developing 'from acts of locating oneself within environmental contexts through daily routines as well as during exceptional circumstances' (Cicognani et al. 2008, p. 34). The symbolic meanings that individuals ascribe to settings is called place meaning, often defined by answers to questions such as 'what does this place mean to you?' (Smaldone et al. 2005). Place meaning is a multidimensional construct reflecting an individual's social interactions, culture, politics and economics (Ardoin 2006). In that context, place meanings are attributed to different themes such as 'environment', 'self' and 'others' (Gustafson 2001).

The idea of place and sense of place has been explored in relation to moving from home to university (Chow and Healey 2008). Preferences for urban and rural locations (urbanophilia versus urbanophobia, see Félonneau 2004) also exist with individuals seeing themselves as either city people or country people (Hummon 1992). Lewicka (2010) writes that distinctions between types of place are closely linked to social capital (bonding and bridging) and that people with many distant friends and acquaintances (bridging social capital) tend to have many more close friends and strong family ties (bonding social capital).

The most consistent predictor of place attachment is length of residence but socio-demographic, social, and physical-environmental are others. Since place identity is a closely related construct to attachment, place identity may also be seen as a predictor. Feldman (1990) suggests that with an increase in mobility within society that attachments shift from place to what he calls 'settlement identity', such as 'city person', 'mountain person', 'rural person', etc. The aspects of place theory that little is known about are the processes through which people become attached to places. Lewicka (2010) also presents an alternative perspective to the concept of attachment, discussed by Beckley (2003), who claims that 'magnets' are factors that attract people to a place and 'anchors' are the factors that prevent people from moving from a place. 'Magnet'/'anchors' and 'settlement identity' are particularly relevant concepts in terms of students' decisions about university. Place attachment develops independently of residence and the physical or recreational assets of place. Magnets potentially draw people away from home attachments; social and community factors potentially anchor them to home locations. The consequences for students of such place attachment include decreased mobility and subsequent potential restrictions on life opportunities.

Decision-Making in Higher Education

Participation in science, technology, engineering and mathematics (STEM) is a concern to researchers and policy makers (European Commission 2010; Bøe et al. 2011). In most OECD countries there has been a large increase in the numbers of students attending higher education over the past 15 years; however, despite the absolute numbers of students in STEM fields also experiencing an increase, the proportion of STEM students has steadily decreased during the same period (OECD 2006). This trend is particularly prominent in disciplines such as mathematics and the physical sciences. In the UK, young people are 20 % more likely to go on to higher education than was the case in the mid-1990s (Thompson and Bekjradina 2009) but it is thought that there are not enough young people studying STEM at the higher education level to sustain desired economic growth (HM Treasury 2004). [This notion is contested by Smith and Gorard (2011) who have shown that, following graduation, the majority of science degree holders enter occupations that are not directly related to science.] To address the perceived problem, the UK Economic and Social Science Research Council commissioned research to provide evidence for policy proposals that could boost applications for STEM courses at university and, a year later, in 2008, the Higher Education Funding Council for England launched a £350 million rescue plan to attempt to counteract the decline in the number of students taking science and mathematics (Royal Society 2007).

The literature on student college choice from the 1980s suggests a three-stage process in decision-making: the predisposition phase, the search phase and the choice phase (Jackson 1982; Litten 1982; Chapman 1981; Hossler and Gallagher 1987; Hossler et al. 1998). The choice phase involves deciding which university a student will actually attend and is influenced by educational and occupational aspirations, costs and financial concerns and university courtship activities (Hossler and Gallagher 1987; St. John 1990). According to the National Union of Students' Student Experience report (NUS 2008) a large proportion of students (31 %) are motivated to choose a university close to home, particularly students from lower socio-economic groupings. This figure is, perhaps, not surprising given that universities are able to charge fees of up to £9k per year to most students. These factors do not pay attention to external factors such as fulfilling entry requirements.

In 2005, the UK Higher Education Careers Service Unit (HECSU) launched a major programme of research (longitudinal study, named Futuretrack), designed to explore the process of entry into and through higher education of all applicants to full-time UK higher education courses who applied through the Universities and Colleges Admissions Service (UCAS) in 2006. The project aims to provide robust and comprehensive evidence to clarify the socio-economic and educational factors that determine career choices, and outcomes. A fourth survey, conducted in winter 2011–2012 five years after most respondents embarked on their courses, is currently under analysis.

According to Purcell et al. (2008) once students have made the decision to apply to a university, they start the process by either choosing the institution or location in which they hope to study and the area of study or discipline they are interested in, or have a very precise idea about the exact course they aspire to enter. The most frequent order of choice of 'traditional' students appeared to be broad subject area, followed by institution. For older students or students with a less established tradition of HE participation, most choose location first, thus restricting their options.

The most popular reasons for choosing to study a particular course were interest in the course and employment or career-related reasons (Purcell et al. 2008). Age and social class affected the order of these reasons, with younger applicants more likely to choose subjects they were good at or enjoyed, and less likely than older applicants to give instrumental, employment-related reasons for choosing their course. Applicants from higher social classes were also more likely to choose subjects they enjoyed or were good at, whilst those from lower social classes were more likely to give employment-related reasons. Older applicants generally had clearer reasons for choosing their course. Students coming from homes where progression to HE was regarded as the norm were likely to perceive themselves as having wider choices and greater likelihood of having had encouragement from parents, teachers and friends to apply for HE. However, they may be more likely to progress directly from subjects they enjoyed at school to study of these same subjects in higher education without considering the implications of choices. Purcell et al. (2008) state that 'prior achievement and experience of a subject are good reasons for studying it but there is a danger that other options – and the longer-term implications of choices – may not be considered adequately in cases where students proceed without much thought or guidance to opt for their best school subject' p. 71. Subject choice remained profoundly gendered, and women exhibited greater clarity in ideas about the career they aspired to in the analysis of vocational subject choices (Purcell et al. 2008).

IRIS and Choosing Undergraduate Courses

This chapter reports on one English research strand within the IRIS project, the Choosing Undergraduate Courses (CUC) study. The CUC study examines university students' understandings of choosing undergraduate courses. Based on an analysis of focus groups and biographic interviews, the CUC project explores the way university students make sense of the priorities, considerations, values and experiences on which they base their educational choice. We explore how young people's educational priorities can be interpreted through perspectives on late modern societies as outlined by Bøe and Henriksen in Chap. 2, and identity, outlined by Holmegaard, Ulriksen and Madsen in Chap. 3. Although all students in our focus groups discussed more typical factors of influence such as interest, achievement, family, and teachers, the significance of place to the participants was clear. Consequently, this chapter focuses on students in one university and examines the role of place in the course choice decision-making process using thematic analysis (Braun and Clarke 2006) of focus group narratives.

Methods

The CUC approach is a qualitative study based on 20 focus groups (Vaughn et al. 1996) of STEM and non-STEM first-year undergraduate students (male and female) in four different English universities (one northern, one western, and two based in London) and biographical interviews that explore the critical pertinent features in their lives (social as well as educational) that led them to study STEM (or other courses). Students were invited to participate through their university email system and, following the focus groups, a subset were invited to participate in an in-depth biographical interview. The STEM students were selected from undergraduate programmes in eight subject areas, identified by ISCED codes: Biology; Physics; Chemistry; Mathematics; Statistics; Computer Science; Engineering Mechanics and Metal Work; Engineering Electronics and Automation; and Engineering Electronics and Process. The non-STEM students were selected from undergraduate programmes in six subject areas: Arts; Humanities; Social and behavioural science; Journalism; Business; Law. While the intention of the research was to study the attitudes and experiences of English university students, several foreign national students participated, reflecting the diversity of the student

Table 8.1 Breakdown of the CUC sample

Focus group	Category	Gender	Number of participants (n)
1	STEM	Male	4
2	STEM	Male	7
3	STEM	Female	9
4	Non-STEM	Male	3
5	Non-STEM	Female	10

population and recruitment to English universities. According to Keown (1983, p. 66) 'homogeneous groups . . . are generally more comfortable and open with each other, whereas mixed sex, ethnic, or socioeconomic groups make it more difficult to achieve a high degree of group interaction'. Consequently, the groups were organised by gender and STEM/non-STEM participation.

Students enrolled in STEM and non-STEM undergraduate programmes were asked to share their thoughts, experiences and feelings about educational choice within each focus group. The discussions, which lasted 120 min were audio-recorded, transcribed and analysed (Barbour and Kitzinger 1999; Morgan 1996) facilitating collective sense-making (Wilkinson (1998a, p. 186) and the 'co-construction of meaning in action' (Wilkinson 1998b, p. 338). The sample (see Table 8.1) on which this chapter is based came from one university in the north of England (n = 33) – the total number of focus group participants across all four universities was 119; 70 were female and 39 male. Recruitment to the study was challenging despite the assistance of gatekeepers within each university. As a result, the number of participants in each focus group varied from the ideal sample of 6–8. All focus groups were arranged with a minimum of six participants; however, several did not turn up on the day of the focus group.

Findings

The students at this northern university acknowledged the importance and significance of place in their educational decision-making. The primary focus of the group discussion was on how the students came to the decision to study either their STEM or non-STEM course. However, what continually surfaced across all the groups were the meanings that students attached to the location/place of the university. Their discourse reveals thinking about the attractions of that place, engagement or estrangement with it and meaningful places. Both the STEM and non-STEM groups offered similar perspectives on how they were influenced in their degree and university choice by both engagement and estrangement with places. Students' discussions around educational choice and decision-making resulted in a diverse range of perceptions and experiences both within and between the groups in this case study. However, several overarching issues, and also common themes, were identified. Firstly, overarching issues included students' reflections that the decisions that they made on the basis of place were superficial and secondly their sense of place as a part of their self identity. Common themes for place meanings include the reputation of the university, the promise of a social life and the influence of friends and family.

In the exploration of the findings that follow, themes common across both groups are presented. The central difference between the groups is where we will begin: the non-STEM students' perception that an acknowledgement of 'place' considerations equated to the superficial nature of choice, whereas STEM students considered place to be key in terms of identifying with their chosen location of study.

Place, Important Yet Superficial

International literature highlights that there are multiple influences on student choice (James 2000; Sjaastad 2011) and the range of factors deemed to be most influential. From the opening of the 'place' discussion with the male non-STEM group, it became clear that for this group of students the choice, in their opinion, was possibly careless, frivolous and perhaps lacking in thorough considerations. Jim, a psychology student, introduces himself to the group and claims immediately 'X is a good city for going to concerts and shows and it's one of the main reasons I chose to come here, I think'. Connor is a first-year student enrolled on geography with transport planning course with varied interests ranging from sport to music. He comments on his reflections on educational choice that he had prior to attending the focus group believing that, overall, despite considerations of whether he would be accepted with the grades he had, that the choice he (centered on place) and many others make, is highly superficial. The students (all male) acknowledge the importance of place as a component of choosing their undergraduate course but view such considerations in a negative light. While discussing their choice process, students reflected on the shallowness of their decisions, such as giving considerations to the city as a social scene rather than considering the programme of study.

> One of the thoughts I had is that when people are choosing their courses its relatively superficial. We've mentioned nightlife, everyone mentioned music and night life. Obviously some universities only offer certain courses and when you're applying it can narrow it down massively, but actually the choices people make can be highly superficial.

These students are, perhaps, being a little over-critical of their decisions given that, in most cases, their thoughts of place have come after considering the A-levels they possess, their options with the grades they are likely to achieve, and choosing an area of study. The decisions made are not inconsequential but are certainly not what one (or one's parents) might think is important such as the quality of teaching.

A similar thread was observed in the female non-STEM group with Norah's primary concern being location over the course choice.

> I didn't choose my course first, everyone, like, chose their uni based on where it was, I wanted to be in the city, in a lively area.

Elizabeth, a geography student with strong musical interests, discussed her choice process. Elizabeth decided to base her choice on whether she would like the university and the place of study because in her view, the degree programmes would be largely similar:

> I thought that everywhere is going to have a bit of the same stuff that you've done before and I'm going to have no idea whether I'm going to enjoy it, so I better concentrate more on whether I actually liked the uni, rather than the actual modules that we would be doing.

Sense of Self

A primary rationale for the choice of degree course by a majority of students in this case study was related to the experiences associated with the place and how that would contribute to their sense of self or evolving self. For both Emily and Max the fact that their chosen university offered work experience and placements as part of their degree programme was a particular draw for them. Both work placements and field-work were seen as important group and socialising experiences whereby students could build attachments and friendships while gaining valuable experience to take forward to their future job applications. Placement and study abroad options illustrate aspects of their evolving sense of self and how they see themselves both currently and in the future.

Feldman (1990) found that the majority of his sample identified as being a 'city person', a 'suburban person', a 'small town person', or a 'country/mountain person' in his study of the identification of self with settlement type in Denver. Clear from the students' discussions in our focus groups was students' identification of self with a type of settlement, in this case, with a city. Although some students originated in smaller, rural areas, the magnet of the city was strong. A vibrant city with a vibrant nightlife was more appealing than quiet country settings for some students. Furthermore, these identifications with city extended to participants' view of their future selves and aligned with their evolving sense of self. In Jackie's case, a mathematics student, that meant studying in the city now and working in a city in the future. The importance of place is evident in her views of her life trajectory, including career and family:

> For me, I know I definitely want to work in the city and I think that's something that I've had to think about. I mean, obviously, it's not a firm decision yet, but I know that I'm going to have to wait a good 10 years to get my career, 'cause I want to aim high, so I want to be very focused for 10 years or so before having a family.

Similarly, chemistry student Cathy's future plans include a life in the city:

> I don't know, like I do want to do something to do with chemistry but I'm not sure what, but I know that I do want to move to London after I've done my degree and be like, not a business woman but like, you know, just like a 'London gal'.

For some students, the thought of moving from home was difficult, mainly on financial grounds. However, the draw to the university experience was enough of a pull to overcome this barrier. Although cost is not often seen as an influential factor (Brennan 2001) it often limits choice to particular locations (Connor et al. 2001). Patiniotis and Holdsworth (2005) report that many students from lower income families live at home, as a debt avoidance strategy. Caroline considered living at home as an option, exploring training as an accountant and earning a salary. Her aspirations aligned more with middle-class families completing the first step towards independence by moving away to university. Her desire for independence resulted in her decision to study business at university because ultimately, she couldn't '*see myself moving on with my life enough*':

> I know it's a big price tag and I'll probably still end up as an accountant in the long run but I couldn't see myself moving out of my house. If I was to be an accountant at home, I'd be living with my parents still and I'm almost 20, I couldn't see myself gaining any independence and so that's a big factor that made me decide.

Caroline was not alone in these types of deliberations. Emily, a chemistry and Spanish student, also shows the importance placed on the spatial practices of young people and their transition to university (see Holdsworth 2009; Cicognani, Menezes, & Nata 2010):

> The main issue was finance really, that was the only concern. My problem was either to move away from home or stay at home. I needed to be aware that if I do move away, I'm going to have X amount more debt than you would have at home. But at the end of the day, I wanted the whole university experience, so I think moving away has helped me a lot more and I was willing to risk the debt kind of thing.

Place Meaning – The Attractions to Place

Proshansky (1978) proposed that people's psychological bonds with places transcend a relationship to a specific locale and are influenced by an individual's unique environmental experiences as well as those experiences common to all individuals living in similar or particular settings. The students discuss attractions to place and the meanings ascribed, and descriptions of why place was important in their decision to choose their undergraduate courses. The students discussed some key attractions to place, notably the reputation of the university, the social life of the city-based university, and the prospect of friendships. These articulations of place and decision-making show that the students had generalised conceptions of places developed by direct experience and through more informal communications.

Reputation

During the focus group discussions it became clear that the prestige, reputation or ranking of the university was an important consideration for most students. In line with Hinton (2011) and Briggs and Wilson (2007), the students in this study were driven by a desire to attend what they perceived to be a 'good university'. Max is studying geological sciences, he explained: 'It's getting a reputable degree from a reputable university that'll get you a job at the end of the day, with a good, reputable company'. This resonates with Soutar and Turner's (2002) investigation of students' preference for university selection: academic reputation and employment prospects are key factors. Similarly, Jessica, an economics student, struggled with her degree choice since she had not realised that her choices at 16 would influence her so much. Jessica studied economics, human biology, and philosophy to A-level and mathematics to AS level. Consequently, though she could have opted to do a STEM degree, she choose to do economics in order to attend a more prestigious university that would accept her with her subject combinations. As outlined in Pearson's (1997) study, students are often poorly informed while making decisions about where to study, which in Jessica's case stems from earlier decisions about A-levels.

> The only uni that would have taken me for a science-based degree without the chemistry would have been the polytechnic kind of uni and I wanted to go to a Russell Group [higher status university]. I wanted to come out of a uni that had a good reputation, I didn't want to come out of a uni that didn't.

Social Life and City

For chemistry student Cathy, the social life expectations of her choice were strong, a factor which was linked to her sense of self and the rewards ('good on your CV') that would result from being involved in social activities:

> I think social life is quite big, as in I chose X because I knew it was very social, there was a lot going on and there's a lot of societies you can join and get involved in, which is good on your CV and stuff like that.

For many students, this aspect of university life was central to their decision-making process, together with the affordability of the scene. Students took advice on locations and social life from siblings and friends that were attending various universities and surprisingly, the weather at the time of visits also proved influential. Similarly to Hinton (2011), where moving away from home was part of achieving the 'best' experience, spatial mobility played a central role in our students' discussion, particularly in light of the importance of enjoying the 'best' university experiences.

Influence of Family and Friends in Choosing Place

For some students, the influence of the city extended to a comparison to the capital city, particularly in relation to concerns about finance. Parental influence, or extended families, has been noted previously as an influence on student choice

(Cole and Thompson 1999) and Max's parents expressed concerns about a potential move away to university, primarily on cost grounds. Connor's parents were also influential.

> Your parents and friends get involved with it, my parents and my mum especially tried to influence me, by making sure I chose the right course, so it varies on who you're talking to.

Anna however reflects on the role played by her brother, who also attends the university she has chosen, and her friends in her decision.

> My brother's here and so I came up for weekends, to go out and everything like that. I didn't do an open day, but what I did do was better because, you know, you get to experience how cheap a taxi is, how much a dinner is, what the halls are like. A lot of my friends from the year above, what they said also helped me decide.

For Connor, the attraction of university as a new place in his life involved the development of new social and cultural experiences. He and Sean discussed the merits of attending a new place and breaking old ties and attachments to existing friends. This process would ensure that their home-town and home ties would no longer symbolise the self:

> Connor: A part of going to university for me was actually making new friends. From experience, I think that it's good that you go to university where you haven't got a lot of friends that you know from home, because it means that you're a lot more, I don't know how's the best way to...
>
> Sean: You're forced into making friends.
>
> Connor: Yeah, you're forced into making new friends and I know lots of people from where I live go to [another university] and friends have sort of stuck together with each other, which is good and nice, but haven't necessarily made the same new friends that they would have made.

Conclusions

As Cresswell (2004) has noted, place is not just a thing in the world but a way of understanding the world, and students' relationships to places are complex (Chow and Healey 2008). In this chapter we have explored how students utilise place as a way of seeing, knowing and understanding the world of educational choice. This study has shown how young people invest their sense of self in their educational choices and the meanings attached to place that attract them. Similar to Purcell et al. (2008) our participants discussed the process of either choosing the institution or location in which they hoped to study through place attachments and attractions. This process followed their initial decision to apply to a university, or the area of study or discipline they were interested in, highlighting the social and the academic factors that influence students' aspirations, goals and intentions. Unlike Chow and Healey's participants (2008), our students did not discuss a loss of sense of place or belonging and detachment from their hometown. Instead, their articulations of place centred on experiences that affected their sense of self and evolving self, and the attractions to new places of study and residence based on reputation, social life and the city, and how significant persons affected their choices. Deciding on an undergraduate course and a university programme marks a significant transition in the lives of students as they contemplate the move from home and school to university. Place attachments held and their sense of self or place identity are shaped during this transition.

References

Ardoin, N. (2006). Toward an interdisciplinary understanding of place: Lessons for environmental education. *Canadian Journal of Environmental Education, 11*(1), 112–126.

Barbour, R., & Kitzinger, J. (1999). *Developing focus group research: Politics, theory and practice*. London: Sage.

Beckley, T. M. (2003). The relative importance of sociocultural and ecological factors in attachment to place. In L. E. Kruger (Ed.), *Understanding community-forest relations* (pp. 105–126). Portland: U.S. Department of Agriculture, Forest Service, Pacific Northwest Research Station. Gen. Tech. Rep. PNW-GTR-566.

Bøe, M. V., Henriksen, E. K., Lyons, T., & Schreiner, C. (2011). Participation in science and technology: Young people's achievement-related choices in late-modern societies. *Studies in Science Education, 47*(1), 37–72. doi:10.1080/03057267.2011.549621.

Braun, V., & Clarke, V. (2006). Using thematic analysis in psychology. *Qualitative Research in Psychology, 3*(2), 77–101.

Brennan, L. (2001). *How prospective students choose universities: A buyer behaviour perspective*. Melbourne: Centre for the Study of Higher Education, University of Melbourne.

Briggs, S., & Wilson, A. (2007). Which university? A study of the influence of cost and information factors on Scottish undergraduate choice. *Journal of Higher Education Policy and Management, 29*(1), 57–72. doi:10.1080/13600800601175789.

Burdge, R. J., & Ludtke, R. L. (1972). Social separation among displaced rural families: The case of flood control reservoirs. In W. R. Burch, N. H. Cheek, & L. Taylor (Eds.), *Social behavior, natural resources, and the environment*. New York: Harper & Row.

Canter, D. (1977). *Psychology of place*. New York: St. Martin's Press.

Cassidy, C., & Trew, K. (2001). Assessing identity change: A longitudinal study of the transition from school to college. *Group Processes & Intergroup Relations, 4*(1), 49–60. doi:10.1177/1368430201041004.

Chapman, D. W. (1981). A model of student college choice. *The Journal of Higher Education, 52*(5), 490–505.

Chow, K., & Healey, M. (2008). Place attachment and place identity: First-year undergraduates making the transition from home to university. *Journal of Environmental Psychology, 28(4)*, 362–372. doi: http://dx.doi.org/10.1016/j.jenvp.2008.02.011.

Cicognani, E., Pirini, C., Keyes, C., Joshanloo, M., Rostami, R., & Nosratabadi, M. (2008). Social participation, sense of community and social well being: A study on American, Italian and Iranian university students. *Social Indicators Research, 89*(1), 97–112. doi:10.1007/s11205-007-9222-3.

Cicognani, E., Menezes, I., & Nata, G. (2010). University students' sense of belonging to the home town: The role of residential mobility. *Social Indicators Research, 104*(1), 33–45. doi:10.1007/s11205-010-9716-2.

Cole, L., & Thompson, G. W. (1999). Survey of current students: Implications for recruitment and retention. *Nacta Journal, 43*, 15–19.

Connor, H., Pearson, R., Pollard, E., Tyers, C., & Willison, R. (2001). *The right choice? A follow-up to making the right choice*. London: Universities UK.

Cresswell, T. (2004). *Place: A short introduction*. Oxford: Blackwell.

Eijck, M., & Roth, W.-M. (2010). Towards a chronotropic theory of "place" in place-based education. *Cultural Studies of Science Education, 5*(4), 869–898. doi:10.1007/s11422-010-9278-2.

European Commission. (2010). *Europe 2020. A European strategy for smart, sustainable and inclusive growth*. Brussels: European Commission.

Feldman, R. M. (1990). Settlement-identity: Psychological bonds with home places in a mobile society. *Environment and Behavior, 22*(2), 183–229. doi:10.1177/0013916590222002.

Félonneau, M.-L. (2004). Love and loathing of the city: Urbanophilia and urbanophobia, topological identity and perceived incivilities. *Journal of Environmental Psychology, 24*(1), 43–52. doi:10.1016/s0272-4944(03)00049-5.

Fisher, S., & Hood, B. (1987). The stress of the transition to university: A longitudinal study of psychological disturbance, absent-mindedness and vulnerability to homesickness. *British Journal of Psychology, 78*(4), 425–441.

Gieryn, F. (2000). A space for place in sociology. *Annual Review of Sociology, 26*, 463–496. doi: 10.1146/annurev.soc.26.1.463.

Gustafson, P. E. R. (2001). Meanings of place: Everyday experience and theoretical conceptualizations. *Journal of Environmental Psychology, 21*(1), 5–16. doi:10.1006/jevp.2000.0185.

Halpenny, E. A. (2010). Pro-environmental behaviours and park visitors: The effect of place attachment. *Journal of Environmental Psychology, 30*(4), 409–421. doi:10.1016/j.jenvp.2010.04.006.

Hauge, Å. L. (2007). Identity and place: A critical comparison of three identity theories. *Architectural Science Review, 50*(1), 44–51. doi:10.3763/asre.2007.5007.

Hinton, D. (2011). 'Wales is my home': Higher education aspirations and student mobilities in Wales. *Children's Geographies, 9*(1), 23–34. doi:10.1080/14733285.2011.540436.

Holdsworth, C. (2009). Going away to uni': Mobility, modernity, and independence of English higher education students. *Environment and Planning, 41*(8), 1849. A.

Hossler, D., & Gallagher, K. S. (1987). Studying student college choice: A three-phase model and the implications for policymakers. *College and University, 62*(3), 207–221.

Hossler, D., Schmit, J., & Vesper, N. (1998). *Going to college: How social, economic, and educational factors influence the decisions students make*. Philadelphia: Johns Hopkins University Press.

Hummon, D. (1992). Community attachment. Local sentiment and sense of place. In I. Altman & S. M. Low (Eds.), *Place attachment* (pp. 253–277). New York/London: Plenum Press.

Jackson, G. A. (1982). Public efficiency and private choice in higher education. *Educational Evaluation and Policy Analysis, 4*(2), 237–247.

James, R. (2000). Non-traditional students and their university participation: An Australian perspective on persistent inequities and the new ideology of student choice. *Tertiary Education and Management, 6*(2), 105–118.

John, E. P. (1990). Price response in enrollment decisions: An analysis of the high school and beyond sophomore cohort. *Research in Higher Education, 31*(2), 161–176.

Jorgensen, B. S., & Stedman, R. C. (2001). Sense of place as an attitude: Lakeshore owners' attitudes toward their properties. *Journal of Environmental Psychology, 21*(3), 233–248. doi:10.1006/jevp.2001.0226.

Keown, C. (1983). Focus group research: Tool for the retailer. In T. J. Hayes & C. B. Tathum (Eds.), *Focus group interviews: A reader* (pp. 64–70). Chicago: American Marketing Association.

Korpela, K. M. (1989). Place-identity as a product of environmental self-regulation. *Journal of Environmental Psychology, 9(3)*, 241–256. doi: http://dx.doi.org/10.1016/S0272-4944(89)80038–6.

Kudryavtsev, A., Stedman, R. C., & Krasny, M. E. (2012). Sense of place in environmental education. *Environmental Education Research, 18*(2), 229–250. doi:10.1080/13504622.2011.609615.

Kyle, G., Graefe, A., Manning, R., & Bacon, J. (2003). An examination of the relationship between leisure activity involvement and place attachment among hikers along the Appalachian Trail. *Journal of Leisure Research, 35*(3), 249–273.

Lee, T. H. (2011). How recreation involvement, place attachment and conservation commitment affect environmentally responsible behavior. *Journal of Sustainable Tourism, 19*(7), 895–915. doi:10.1080/09669582.2011.570345.

Lewicka, M. (2010). What makes neighborhood different from home and city? Effects of place scale on place attachment. *Journal of Environmental Psychology, 30*(1), 35–51. doi:10.1016/j.jenvp.2009.05.004.

Lewicka, M. (2011). Place attachment: How far have we come in the last 40 years? *Journal of Environmental Psychology, 31*(3), 207–230. doi:10.1016/j.jenvp.2010.10.001.

Litten, L. H. (1982). Different strokes in the applicant pool: Some refinements in a model of student college choice. *The Journal of Higher Education, 53*(4), 383–402.

Low, S. M., & Altman, I. (1992). Place attachment. In I. Altman & S. Low (Eds.), *Place attachment* (Vol. 12, pp. 1–12). New York: Plenum.

Malpas, J. (2008). New media, cultural heritage and the sense of place: Mapping the conceptual ground. *International Journal of Heritage Studies, 14*(3), 197–209. doi:10.1080/13527250801953652.

Marcouyeux, J., & Fleury-Bahi, G. (2011). Place-identity in a school setting: Effects of the place image. *Environment and Behavior, 43*(3), 344–362.

Massey, D. (1995). The conceptualization of place. In D. Massey & P. Jess (Eds.), *A place in the world?* (pp. 45–85). Oxford: Oxford University Press/The Open University.

Morgan, D. L. (1996). Focus groups. *Annual Review of Sociology, 22*, 129–152.

National Union of Students (NUS). (2008). *NUS student experience report*. London: National Union of Students.

Organisation for Economic Co-Operation and Development (OECD). (2006). *Evolution of student interest in science and technology studies policy report*. Paris: OECD Global Science Forum.

Patiniotis, J., & Holdsworth, C. (2005). 'Seize that chance!' Leaving home and transitions to higher education. *Journal of Youth Studies, 8*(1), 81–95. doi:10.1080/13676260500063710.

Pearson, R. (1997). *As the cost rises so does the need for students to have better information to make better choices* (House of Commons Select Committee: The Dearing Report). London: HMSO.

Pretty, G. H., Chipeur, H. M., & Bramston, P. (2003). Sense of place amongst adolescents and adults in two rural Australian towns: The discriminating features of place attachment, sense of community and place dependence in relation to place identity. *Journal of Environmental Psychology, 23*(3), 273–287.

Proshansky, H. M. (1978). The city and self-identity. *Environment and Behavior, 10*, 147–170.

Proshansky, H. M., Fabian, A. K., & Kaminoff, R. (1983). Place-identity: Physical world socialization of the self. *Journal of Environmental Psychology, 3*(1), 57–83.

Purcell, K., Elias, P., & Ellison, R. (2008). *Futuretrack: Applying for higher education – The diversity of career choices, plans and expectations*. Warwick: Warwick Institute for Employment Research, University of Warwick.

Royal Society. (2007). *A state of the nation report on science and mathematics education*. London: The Royal Society.

Sack, R. (1997). *Homo geographicus*. Baltimore: Johns Hopkins University Press.

Sampson, K. A., & Goodrich, C. G. (2009). Making place: Identity construction and community formation through "sense of place" in Westland, New Zealand. *Society & Natural Resources, 22*(10), 901–915. doi:10.1080/08941920802178172.

Scannell, L., & Gifford, R. (2010). Defining place attachment: A tripartite organizing framework. *Journal of Environmental Psychology, 30*(1), 1–10. doi:10.1016/j.jenvp.2009.09.006.

Sjaastad, J. (2011). Sources of inspiration: The role of significant persons in young people's choice of science in higher education. *International Journal of Science Education, 34*(10), 1615–1636. doi:10.1080/09500693.2011.590543.

Smaldone, D., Harris, C., & Sanyal, N. (2005). An exploration of place as a process: The case of Jackson Hole, WY. *Journal of Environmental Psychology, 25*(4), 397–414. doi:10.1016/j.jenvp.2005.12.003.

Smith, E., & Gorard, S. (2011). Is there a shortage of scientists? A re-analysis of supply for the UK. *British Journal of Educational Studies, 59*(2), 159–177. doi:10.1080/00071005.2011.578567.

Soutar, G. N., & Turner, J. P. (2002). Students' preferences for university: A conjoint analysis. *International Journal of Educational Management, 16*(1), 40–45.

Stedman, R. C. (2002). Toward a social psychology of place: Predicting behavior from place-based cognitions, attitude, and identity. *Environment and Behavior, 34*(5), 561–581.

Thompson, J., & Bekjradina, B. (2009). *Male and female participation and progression in higher education.* Oxford: Higher Education Policy Institute (HEPI).

Treasury, H. (2004). *Science & innovation investment framework 2004–2014*. London: HM Treasury.

Trentelman, C. K. (2009). Place attachment and community attachment: A primer grounded in the lived experience of a community sociologist. *Society & Natural Resources, 22*(3), 191–210. doi:10.1080/08941920802191712.

Vaske, J. J., & Kobrin, K. C. (2001). Place attachment and environmentally responsible behavior. *The Journal of Environmental Education, 32*(4), 16–21. doi:10.1080/00958960109598658.

Vaughn, S., Schumm, J. S., & Sinagub, J. (1996). *Focus group interviews in education and psychology*. London: Sage.

Wilkinson, S. (1998a). Focus group methodology: A review. *International Journal of Social Research Methodology, 1*(3), 181–203.

Wilkinson, S. (1998b). Focus groups in health research exploring the meanings of health and illness. *Journal of Health Psychology, 3*(3), 329–348.

Chapter 9
Short Stories of Educational Choice: In the Words of Science and Technology Students

Fredrik Jensen and Ellen Karoline Henriksen

Introduction and Aims

How do science and technology[1] students describe their educational choice? In this chapter we look at responses to an open-ended question in the IRIS Questionnaire (IRIS Q) from first-year students in Norway, Denmark and England. Similar analyses of Italian students' responses to the same question are presented in Chap. 18.

Eccles et al.'s expectancy-value model presented in Chap. 2 describes how young people base their educational choice on their expectation of success and the interest and enjoyment, attainment value, utility value and cost they ascribe to various educational options. Chapter 2 further describes late-modern society as a place and time where non-materialistic values are important, where young people feel they can make their own choices, free from traditions, and where self-realisation and identity development are central in young people's lives. In Chap. 3, Holmegaard, Ulriksen and Madsen use a narrative approach to look at how students negotiate their educational choice over time, gradually reworking their narrative of the choice to make it convincing to the people it is shared with and to themselves.

[1] Most chapters in this book concern participation in science, technology, engineering and mathematics (STEM) education. The present chapter presents results from a subsample of STEM and does not include engineering. The term "STEM" is nonetheless used in some instances in the Discussion, assuming that the perspectives discussed will be generally relevant for all STEM disciplines.

F. Jensen (✉)
Norwegian Centre for Science Education, University of Oslo,
P.O. Box 1106, Blindern, N-0317 Oslo, Norway
e-mail: fredrik.jensen@ils.uio.no

E.K. Henriksen
Department of Physics, University of Oslo, P.O. Box 1048, Blindern, N-0316 Oslo, Norway
e-mail: e.k.henriksen@fys.uio.no

E.K. Henriksen et al. (eds.), *Understanding Student Participation and Choice in Science and Technology Education*, DOI 10.1007/978-94-007-7793-4_9

In line with late modern ideals, it is important for the students that the choice is personal and unique, and that it fits with their conception of who they are.

In the Eccles et al. model, interest-enjoyment value, along with three other value categories and expectation of success, directly influence educational choice, whereas these values are in turn related to several psychological and social/cultural variables such as self-concept, affective reactions and memories, cultural *milieu* and socializers' behaviour (Eccles and Wigfield 2002).

Interest has been found in a number of studies to be important in young people's description of their educational choice (Eccles and Wigfield 2002; Hazari et al. 2010; Maltese and Tai 2009). In particular, choices of physics seem to be largely interest-motivated (Reid and Skryabina 2002; Rødseth and Bungum 2010; Bøe and Henriksen 2013). According to Hazari et al. (2010, p. 2), "when studying students' choice of field, the development of their interests is of critical relevance. It is likely that the link between the development of interest and career choice is mediated by changes in self-perceptions (and identity)".

Krapp and Prenzel (2011), who have reviewed research and theoretical models concerning interest in science, described interest as "a phenomenon that emerges from an individual's interaction with his or her environment" (ibid., p. 31) and emphasised its dependence on cognitive as well as emotional aspects. Several chapters in this volume, and also previous research (Sjaastad 2011; Bøe and Henriksen 2013) indicate that interest in science and mathematics is related to significant persons and to engagement in out-of-school activities such as popular science, science centres or science camps.

In this chapter, we present an analysis of students' brief accounts of their educational choice, focusing on the values and expectations which emerge as important factors. In particular, we will look at respondents' expressions of subject interest and at clues concerning how this interest has arisen and developed through school and family experiences, leisure activities, science outreach and exposure to popular culture. We aim to identify the discourses that young people draw upon when they describe their choices and to relate these to expectancy-value theory as well as choice narratives and late modern ideals. Silverman (2006) describes how research can identify the broad discourses which people use to define their identities. For instance, what vocabularies and arguments do individuals use in different contexts, and how do these repertoires get invoked?

Methods

The open-ended request in the IRIS questionnaire (Appendix), "Please describe how you came to choose this course", prompts responses concerning all sorts of influences, priorities and prospects in the decision-making process. Written responses from Norwegian, Danish and English students from five disciplines – biology, physics, chemistry, mathematics, and computer science – have been included in this analysis. In total, the data set comprises 2,146 responses.

Table 9.1 Number of respondents in the data material analysed here, by nationality and discipline, and total number of IRIS Q respondents in the same categories (in *parentheses*)

	Norway	Denmark	UK
Biology	57 (234)	47 (260)	46 (300)
Physics	48 (137)	45 (95)	69 (69)
Chemistry	35 (35)	31 (31)	72 (72)
Mathematics	37 (37)	66 (103)	50 (135)
Computer science	53 (326)	57 (241)	71 (71)
Total	230 (769)	246 (730)	308 (647)
Total (all countries)	**784 (2,146)**		

Using the random number function in Excel, we reduced the amount of data, closing in on around 50 responses for each of the five disciplines for each country. This procedure was carried out partly in order to make the analysis manageable and partly because theoretical saturation was expected to be reached with a far smaller material than the full data set. In total we used 784 responses (Table 9.1).

Responses were analysed qualitatively using the NVivo 9 software to code and retrieve quotations. A thematic analysis was performed, following the recommended approach of Braun and Clarke (2006), with the aim of finding repeated patterns of meaning. An open coding process was adopted (Strauss and Corbin 1990) where codes were mainly created inductively, based on the respondents' actual expressions. However, the coding process was guided by the theoretical frameworks outlined above, notably Eccles et al's expectancy-value model, narrative psychology and perspectives on youth in late-modern societies. Several codes were assigned to the same passage where appropriate; for instance, the response "Liked the subjects. Got good results. Good job opportunities" was assigned codes "Interest and enjoyment", "Expectation of success", "Utility value", and "School experiences". Responses were reviewed in several cycles. Coding and interpretation were developed and refined as a result of this process and following discussions with colleagues until we had a coding framework and an account that were judged by the research group to be a valid representation of the students' responses. In the Results section we indicate the frequencies of the different codes assigned, in line with the recommendations of Onwuegbuzie and Daniel (2003); however, frequencies in themselves cannot be interpreted as directly indicative of the relative *importance* of various values or considerations for educational choice. Quotes given in the next sections have been translated into English (where necessary) by the authors.

Students' Interpretation of the Open-Ended Question

To help us interpret the written responses given to the open question, short, individual interviews were performed with six first-year physics students at the University of Oslo during spring 2012 (Pettersen 2012). The interviews were conducted immediately after the students had completed the questionnaire.

Such interviews can give more detailed information about how respondents interpret and respond to the question they are asked, and thus provide evidence to help establish the validity of the study (Wilson 2005). Audio-recordings from the interviews were transcribed and analysed using a similar approach as for the open-ended question. These interviews, and the results reported in the two next paragraphs, are described in more detail by Pettersen (2012).

Since the open-ended question was part of the longer IRIS questionnaire, most answers given to this question were likely to be short and influenced by the contexts and associations evoked by the previous (closed) questions. Concerning respondents' interpretation of the question, the six students who were interviewed indeed interpreted the question as intended by the researchers; for instance, as one respondent put it, "[The question concerns] what made you choose the [educational] strand that you chose (...) were there any particular driving forces in or around you that led to the choice".

In the interviews several students expressed that when responding to the open-ended question, they gave the first answer that came to their mind. Some said that they gave an answer resembling responses they had given to similar questions on previous occasions, when asked by friends, family, or others – they had a ready-made educational choice narrative that was recalled when prompted by the IRIS Q open question.

There are good reasons to believe that the amount of thought and reflection that was put into the answers varied a lot. In one of the interviews, the student insisted that his subject interest, and nothing else, explained his choice. However, later in the interview, he said that he enjoyed watching science television shows when he was young, "... but I have not really given that much thought before now". During the interview he recalled experiences that could have contributed to his present science interest. Several of the students expressed that their response to the open-ended question reflected the experiences or priorities they considered *most important* for their choice. To what extent the factors which students rate as most important, are actually the ones that best explain their choice will be discussed in the last part of the chapter.

Descriptions Related to Expectancies and Values

In the following, we present results from the thematic analysis of the 784 written responses to the open-ended question about the educational choice process. In the present section, results related to expectancies and values as described in the Eccles et al. model are presented, whereas the next section describes findings related to influence factors that shape expectancies and values. For each quote given, the nationality, gender and chosen discipline of the respondent is indicated in parentheses.

Interest and Enjoyment

On the open-ended question "Please describe how you came to choose this course", interest and enjoyment was the most frequent response given. In total, 533 of the 784 answers referred to this factor. Many responses were short, just stating that the choice was guided by interest:

> The interest for biology and chemistry (Norway, Female, Biology)
> Chose biology because of interests in animals and nature (Norway, Male, Biology)
> Chose out of interest (Denmark, Female, Mathematics)

A large proportion of the students wrote that they made their choice based on a long-lasting interest for the subject, often with roots in early adolescence. Several mentioned popular science and leisure activities in childhood as starting points for their interest in science and mathematics.

> Used to read up on astronomy (just stars and planets mainly) when I was younger. This eventually developed into a love of science particularly physics. (England, Male, Physics)
> ... I have always since I was young had an interest for chemistry, and have among other things in my childhood years experimented with chemistry, including fireworks, etc. that have enhanced my interest for the subject. (Norway, Male, Chemistry)

Some of the students wrote that their own interest or enjoyment of the subject (and nothing else) determined their decision.

> My own interests (Norway, Male, Chemistry)
> Only my own well-being. Have already been through a hard education, now I want to do something I LIKE doing (Norway, Female, Mathematics)

These quotes might indicate that these students rate interest as the most important factor in the choice process, more important than for instance job safety, income or opportunities to work or study abroad. It might also be a way to express that it is important that they made their own choice, free from other people's opinions or expectations.

Utility Value

Of the 784 responses, 135 were coded with utility value, indicating that the expected extrinsic outcomes of the choice were important for many. In general, three types of utility value were identified: high income, safe job, and (a wide variety of) study or job opportunities.

> (...) much because of the good wages, and that one is «guaranteed» a job. (Norway, Male, Biology)
> I chose maths as it has applications in nearly everything and would give me a wide range of career paths upon leaving university. (England, Male, Mathematics)

Expectation of Success and Cost

As many as 124 responses were coded with 'expectation of success'. Here, many students stated that they chose a course that they expected they would master, often based on previous successes in the subject, as we shall see in the next section.

> I have always been good at physics and enjoyed it, I didn't want to do anything else. (England, Female, Physics)
>
> Maths is my best subject, which I find interesting (England, Male, Mathematics)

Cost, as defined in the Eccles et al. model, concerns the potentially negative aspects of the choice, for instance what leisure activities must be limited in order to follow a specific course. This category was not identified in any of the responses to the open-ended question, which was to be expected since the open-ended question asked for what made respondents choose their course, and not for the arguments against the choice.

Attainment Value

Attainment value in the Eccles et al. model is closely related to identity and to the prospect of attaining the goals (in this case a STEM education) one has set for oneself. Attainment value is not as readily identifiable in students' answers as are, for example, interest-enjoyment or utility value; it is often expressed more indirectly, but may be identified in several of the responses analysed here. Altogether, 70 responses received codes related to attainment value. For instance, some students wrote that their choice fitted well with how they perceived themselves.

> (…) at the 'open house' at [Copenhagen University], I thought chemistry sounded exciting and like something I could see myself studying (Denmark, Female, Chemistry)
>
> Because I was the nerd of the family (Denmark, Female, Computer science)

Responses expressing values such as idealism, prestige, and self-development have also been interpreted as expressions of attainment value in our analysis.

> Wish to contribute to more equitable distribution between rich and poor, and/or work with climate issues. (Norway, Female, Biology)
>
> I wanted to study something that I am personally interested in and that would impress people, and just generally sound good on a CV (England, Male, Physics)
>
> I felt that by choosing this course I would face increasing challenges and improve myself as a person. (England, Male, Physics)

The last quote is an example of students describing their chosen subject as offering positive opportunities for self-development through challenges, and illustrates the close connection between attainment value, interest-enjoyment value and expectation of success.

Influence Factors Shaping Expectancies and Values

In the Eccles et al. model, interest/enjoyment, attainment value, utility value and expectation of success are considered to directly influence educational choice, whereas a range of personal and cultural influence factors in turn shape expectations and values. Do IRIS respondents' brief accounts of their educational choice tell us something about these influence factors that lie behind expectations and values; notably, do they tell us what brought about and supported the strong interest that many respondents cite as the primary driving force behind their choice? In this section we present findings related to influence factors that shape expectancies and values.

School Experiences and Teachers

School experience was the source of influence most widely referred to. In total, 120 responses described school science and mathematics as an important influence and 99 of these responses were also coded *interest and enjoyment*, indicating that many relate their subject interest to experiences from school.

> This was what I found most interesting in upper secondary school, and what sounded most exciting to work with in the future. Biotechnology is the new oil! (Norway, Female, Biology)
>
> I liked the subject in school, and thought the programme appeared good at the [university]. (Norway, Female, Mathematics)

School experiences were related to expectation of success by some students. Thirty-five responses were coded into both the 'school experience' and the 'expectation of success' category. The students typically wrote that they chose a study programme containing subjects they previously had success in and thus expected to master in higher education.

> It was my favourite subject at high school and college, and the area where I did the best. (England, Male, Chemistry)

Perhaps unsurprisingly, many of the students who described school experiences as important also mentioned good teachers. Teachers are mentioned in 56 of the 784 responses analysed, which makes them the most frequently mentioned group of persons. They are described as skilled in the subject matter, good at teaching, they engage in discussions with the students, praise students' efforts, and provide advice about educational choice. Most frequently, teachers are related to subject interest and enjoyment (41 responses), and expectation of success (15 responses).

> Was fascinated with the concepts at A-Level, and had very good teachers who loved their subject. (England, Female, Chemistry)
>
> (...) was praised by my teachers a lot so gave me the courage and motivation to continue with it. (England, Female, Biology)

> I was much in doubt about whether I should start this year at all, and whether [it should be] mathematics. Upon leaving upper secondary, I talked to my mathematics teacher about it, and the next morning I woke up and knew that this was my calling. (Denmark, Female, Mathematics)

People's Influence on Interest and Enjoyment

Family was mentioned in 27 of the open responses, and 20 of these also received the *interest and enjoyment* code. Family members were said to have influenced respondents in three slightly overlapping ways: through having a STEM degree or working in the field; by giving recommendations, encouraging or discussing the choice with the respondent; or by having done (or encouraged) science and mathematics activities with the respondent.

> Had a long running interest in the natural world since I was young, maybe influenced by my mum who has a biology degree and taught biology for a while. (England, Female, Biology)
>
> (...) discussed it with parents and teachers and decided to go for it. (England, Male, Chemistry)
>
> (...) not to mention that I've always been encouraged at home and school to attempt to solve problems. (England, Male, Physics)

Like family members, *friends* were also described as having given recommendations, discussed the choice, and engaged in science or technology activities. Also, some respondents wrote that they chose where to study based on where their boy/girl-friend or friends studied, in some cases in order to study at a place where they already had friends, but also based on the information that the peers provided. Compared to parents, friends might be in a better position to provide up-to-date reports and information about the educational institutions and student life.

> I have a friend who has studied in this programme before, and therefore knew how it was built up. (Norway, Female, Biology)
>
> (...) boyfriend who was going to study in the same city (Norway, Female, Physics)

One student was captivated by her boyfriend's mathematics exercises in upper secondary school. Originally, language was her chosen subject of specialization; however, pondering on these exercises resulted in her rating mathematics as more fun than language. Another student wrote that many social and fun group discussions when doing chemistry homework developed into an interest for the subject.

A few students wrote that they had received information and recommendations from STEM students or researchers.

> Talked to a Master student from the [university] who presented what she was doing in a very interesting way. Found out that this must be something for me, since I have always liked biology... (Norway, Female, Biology)

Only three students wrote that they had received guidance from career advisors. In all three responses it was visible that the students had made their own assessment of how well the option fitted with their own values and priorities, after having received information and suggestions from the career advisor.

> I was made aware of it by a career advisor. I had never heard of it before, but all this about being creative and working in groups sounded quite exciting, so therefore I applied. (Denmark, Male, Computer science)

Popular Science and Leisure Activities Influencing Interest and Enjoyment

Articles, books, TV programmes, magazines, films and documentaries were all mentioned as sources of inspiration by IRIS respondents. Popular science was usually related to subject interest, and was often referred to as a trigger of interest. For many, popular science books or TV programmes (including fiction with science and/or technology content) functioned as a starting point for their interest in the subject. In several cases popular science was mentioned alongside having a long-lasting interest for the subject, often from an early age.

> I am a die-hard Star Wars-fan and have seen all 210 episodes of the 10 year long Stargate SG1 series . . . To say that this has determined my choice would be to exaggerate – but it has kindled dreams. (Denmark, Female, Physics)
>
> Ever since I was very small I've been in love with astronomy. I first started reading simple astronomy textbooks at age 8. By the time I finished school I had three different telescopes, had read hundreds of books on the subject and had even gotten excellent grades in 2 astronomy classes at a local community college . . . Last summer I found myself in the position to start at [University] and took it. (England, Female, Physics)

One student expressed that an experience with a particularly difficult popular science text propelled his wish to understand the subject:

> I started reading about physics, first popular science literature and then more advanced material. Unavoidably, I came across topics that I had no preparation to understand, and it was the desire to understand that made me choose this study (Norway, Male, Physics)

For some, popular science provided motivation for understanding how nature works:

> When I was around 15, I got the idea (from a colourful Hawking book) to become an astrophysicist. It stuck through upper secondary school, and I think I reached a point where I could not live without getting the answers to how the world had come to look like it does = how things work, which forces have created our universe – and not the least, how can we use it? That is why I chose physics, even if I think it surprised many. (Denmark, Female, Physics)

Popular science books, magazines and TV programmes were broadly represented, particularly in the quotes from physics students. On the other hand, computer science students in particular referred to leisure activities. These students referred to programming and other computer usage as a hobby that contributed to their interest for computer science.

> Spend my leisure time on computers and that has created an interest for it (Denmark, Male, Computer science)

A few of the chemistry students referred to experiments they had performed at home with friends or family as adolescents. A handful of the biology students also referred to leisure activities, particularly outdoor activities, as important for boosting their interest in biology.

> An interest from when I was a child. All sailing trips and hiking trips that were filled with experiences of animals and nature, that I was very curious about. The fish books on the shelf. The birds on the window sill. Nature in general. (Norway, Female, Biology)

Outreach Activities Related to Interest and Enjoyment

Popular science was mentioned by respondents as stimulating interest across all age periods. In upper secondary school it was also described as contributing to the decision of what to study by confirming that the education would be interesting.

> Of the A-levels I did the physics lessons were the most inspiring, thought provoking and ultimately the most enjoyable. I was also fascinated by developments within physics that I read about in New Scientist. So I choose the course because it was the area of my lessons and additional reading I found most interesting and enjoyable. (England, Female, Physics).

The outreach activities mentioned in response to the open question – competitions, educational fairs, open day events, information brochures, etc. – were mostly mentioned as important in the year or so before applying for tertiary education. Although some students reported that outreach activities had stimulated their interests, they were most often referred to as helpful in making sure that the respondent was making the right educational decision.

> Interest from an early age, Kennedy Space Centre in Florida and other Science museums e.g. Air & Space Museum in Washington. Good teachers in High School and College kept me interested and open days at Universities cemented my choice and moved me away from Forensic Science. Also the promise of the skills being transferable to the workplace was very important. (England, Male, Physics)

Six students wrote that they gained inspiration from science and mathematics competitions. One participated in a Danish mathematics competition, five participated in the physics or chemistry Olympiads.

> I master chemistry quite well and have found out that (ha-ha) the chemistry is right. Since the start of secondary school I have known that it should be a STEM subject, much thanks to a very skilled teacher. During my participation in the chemistry Olympiads I decided to study chemistry. (Norway, Male, Chemistry)

Guttersrud and Angell (2002) investigated the career paths of Norwegian students who had participated in the physics, chemistry or mathematics Olympiads finals. They found that the majority later chose tertiary STEM education at universities and university colleges, and approximately 60 % of the participants answered that the Olympiads had (some) influence on their choice of education.

Open day events at universities were mentioned by some students. These events give useful information, making it easier to narrow down and settle on a decision.

In particular, a Danish initiative called "Student for a day" was mentioned in a handful of responses. As the title suggests, secondary students can follow a tertiary STEM student for a day, attending lectures, labs, etc. Some students referred to lectures they had attended in open days and similar events. Typically, these students wrote that these lectures confirmed their interest for the subject. Two such outreach events, CERN master class and "The girls' day" at a technological university, are described in some detail in Chap. 12 in the present volume.

> My Love and interest in Mathematics was confirmed after a mathematics lecture and my ability to analyse and solve any mathematical problem easily during my A levels. (England, Male, Mathematics)

Expositions, fairs, brochures and web pages provide information about education and career opportunities. Five students mentioned expositions. Here, they found information that guided them to courses resonating with their wishes, such as interest and enjoyment, an appealing student life or guaranteed employability.

> I went to a fair and had explained to me which educations suited my interests. And then I chose this one. (Denmark, Male, Computer science)

The Internet in general, and in particular websites provided by educational institutions as well as a few specific pages provided by ministries of education, were the most frequently mentioned sources of information.

> I was certain that I wanted to study something to do with science, but was very uncertain about what and where. Spent much time on the Internet, and I would say that I made my final decision based on what I read on the Net. (Norway, Female, Biology)

In the Norwegian *Lily* study, Schreiner et al. (2010) found that first year tertiary students primarily had visited the higher education institutions' own web pages, whereas a range of campaign websites (provided by industry and other stakeholder organisations) were far less visited.

Discussion

In the descriptions of educational choice analysed in this chapter, expressions of interest and intrinsic motivation dominated the responses, but utility value, expectation of success and attainment value were also among the motivations described. Concerning the sources of influence on the choice, respondents referred to school experiences, family and friend influence, popular science and outreach. These sources of inspiration were more often linked to subject interest and enjoyment than to utility value or expectation of success. The Eccles et al. model of educational choice (Chap. 2) emphasises the mutual interaction between interest and enjoyment value, expectation of success, and a range of factors such as cultural setting, childhood experiences, socialisers such as parents and teachers, and so on, in line with what we have seen in the present chapter.

The strong predominance of references to personal interest in responses to the IRIS Q open-ended question was also found for Australian IRIS Q respondents (Lyons et al. 2012), and is also in line with a number of previous studies (Osborne and Collins 2001; Bøe and Henriksen 2013; Maltese and Tai 2009). The analysis of Italian IRIS Q results presented in Chap. 18 has many similarities with the frameworks and results presented here: Also for the Italian data set it was found that intrinsic value (interest and enjoyment) dominated the responses. Chapter 18's category "cultural features" largely covers the findings presented here under the heading "Influence factors shaping expectancies and values" (influence from school, family members, popular science and outreach, etc.). An interesting difference is the category "Innate (natural) features" in the analysis of the Italian data. This category expresses students' belief that a predisposition or innate aptitude is necessary to study STEM. This perception may be related to the idea discussed in Chap. 6 and elsewhere in this volume that only the most dedicated (the brightest and most interested) can study STEM.

Interest, Enjoyment, Identity and Late Modern Ideals

An aim of the present chapter was to identify the discourses – the shared repertoire of common arguments and vocabularies – which students draw on when describing their educational choice.

The students' accounts indicate that there is not a single story about STEM choice; a broad range of direct and indirect influences on the decision are described. However, interest and enjoyment is undoubtedly the single most referred to priority. In the late-modern youth culture of which IRIS Q respondents are arguably a part, the ideal is to choose an education that is rewarding and fulfilling. Students expect to be passionate about their chosen education; tediousness is perceived as betraying their identity (Illeris et al. 2002; Ulriksen 2003). Many respondents wrote that they chose their course according to a subject interest they have had for a long time, often from early adolescence, and thus as something that is part of who they are. As described in Chap. 3 in this volume, each individual student has to find the criteria for what is a right educational choice within themselves, and the choice needs to appear as "true" to their identity. Many of the respondents in the study by Holmegaard et al. (2012) held the idea that there is a "right choice", and that their choice of education should therefore be *authentic* in the sense that it should correspond with who they really are. The choice should be individual, personal and special. In line with these perspectives, it makes sense to choose a subject based on a long-lived and well matured interest. In the present chapter, we have seen that in students' accounts of their choice, interest is linked to external influences on the choice – school experiences, leisure activities, etc. – in such a way that these influences contribute to a coherent, authentic and convincing choice description that fits with the student's identity.

The findings in this chapter indicate that choosing something that one is interested in and expects to enjoy is perceived by many to be "the right answer" to the question of which education to choose, and is thus central in young people's discourses about educational choice. It is important to bear in mind that the findings reported here emerge from students' self-reports and that these short narratives are constructed retrospectively by the students in a process of constructing a coherent choice-narrative (see Chap. 3). In order to understand the mechanisms underlying educational choices, there may be other factors that are less visible in students' accounts, but that are nonetheless powerful explanatory factors (for instance influence from parents; see below). However, the present study shows that such impact of parental background is only to a limited extent visible in students' own account of their educational choice. The students themselves are likely to report that they made their own decision, free from the expectations and opinions of others.

Sources of Inspiration for Educational Choice

Family members (notably parents) were described by some IRIS Q respondents as having influenced choice of a science or technology education either directly (through discussing the choice) or indirectly (for instance through having engaged in science-related activities). Family influence on educational choice has been thoroughly documented. For example, several studies suggest that the family's socio-economic background is, to some extent, reproduced by their children and thus predicts educational attainment and choices (e.g. Dustmann 2004; Schnabel et al. 2002; Werfhorst et al. 2003). Based on data from the first 7 years of the British Household Panel Study, Ermisch and Francesconi (2001) wrote that "Parents' educational attainments are found to be very strongly associated with their children's educational attainments" (p. 137). The impact of parent educational level on students' risk of leaving STEM education without graduating is described in Chap. 14.

Teachers were mentioned more frequently than family members in the responses analysed here, in line with a large body of literature describing teachers' role in forming attitudes to science and contributing to educational choice (Cerinsek et al. 2012; Bøe et al. 2011; Hazari et al. 2010). Drawing upon questionnaire responses from 5,007 Norwegian science, technology, engineering and mathematics (STEM) students in their first year of higher education, Sjaastad (2011) investigated significant persons' influence on educational choices. He suggested that teachers are key factors in inspiring and motivating STEM choices. In his study, teachers were most frequently mentioned by girls, and he suggested that this may be because girls, more than boys, depend on other people to build self-efficacy (Zeldin et al. 2008; Zeldin and Pajares 2000). Furthermore, teachers were more frequently mentioned by students in theoretical STEM disciplines such as mathematics or physics than by students in more applied disciplines such as engineering. This may indicate that teachers are best at inspiring choices of higher education in the subject they teach, and not as good at showing how a mathematics or science interest may be pursued in more practice-oriented educations and professions.

Sjaastad (2011) suggested that persons who have a personal relationship to young people (notably teachers and parents) are in a particular position to support young people in choosing a STEM education through displaying and defining the subjects and their applications, through modelling a STEM identity, and through helping young people in their identity work and with identifying their own interests and abilities. Interventions aimed at helping parents and teachers support young people's educational choice process thus appears as a promising way of improving[2] STEM participation.

Popular science was mentioned in the responses analysed here as stimulating interest across all age periods. In upper secondary school it was also described as contributing to the decision of what to study by confirming that the education will be interesting. Based on their descriptions, it seems that students relate differently to popular science and outreach activities during different age periods. Maltese and Tai (2009) found that the majority of their sample of chemistry and physics graduate students and scientists reported that their interest in science began before middle school. Moreover, 45 % reported that 'intrinsic self-interest' was the source of their interest, whereas an additional 40 % related their interest to a school or education-based experience such as a science competition or science camp. The remaining 15 % referred to a family member as having initiated their interest. In early adolescence, most people are open-minded and show interest in a wide variety of subjects (Krapp and Prenzel 2011), while in the last part of upper secondary school, many make educational choices based on interests that are narrowed down and well matured. In this period of time, they are perhaps more receptive to information or input that can support or help them make their decision (school visits, educational fairs, universities' websites etc.).

Concerning the impact of popular science on interest development and educational choice, Jidesjö (2012) pointed to media's influence on youth's identity project and documented similarities between young people's interest in science (as expressed through the ROSE questionnaire study) and the programmes broadcast on an international popularising TV channel. Astronomy has been found to be high on young people's list of science-related interests (Angell et al. 2004; Osborne and Collins 2001; Schreiner and Sjøberg 2007), and this is visible also in the stories about popular science inspiration in the present chapter. The influence of television programmes on young people's image of science and scientists has also been discussed by Dhingra (2003) who wrote that "Particularly for students who do not have personal knowledge about science through family members or friends, such television role models may be especially significant."

[2] For a definition of what we mean by "improved participation", please refer to Chap. 1, Introduction.

Conclusion

Interest and enjoyment dominates as an explanation for educational choice in students' own accounts. This may be interpreted as an expression of the late-modern ideal of making an authentic educational choice, true to one's identity. Inspiration and influence from school, family influence, popular science and outreach was also described by respondents, mainly in terms of having created interest. In order to improve participation in STEM tertiary education, it is important to kindle and maintain interest and create opportunities for students to develop "STEM identities" through school as well as out-of-school settings. This is in line with Osborne et al.'s (2003) emphasis on providing positive STEM experiences from an early age, for instance through popular science and outreach activities. In later school years, it becomes important to maintain students' interests by providing contents and contexts that are perceived as interesting and relevant. In late adolescence, popular science and, in particular, outreach activities and recruitment efforts may work as important influences in connection with educational decision points. It might be worth encouraging cooperation between STEM participation stakeholders and popular media in order to strengthen the public image of STEM and its practitioners. Parents and teachers might be included in the target groups for efforts to improve STEM participation because of their opportunity to support young people in their educational decision-making process.

References

Angell, C., Guttersrud, Ø., Henriksen, E. K., & Isnes, A. (2004). Physics: Frightful but fun. Pupils' and teachers' views of physics and physics teaching. *Science Education, 88*(5), 683–706.

Braun, V., & Clarke, V. (2006). Using thematic analysis in psychology. *Qualitative Research in Psychology, 3*(2), 77.

Bøe, M. V., & Henriksen, E. K. (2013). Love it or leave it: Norwegian students' motivations and expectations for postcompulsory physics. *Science Education, 97*(4), 550–573. doi:10.1002/sce.21068.

Bøe, M. V., Henriksen, E. K., Lyons, T., & Schreiner, C. (2011). Participation in science and technology: Young people's achievement-related choices in late modern societies. *Studies in Science Education, 47*(1), 37–72.

Cerinsek, G., Hribar, T., Glodez, N., & Dolinsek, S. (2012). Which are my future career priorities and what influenced my choice of studying science, technology, engineering or mathematics? Some insights on educational choice—case of Slovenia. *International Journal of Science Education*, 1–27, doi:10.1080/09500693.2012.681813.

Dhingra, K. (2003). Thinking about television science: How students understand the nature of science from different program genres. *Journal of Research in Science Teaching, 40*(2), 234–256.

Dustmann, C. (2004). Parental background, secondary school track choice, and wages. *Oxford Economic Papers, 56*(2), 209–230. doi:10.1093/oep/gpf048.

Eccles, J., & Wigfield, A. (2002). Motivational beliefs, values, and goals. *Annual Reviews Psychology, 53*, 109–132.

Ermisch, J., & Francesconi, M. (2001). Family matters: Impacts of family background on educational attainments. *Economica, 68*(270), 137–156. doi:10.1111/1468-0335.00239.

Guttersrud, Ø., & Angell, C. (2002). *Fagolympiadenes finaledeltakere – hvor blir de av?* [Participants in physics-, chemistry and mathematics olympiads – Where do they go?]. Oslo: Department of Physics, University of Oslo.

Hazari, Z., Sonnert, G., Sadler, P. M., & Shanahan, M.-C. (2010). Connecting high school physics experiences, outcome expectations, physics identity, and physics career choice: A gender study. *Journal of Research in Science Teaching, 47*(8), 978–1003.

Holmegaard, H.T., Ulriksen, L.M., Madsen, L.M. (2012). The process of choosing what to study: A longitudinal study of upper secondary students' identity work when choosing higher education. *Scandinavian Journal of Educational Research*, 1–20 doi:10.1080/00313831.2012.696212.

Illeris, K., Katznelson, N., Simonsen, B., & Ulriksen, L. (2002). *Ungdom, identitet og uddannelse* [Youth, identity and education]. Frederiksberg: Roskilde universitetsforlag.

Jidesjö, A (2012). *En problematisering av ungdomars intresse för naturvetenskap och teknik i skola och samhälle – Innehåll, medierna och utbildningens funktion.* PhD thesis, Linköping University.

Krapp, A., & Prenzel, M. (2011). Research on interest in science: Theories, methods, and findings. *International Journal of Science Education, 33*(1), 27–50.

Lyons, T., Quinn, F., Rizk, N., Anderson, N., Hubber, P., Kenny, J., et al. (2012). *Starting out in STEM. A study of young men nd women in first year science, technology, engineering and mathematics courses.* University of New England: SiMERR National Research Centre.

Maltese, A. V., & Tai, R. H. (2009). Eyeballs in the fridge: Sources of early interest in science. *International Journal of Science Education, 32*(5), 669–685.

Onwuegbuzie, A. J., & Daniel, L. G. (2003). Typology of analytical and interpretational errors in quantitative and qualitative educational research. *Current Issues in Education, 6*(2), 1–29.

Osborne, J., & Collins, S. (2001). Pupils' views of the role and value of the science curriculum: A focus-group study. *International Journal of Science Education, 23*(5), 441–467.

Osborne, J., Simon, S., & Collins, S. (2003). Attitudes towards science: A review of the literature and its implications. *International Journal of Science Education, 25*(9), 1049–1079.

Pettersen, A. (2012). *Hvordan beskriver norske fysikkstudenter bakgrunnen for sitt utdanningsvalg* [How do Norwegian physics students describe the background for their choice of education?]. Oslo: University of Oslo.

Reid, N., & Skryabina, E. A. (2002). Attitudes towards physics. *Research in Science & Technological Education, 20*(1), 67–81.

Rødseth, S., & Bungum, B. (2010). Hva inspirerer til fysikkstudier? En undersøkelse av begynnerstudenter på fysikk. *NorDiNa, 2010*(1).

Schnabel, K. U., Alfeld, C., Eccles, J. S., Köller, O., & Baumert, J. (2002). Parental influence on students' educational choices in the United States and Germany: Different ramifications—Same effect? *Journal of Vocational Behavior, 60*(2), 178–198. doi:10.1006/jvbe.2001.1863.

Schreiner, C., & Sjøberg, S. (2007). Science education and youth's identity construction – Two incompatible projects? In D. Corrigan, J. Dillon, & R. Gunstone (Eds.), *The re-emergence of values in science education* (pp. 231–247). Rotterdam: Sense Publishers.

Schreiner, C., Henriksen, E. K., Sjaastad, J., Jensen, F., & Løken, M (2010). *Vilje-con-valg: valg og bortvalg av realfag i høyere utdanning* [Choosing – Or not choosing – Science and technology in higher education]. *KIMEN, 2010* (2), http://www.naturfagsenteret.no/tidsskrift/kimen.html

Silverman, D. (2006). *Interpreting qualitative data: Methods for analyzing talk, text and interaction.* Los Angeles: Sage.

Sjaastad, J. (2011). Sources of inspiration: The role of significant persons in young people's choice of science in higher education. *International Journal of Science Education, 33*, 1–22.

Strauss, A. L., & Corbin, J. M. (1990). *Basics of qualitative research: Grounded theory procedures and techniques.* Newbury Park: Sage.

Ulriksen, L. (2003). Børne- og ungdomskultur og naturfaglige uddannelser [Child- and youth culture and science education]. In H. Busch, S. Horst, & R. Troelsen (Eds.), *Inspiration til fremtidens naturfaglige uddannelser*. Copenhagen: Undervisningsministeriets forlag.

Werfhorst, H. G. V. D., Sullivan, A., & Cheung, S. Y. (2003). Social class, ability and choice of subject in secondary and tertiary education in Britain. *British Educational Research Journal, 29*(1), 41–62.

Wilson, M. (2005). *Constructing measures: An item response modeling approach*. Mahwah: Erlbaum.

Zeldin, A. L., & Pajares, F. (2000). Against the odds: Self-efficacy beliefs of women in mathematical, scientific, and technological careers. *American Educational Research Journal, 37*, 215–246.

Zeldin, A. L., Britner, S. L., & Pajares, F. (2008). A comparative study of the self-efficacy beliefs of successful men and women in mathematics, science and technology careers. *Journal of Research in Science Teaching, 45*(9), 1036–1058.

Chapter 10
Understanding Declining Science Participation in Australia: A Systemic Perspective

Terry Lyons and Frances Quinn

Introduction

Trends in Participation in High School Science

Concerns about young people's participation in STEM courses and careers in Australia have seldom been far from the headlines over the last two decades. However, they have become more strident recently due to the rapid growth in mining and allied engineering industries, the expansion of medical research and the impending retirements of many in the STEM workforce. Claims that supply will not meet these demands are often accompanied by figures showing a continuing fall in the proportions of senior high school students taking physics, chemistry and biology. For example, a 2012 report by the Chief Scientist of Australia indicated that participation rates in these three subjects among final year (Year 12) high school students fell 32 %, 25 % and 32 % respectively between 1992 and 2010 (see Fig. 10.1). Over this period the Year 12 cohort increased by around 9 %, so these proportional declines represent declines in actual numbers; around 10,000 fewer physics students, 7,000 fewer chemistry students and 16,000 fewer biology students (Office of the Chief Scientist 2012a).

There has been a great deal of speculation in Australia about the underlying causes of these declines, though surprisingly little empirical research. This chapter presents three contributions from the *Choosing Science* project to our understanding of this problem. *Choosing Science* (Lyons and Quinn 2010) was a large-scale national study of the influences on Year 10 (15–16 year old) students' decisions

T. Lyons (✉)
School of Curriculum, Queensland University of Technology, Brisbane, QLD 4059, Australia
e-mail: terry.lyons@qut.edu.au

F. Quinn
School of Education, University of New England, Armidale, NSW 2351, Australia

E.K. Henriksen et al. (eds.), *Understanding Student Participation and Choice in Science and Technology Education*, DOI 10.1007/978-94-007-7793-4_10

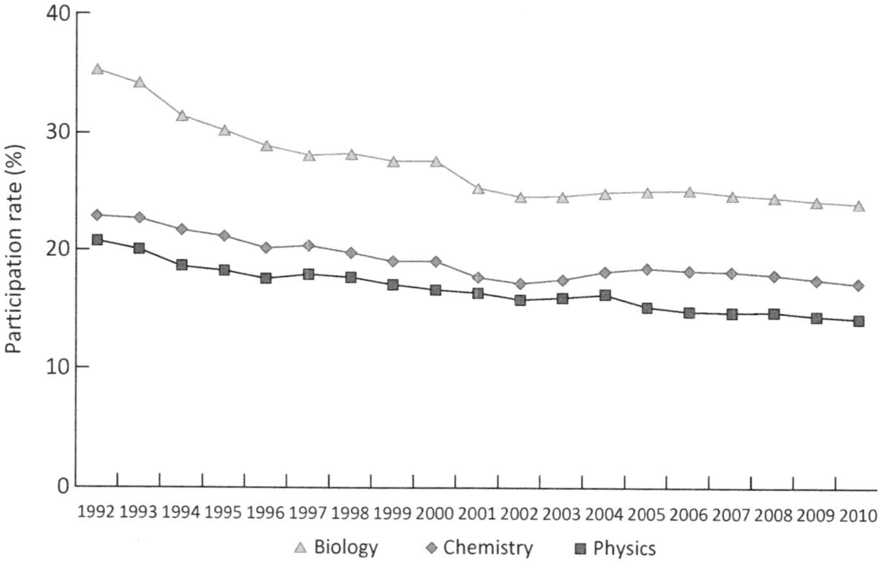

Fig. 10.1 Proportions of Australian Year 12 students enrolled in physics, chemistry and biology subjects, 1992–2010 (Prepared by Kennedy for the Health of Australian Science report (Office of the Chief Scientist 2012a))

about whether or not to take science in their final two years of school. The chapter begins with an overview of school science education in Australia, before focusing on some of the most common explanations for declining participation rates. We introduce the *Choosing Science* study, outlining our research approach and presenting our findings with respect to the two hypotheses. We then draw on additional evidence to argue that the declines are unlikely to be related to changes in attitudes towards science or declines in the quality of science teaching and curricula. Rather, we contend that they are associated with structural changes in school and university curricula and the interrelationship between students' developing identities, the expanded curriculum marketplace and broader developments in the education landscape.

School Science Education in Australia

Science is a compulsory subject from Foundation (age 4–5) to Year 10 (age 15) in all eight Australian states and territories. During these years it is taught as an integrated subject rather than as separate discipline strands. At the end of Year 10, students have the opportunity to select their subjects for Years 11 and 12. English is the only compulsory subject for these years and while there is some variation between states and territories, in general the sciences on offer include physics, chemistry, biology, earth and environmental science and integrated science. Students can choose one or more of these subjects, with the most common combinations traditionally being physics and chemistry, or chemistry and biology. They can also decide not to take any science subjects; a decision which is becoming more common. In 1993, around 32 % of Year 12 students took no science subjects. In 2011, estimates put this figure at around 50 %. Quite apart from any concerns about meeting future demand for scientists, this trend has serious long term implications for Australian society, raising the question of how well-equipped we will be to make informed decisions about critical science-related issues when only half our citizens studied science in any formal way beyond age 16.

As other chapters in this book attest, concerns about declines in young people's participation in science are certainly not confined to Australia and our discussion will draw upon international literature where relevant. However, as our findings arose from evidence collected from Australian students we make no assumptions about the extent to which our conclusions apply to other countries or contexts.

Hypotheses About the Causes of Declining Participation

Attempts by researchers, policy makers or commentators to address the issue of declining participation in science tend to fall into four categories depending on their hypotheses about the locus of the problem. These categories are: uninspiring pedagogy, unengaging curriculum, poor student attitudes towards science and changes in curriculum and education policy.

1. *Ineffective or uninspiring science pedagogy*
   A number of Australian and international researchers have pointed to unengaging pedagogy as a principal cause of declining participation in school science. Certainly the science education literature provides plenty of examples of widespread teacher-centred or exam-focused pedagogies which many students find uninspiring (e.g. Bennett and Hogarth 2009; Jackson et al. 2010; Lindahl 2007; Lyons 2006). Indeed, concern about the prevalence and impact of such pedagogies has been the driver of extensive European efforts to promote inquiry-based science education (European Commission 2007). However, as an explanation for declining science participation in Australia (and perhaps elsewhere), this hypothesis falls short in some respects. First, as DeWitt et al. (2011) point out, a number of studies have raised questions about the robustness of links between students' experiences in science classrooms and their intentions to participate in science. In particular, the levels of family cultural capital and social capital have been shown to strongly mediate the influence of school experiences on students' decisions (Leach and Zepke 2005; Lyons 2006). Second, in order to establish the existence of a causal relationship between the quality of science pedagogy and enrolment declines, one would need to demonstrate that the former had waned over time in parallel with the latter. Yet there is little if any evidence that high school science teaching in Australia is worse today than it was 20 years ago, or even earlier. Indeed, criticisms of the quality of science teaching in Australia have long been part of the education landscape and

certainly predate the steepest declines (Tobin and Gallagher 1987). Likewise, critiques of prevailing school science pedagogies in the UK and the US are certainly nothing new (Gardner 1975; Skinner 1967).

In a similar vein, some Australian studies (e.g. Harris et al. 2005) have attempted to draw a connection between declining participation rates and sub-optimal science teacher qualifications. While it is undoubtedly important to ensure that science teachers have sound subject and pedagogical content knowledge, there is no evidence that recent generations of Australian science teachers are less qualified than those who taught them (see for example Tisher 1971; Rosier 1973). So while there is little doubt that science pedagogy and teacher preparation can be improved, attempts to attribute the declines in student participation over the last 20 years to declining teacher quality have little basis. Indeed, suggestions that Australian students have increasingly been exposed to poor quality science teaching are inconsistent with the world-class performances of Australian 15 year olds in the Programme for International Assessment (PISA) studies (OECD 2007, 2010).

2. *Irrelevant or unengaging science curriculum*
   A second and related target of criticism is the traditional science curriculum, with a number of studies revealing that many students find much of the content personally irrelevant (Osborne and Collins 2001). Certainly this has been a common complaint among Australian Year 10 students (Lyons 2006). However, much like the arguments around pedagogy above, it is one thing to identify problems in recent or current curricula, but quite another to argue that these have had a direct impact on participation rates. Historical examinations of science curricula (e.g. Smith and Gunstone 2009) suggest that dry, content-focused and exam-oriented curricula were around long before the enrolment declines of the last 20 years.
3. *Changes in attitudes towards science and science careers*
   This third hypothesis shifts the focus from teachers and curricula to the students themselves; more specifically, to changes in their attitudes towards school science and/or science careers. The argument is generally framed in terms of today's students holding different values and expectations to previous generations and responding less positively towards the representations of science and science careers they see in school and society. Certainly there is no shortage of Australian and international literature advocating room for improvement in students' attitudes towards school science (e.g. Adamuti-Trache 2007; George 2006; Lindahl 2007; Ramsden 1998). This argument is supported by studies identifying links between attitudes towards science and intentions regarding future engagement (e.g. Ainley and Ainley 2011; Osborne et al. 2009), and is certainly worthy of further investigation.
4. *Changes in education structures and policies*
   Like many countries, Australia has experienced extensive education reforms at the school and tertiary levels over the last two decades. The most pertinent change at the high school level has been the increased retention of students to Year 12. For example, in the early 1980s, only around 35 % of students remained

in school to Year 12. Those leaving at the end of Year 10 generally took up manual, technical or sales jobs or apprenticeships, while most of those going on to Year 11 intended to go to university. Over the following decade the retention rate to Year 12 more than doubled (Ainley et al. 2008). To cater for the greater diversity of interests and abilities of the student body, the senior school curriculum diversified to include a wider range of subjects, altering not only the ecology of the school system, but also of the tertiary sector into which increasing numbers of high school graduates were transitioning. These systemic changes were also considered in the *Choosing Science* study, and are discussed in more detail later in the chapter. The essence at this point is that these policy and curriculum changes present another promising explanation for the declining science participation rates in Australia.

In summary, the evidence suggests that the hypotheses offering the greatest potential for understanding these declines are the generational change in students' attitudes towards science and science careers, and the impact of developments in education policy and curriculum diversity. It is the investigation of these two factors to which we now turn.

Investigating Generational Change in Student Attitudes Towards Science and Science Careers

With respect to enrolment declines, the point was made earlier that knowing the nature of students' attitudes at a single point in time is less pertinent than knowing whether these attitudes have changed over time. Even though much of the recent literature on attitudes to school science has been motivated by concerns about declining participation rates over a long period, most studies have either examined students' attitudes at a single point in time, or explored changes in attitudes between different points in their schooling. While such studies make valuable contributions to the field, they do not establish whether current students have different attitudes to earlier cohorts.

To address this issue, Lyons and Quinn (2010) sought to compare the attitudes of contemporary Australian students with those of students a generation earlier when enrolments were proportionally much higher. Such a comparison involved finding reliable benchmark data on attitudes to science collected from an earlier Year 10 cohort. Fortunately the late 1970s and early 1980s saw groundbreaking research undertaken in this field by Fraser (1977, 1978, 1981) using a survey instrument called the Test of Science Related Attitudes (TOSRA) to measure the attitudes of Year 10 Australian school students to school science. The development of TOSRA scales and technical information about its reliability and discriminant validity are described by Fraser (1978) and the full instrument, coding instructions and findings are outlined in Fraser (1981). TOSRA has since been used in many studies in Australia and overseas. It is still considered valid and reliable for gauging students' attitudes to science, with its validity comparing very favourably with other attitudinal scales in an evaluation of attitudinal instruments (Blalock et al. 2008).

Four of the original seven TOSRA scales were used in the *Choosing Science* study. Scale definitions and sample items from Fraser (1978, 1981) provide some indication of the construct being investigated via the five positively and five negatively worded items in each scale. The *Career Interest in Science* scale gauges respondents' interest in pursuing a career in science (for example, 'I would like to be a scientist when I leave school'). The *Social Implications of Science* scale measures attitudes about the benefits and problems accompanying scientific progress (for example, 'Scientific discoveries are doing more harm than good'). The *Normality of Scientists* scale gauges the extent to which respondents perceive scientists as normal people rather than the eccentric stereotype so often portrayed in the media (for example, 'Scientists like sport as much as other people do'). The *Enjoyment of Science Lessons* scale explicitly focuses on students' enjoyment of school science learning experiences (for example, 'I dislike science lessons').

Our 1977 benchmark data came from Fraser's initial survey of Year 10 students in Sydney, New South Wales (NSW) (Fraser 1978). At the time of Fraser's study the proportions of Australian Year 12 students taking physics, chemistry and biology were around 28 %, 31 % and 58 % respectively (Ainley et al. 2008); far above any of the proportions shown in Fig. 10.1. We identified a sub-sample (N = 308) of the *Choosing Science* cohort from Sydney closely matching Fraser's 1977 sample with respect to size, sex breakdown and school characteristics, and compared mean ratings on each scale. We also calculated mean ratings from the full *Choosing Science* cohort of 3,795 for additional comparison. Further details relating to the samples and methodology are provided in the full *Choosing Science* report.

Results of TOSRA Comparisons

Table 10.1 shows means and alpha reliability levels from Fraser's 1977 study and the *Choosing Science* sub-sample data, supporting the high internal reliability of the

Table 10.1 Scale reliabilities, means and standard deviations from the comparable 1977 and 2007 TOSRA samples

TOSRA Scale	Scale reliabilities, means and standard deviations for 1977 and 2007 data with associated effect size				
	Scale α reliability 1977	Scale α reliability 2007	Mean (sd) 1977	Mean (sd) 2007	Effect size (Cohen's d)
Social implications of science	0.82	0.86	37.3 (5.2)	36.0 (6.9)	0.19
Enjoyment of science lessons	0.93	0.93	33.5 (8.6)	31.3 (9.5)	0.23
Normality of scientists	0.78	0.82	36.3 (4.9)	34.7 (6.6)	0.24
Career interest in science	0.91	0.90	28.8 (8.4)	29.1 (8.8)	−0.04

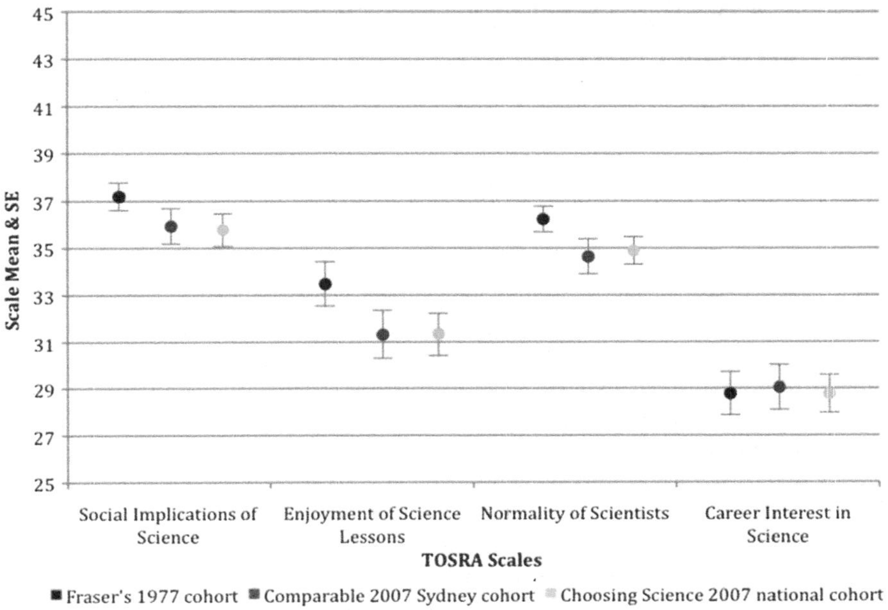

Fig. 10.2 Comparison of mean scores (+/− 2 *SEs*) of all three cohorts on the four TOSRA scales

instrument. The table also shows the Effect Sizes (Cohen's *d*) of any differences in means. The mean scores for each of the four TOSRA scales for all three cohorts are depicted in Fig. 10.2.

As is indicated in Fig. 10.2, there was very little difference between the mean scores of the cohorts on the Career Interest in Science scale, and the effect size of any difference was negligible. Mean scores on the Social Implications of Science, Enjoyment of Science Lessons and Normality of Scientists scales were marginally lower among the 2007 samples, suggesting that these aspects of students' attitudes might be associated with falling participation rates. However, the differences were only significant for the latter two scales and, more importantly, the effect sizes of differences in mean scores for the three scales were only small at less than 0.25. Given that the two studies were 30 years apart, such small effect sizes suggest that the differences are not educationally very meaningful (Cohen 1988; Coe 2002).

These results do not therefore support the contention that contemporary Year 10 students have less interest in science careers or less positive attitudes to the social implications of science than did their counterparts 30 years previously. Given the relative stability of attitudes towards science suggested by this comparison, it appears unlikely that declining science participation rates in Australia are strongly causally related to changed attitudes towards science. This finding signalled a need to look elsewhere for other potential influences on enrolment declines.

Impact of Developments in Education Policy and Curriculum Diversity

In Chap. 7, Ametller and Ryder discussed the impact on subject choice of a specific element of the school science curriculum (socio-scientific issues) while at the same time recognising curriculum as one of many influences on student choice. In view of the results reported above we subsequently explored some of these other, more systemic, influences on subject choice as part of *Choosing Science*. As mentioned earlier, one aspect of the Australian educational landscape that has changed concurrently with the enrolment declines is the curriculum policy context within which students are making their decisions. In this section we report on the results of our review of relevant policy changes and their consequences.

It has been argued by Keating and Walshe (2009, p. 15) that the patterns of youth transition from school to the labour force are ‘largely determined’ by the interaction between education systems, economic conditions and the labour market. We contend here that this interaction between economic conditions and education has contributed to curriculum changes in secondary schools and universities that have very likely impacted on participation rates in senior high school and university science. Downturns in the Australian economy during the early 1980s and 1990s led to sharply increased retention rates from compulsory junior secondary school to the non-compulsory senior years (Years 11 and 12). Apparent retention to Year 12 rose to a peak of 75 % during the severe recession in 1992 (Ainley et al. 2008, p. 21) and has subsequently declined to around 71 % (Clayton et al. 2010). The resultant academic diversity among senior students, together with a number of landmark policy statements, for example, the Dawkins Report (Dawkins and DEET 1988), the Hobart Declaration (Australian Education Council 1989) and the Finn Review (Finn 1991) profoundly influenced curriculum directions in Australian schools. For example, Finn (1991) explicitly recommended that increased student diversity in Years 11 and 12 be accommodated by offering broader curriculum options, including vocationally-oriented options in Years 11 and 12 to articulate with other post-compulsory Vocational Education and Training (VET) courses. All Australian states and territories subsequently reformed their curricula (Keating and Walsh 2009; Dekkers and De Laeter 2001) in part by broadening their subject offerings.

One curriculum change made in response to the changing education context and industry needs over the past few decades has been the inclusion of VET subjects that were traditionally the domain of Colleges of Technical and Further Education (TAFE). Vocational education has since become one of the success stories of recent curriculum reform, and we have seen a national trend of increased participation in VET subjects in schools, with over 40 % of senior students enrolling in at least one VET subject (Clayton et al. 2010). However, students enrolling in VET options are generally less academically inclined and tend not to be enrolled in the biological and especially physical sciences. Hence, although VET participation has risen, VET subjects may be catering primarily for those students who would not otherwise have remained at school for Years 11 and 12 and therefore may not be significantly competing with science for a market share.

However, this same period saw the introduction of many other subjects such as psychology and IT/computer studies which potentially compete for academically inclined students aspiring to university study. In the state of Victoria, for example, psychology is now the third most popular Year 12 subject, being taken by nearly a third of all students (Victorian Curriculum and Assessment Authority 2012). Further, most states and territories now offer religious studies as a Year 12 examination subject. This is often compulsory in the Catholic and other religious denomination schools making up around 35 % of all secondary schools in Australia. In 2009, for example, there were approximately 21,000 Year 12 enrolments in religious studies subjects (unpublished data from the Department of Education, Employment and Workplace Relations).

In addition to the traditional offerings of biology, chemistry, physics and geology, a range of alternative science subjects that also contribute to university entrance requirements were developed, such as environmental science and health/sports/exercise science (Dekkers and De Laeter 1997). The latter have become increasingly popular, with around 69,000 enrolments in 2009 (unpublished data from DEEWR). To appreciate the significance of expanded curriculum offerings one also needs to be aware that Australian states and territories specify that Year 12 students take a minimum number of subjects (or subject units) to qualify for the completion certificate. As the selection of subjects above this minimum in most cases represents additional unnecessary study, most students tend to choose around five subjects for examination. Hence the introduction of new subjects into the senior curriculum tends to decrease participation rates in existing subjects. Between 1993 and 2001, for example, enrolment declines were experienced not only in the key sciences, but also economics, accounting, geography and political/social studies (Australian Council for Educational Research [ACER] 2005). Over this period there was a concomitant increase in enrolments in business studies, secretarial studies, religious studies, hospitality, health/exercise science, computer studies, food and catering, and the arts, among others. In summary, it seems that the context and dynamics of high school subject choice have both changed dramatically, with traditional university-oriented high school subjects facing increased competition for curriculum share.

Interactions with the Tertiary Sector

In addition to the more diverse curriculum *smorgasbord* at the school level, the reciprocal relationship between Australian school and university systems also contributes in several ways to the broader context of high school students' subject choices. First, many school subjects which previously were ineligible for consideration in university entry calculations in Australian universities are now eligible, making them more attractive options for university-oriented students entering Year 11.

Second, universities in the current market-driven tertiary sector have restructured their own curricula to accommodate both new demands from industry and the demand from school leavers with diverse educational backgrounds. New degree programmes have proliferated in such areas as tourism and hospitality management, sport science, sports management and business management, among others, corresponding to Year 12 enrolment trends. These changes have, in turn, given a greater academic legitimacy and status to many school subjects previously considered to be non-university track subjects.

In addition, and perhaps in response to the increased competition in the higher education sector, many Australian universities have relaxed their entry requirements for science courses. School sciences such as physics and chemistry were once considered prerequisites for entry to most undergraduate science courses. However, a review of University Admissions Centre guidebooks over the last two decades reveals that it is now more common to see these subjects listed as 'assumed knowledge' or 'recommended studies', and universities are offering foundation or 'bridging' units in these and other subjects to cater for less-prepared students. This has been an issue of much debate in the community (for example, Belward et al. 2007; Novak 2009; Phillips 2009), with the Australian Academy of Science identifying this as one of the key contributors to declines in mathematics enrolments at the senior high school level (Australian Academy of Science 2006). In terms of the Eccles (2009) model (see also Chap. 2) the utility value of school physics and chemistry has become less tangible. Regardless of the merits or relative difficulty of particular school subjects, it was perhaps inevitable that the market share enjoyed by long-established subjects such as physics, chemistry and biology would decline with the introduction of more Year 12 subject options and changes to university entry requirements.

The Role of Identity in an Extensive Curriculum Marketplace

While we contend that the systemic developments discussed above are most likely behind the declines in science enrolments in Australia, this does not discount the role of identity in student choice. Indeed, a third finding from the *Choosing Science* study highlights the importance of identity in students' negotiations of the expanded curriculum marketplace. Students who had decided to take no science subjects in Year 11 were asked to respond to seven items suggesting reasons for their decisions. Figure 10.3 provides a breakdown of their responses to these items.

The figure shows that the most common reason endorsed by students for not choosing science is that they were unable to picture themselves as scientists. This suggests strongly that their deliberations included an element of cross-referencing between self-image and the images of scientists and science careers, and that perceived incongruence between these images was a considerable influence on

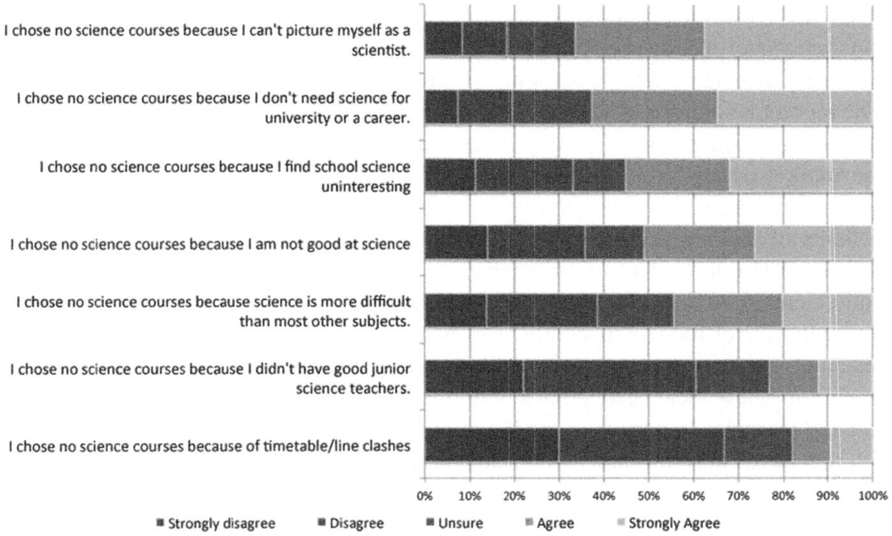

Fig. 10.3 Percentage breakdown of students' responses to items explaining why they chose not to study science in Year 11 (N = 908)

their decisions. One interpretation of this incongruence could be that students are knowledgeable about science careers, but do not see a fit with their own aspirations. Alternatively, students may not have sufficiently well-developed – or sufficiently authentic – images of scientists and science careers to use as reference points when attempting to picture themselves as scientists. Either way, the finding reinforces the importance of identity as a reference point in students' subject deliberations, particularly in the context of increased curriculum choice. Research shows that greater choice can generate greater confusion due to the complexities involved in weighing up advantages and disadvantages of multiple options (White and Hoffrage 2009), leading to an increased reliance on such reference points.

Responding to Declining Science Enrolments

The discussion above highlights the complex interactivity between systemic and personal considerations and the difficulty of teasing apart individual influences on young people's decisions about further participation in science. Borrowing from the systems description of education adopted by Biggs (1993), a student deciding whether or not to choose a science subject is at the core of a nested hierarchy of interacting broader systems including the classroom, the wider school curriculum, the general educational environment (including universities), national educational policies, economic circumstances and political ideologies. Hence arresting the declines in science participation rates will require attention to all components of the system, as well as their interactions.

In the Australian context, much research effort relating to improving STEM outcomes over recent years has been conducted by science educators and focused primarily at teacher and classroom systems (for example, Goodrum et al. 2001). The recommendations of such reports about the need to improve the experiences of students in science classes are as relevant and important as ever, given the reality of the market dynamic and current policy and curriculum settings. The impact on school science of increased competition for market share highlights the potential benefits of making science subjects intrinsically and strategically more attractive and rewarding to students, particularly for girls, who appear to be deserting maths and science courses at a faster rate than boys (Mack and Walsh 2013). Such a classroom level focus will continue to be a productive field of endeavour, as the teaching and learning context is so fluid and there is still so much scope for better integrating new ideas, pedagogies and innovative technologies into science teaching and learning. In particular the burgeoning educational options available via the internet have enormous and as yet scarcely tapped potential to contribute to science education through the use of social media tools, interactive multi-player virtual worlds, remote online experiments and so forth.

A recent report on *Mathematics, Engineering and Science in the National Interest* (Office of the Chief Scientist 2012b) advising the Australian government on "means to encourage greater participation in mathematics, statistics and science courses of study at university" continues this emphasis on teacher and classroom level initiatives. Of the 17 specific recommendations in that report, 11 focused on strategies to promote the first priority of "Inspirational Teaching" through specific preservice and inservice teacher programmes.

However, we argue here and elsewhere (Kennedy et al. 2014) that responses focused at improving the classroom teaching and learning system, while necessary and very welcome, will not be sufficient to address declines in STEM enrolments for two reasons raised earlier in this chapter. The first is the absence of any empirical evidence we are aware of linking declines in enrolments explicitly to the quality of science teaching. The second is the likely impact on students' choices of the system-wide machinations beyond the science classroom, such as curriculum diversification and a market-driven university sector.

The importance of these broader systems has been recognised in a range of other measures addressing STEM participation in senior high school and university. Science and engineering faculties in many Australian universities are offering a variety of outreach programs to connect school students to real science as it is practised, a move in part motivated by the imperative to attract school leavers to their particular institution in an increasingly market-driven tertiary sector. Scientific organisations such as the Commonwealth Scientific and Industrial Research Organisation (CSIRO) offer federally funded outreach programmes such as *Scientists in Schools*. In terms of the importance of identity in students' decision-making it would appear to be crucial that "scientists" are broadly defined and represented in such outreach programs and include a range of the allied or applied sciences, so that students can build up an authentic and inclusive image of what it means to be doing science.

The past two years have seen an upsurge of attention to STEM participation and education, which is taking a wide-angled view of some of the issues across the broader educational and institutional systems. For example, the report on *Mathematics, Engineering and Science in the National Interest* (Office of the Chief Scientist 2012b) partially addresses the issue of university prerequisites, stating that:

> many universities have relaxed the requirement for students to have completed these subjects [physics and chemistry], thus reducing their strategic value. The perceived relative difficulty of these subjects needs to be matched by appropriate rewards (p. 9)

The associated recommendation was for universities to send 'accurate signals' about the value of these school subjects. In the absence of other incentives and in the current tertiary education climate, it appears unlikely that these signals will extend to reinstating prerequisites; it is also possible that reinstating prerequisites might result in fewer university STEM enrolments in any case. Greater clarity on what other meaningful form these 'signals' might take, or what the 'appropriate rewards' might be, could greatly assist this key element of STEM choice.

More generally, several reports referred to in this chapter were only recently commissioned by the Australian Chief Scientist (Dobson 2012; Goodrum et al. 2011; Office of the Chief Scientist 2012a), demonstrating the priority given to science education in the current policy sphere. In response to these initiatives, the Federal government announced a $54 million dollar package to enhance student participation in mathematics and science, with the bulk of the amount supporting partnerships between schools and universities and supporting science and mathematics teachers. Used effectively and aligned with a consistent STEM strategy, this investment has the potential to significantly increase young people's engagement with STEM.

Conclusion

Developing effective policy to increase STEM enrolments depends on an appreciation of the complexity of interrelationships between curriculum, societal, school and student factors associated with the declines. Because declines in STEM enrolments do not appear to correlate to declining attitudes to science, but do correlate to systemic curriculum changes, it cannot be expected that interventions targeting teacher education, science syllabus development or better promotion of science courses and careers will result in these subjects realising the same levels of curriculum market-share they attained in the early 1990s. That target is no longer realistic and it is our view that participation rates in school science will not improve substantially without policies which recognise and address the systemic nature of the problem.

References

Adamuti-Trache, M. (2007). *Who likes science and why? Individual, family and teacher effects* (72 p). Ottawa: Canadian Council on Learning.

Ainley, J., & Ainley, M. (2011). Student engagement with science in early adolescence: The contribution of enjoyment to students' continuing interest in learning about science. *Contemporary Educational Psychology, 36*(1), 4–12.

Ainley, J., Kos, J., & Nicholas, M. (2008). *Participation in science, mathematics and technology in Australian education* (ACER research monograph no 63). Camberwell: Australian Council for Educational Research. Retrieved from http://www.acer.edu.au/documents/Mono63_MathsSciTechSept08.pdf

Australian Academy of Science. (2006). *Mathematics and statistics: Critical skills for Australia's future*. Retrieved from http://www.review.ms.unimelb.edu.au/Report.html

Australian Education Council. (1989). *The Hobart declaration on schooling*. Hobart: Australian Education Council.

Australian Council for Educational Research (2005). *Year 12 subjects and further study* (LSAY Briefing Number 11). ACER. Retrieved May 29, 2007, from http://www.acer.edu.au/lsay/briefs.html

Belward, S., Mullamphy, D., Read, W. & Snedden, G. (2007). Preparation of students for tertiary studies requiring mathematics. *Journal of Australian and New Zealand Industrial and Applied Mathematics (ANZIAM), 47*, C840–C857. Retrieved April 2009, from http://anziamj.austms.org.au/ojs/index.php/ANZIAMJ/article/view/1078

Bennett, J., & Hogarth, S. (2009). Would you want to talk to a scientist at a party? High school students' attitudes to school science and to science. *International Journal of Science Education, 31*(14), 1975–1998.

Biggs, J. (1993). From theory to practice: A cognitive systems approach. *Higher Education Research and Development, 12*(1), 73–85.

Blalock, C., Lichtenstein, M., Owen, S., Pruski, L., Marshall, C., & Topperwein, M. (2008). In pursuit of validity: A comprehensive review of science attitude instruments 1935–2005. *International Journal of Science Education, 30*(7), 961–977.

Clayton, B., Lewanski, R., Pancini, G., & Schutt, S. (2010). *Enhancing the retention of young people to Year 12, especially through vocational skills*. Final Report. Work Based Education Research Centre. Retrieved from http://austcolled.com.au/notepad/article/enhancing-retention-young-people-year-12-especially-through-vocational-skills-project

Coe, R. (2002). *It's the effect size, stupid! What effect size is and why it is important*. Paper presented at the annual conference of the British Educational Research Association, University of Exeter, England, 12–14, September. Retrieved February 3, 2008, from www.leeds.ac.uk/educol/documents/00002182.htm

Cohen, J. (1988). *Statistical power analysis for the behavioural sciences* (2nd ed.). New York: Academic.

Dawkins, J., & Department of Employment, Education and Training. (1988). *Higher education: A policy statement*. Canberra: Australian Government Publishing Service.

Dekkers, J., & De Laeter, J. (1997). The changing nature of upper secondary school science subject enrolments. *Australian Science Teachers' Journal, 43*(4), 35–41.

Dekkers, J., & De Laeter, J. (2001). Enrolment trends in school science education in Australia. *International Journal of Science Education, 23*(5), 487–500.

DeWitt, J., Osborne, J., Archer, L., Dillon, J., Willis, B. & Wong, B. (2011) Young children's aspirations in science: The unequivocal, the uncertain and the unthinkable, *International Journal of Science Education*, 1–27. doi:10.1080/09500693.2011.608197

Dobson, I. (2012). *Unhealthy science? University natural and physical sciences to 2009–10*. University of Helsinki, Monash University & Educational Policy Institute, Melbourne. Retrieved from http://www.chiefscientist.gov.au/2011/10/health-of-australian-science-report/

Eccles, J. (2009). Who am I and What am I going to do with my life? Personal and collective identities as motivators of action. *Educational Psychologist, 44*(2), 78–89.

European Commission. (2007). *Science education NOW: A renewed pedagogy for the future of Europe*. Luxembourg: Office for Official Publications of the European Communities. Retrieved from http://ec.europa.eu/research/science-society/document_library/pdf_06/report-rocard-on-science-education_en.pdf

Finn, B. (1991). *Young people's participation in post-compulsory education and training* [Finn Review]. Report of the Australian Education Council Review Committee. Canberra: Australian Government Publishing Service. Retrieved from http://hdl.voced.edu.au/10707/34856

Fraser, B. (1977). Selection and validation of attitude scales for curriculum evaluation. *Science Education, 61*(3), 317–329.

Fraser, B. (1978). Development of a test of science-related attitudes. *Science Education, 62*(4), 509–515.

Fraser, B. (1981). *TOSRA: Test of science-related attitudes*. Hawthorn: Australian Council for Educational Research.

Gardner, P. (1975). Attitudes to science: A review. *Studies in Science Education, 2*(1), 1–41.

George, R. (2006). A cross-domain analysis of change in students' attitudes toward science and attitudes about the utility of science. *International Journal of Science Education, 28*(6), 571–589.

Goodrum, D., Hackling, M., & Rennie, L. (2001). *The status and quality of teaching and learning of science in Australian schools*. Canberra: Department of Education, Training and Youth Affairs.

Goodrum, D., Druhan, A., & Abbs, J. (2011). *The status and quality of year 11 and 12 science in Australian schools*. Retrieved from http://www.science.org.au/reports/documents/Year-1112-Report-Final.pdf

Harris, K., Jensz, F., & Baldwin, G. (2005). *Who's teaching Science? Meeting the demand for qualified science teachers in Australian secondary schools*. Melbourne: Centre for the Study of Higher Education, University of Melbourne.

Jackson, C., Lasen, M., Callin, R., & Tsey, K. (2010, November 26–27) Enhancing year 9 performance and engagement in science. In *Proceedings of STEM in education conference*, Brisbane, QLD, Australia.

Keating, J., & Walsh, L. (2009). *Submission by the foundation for young Australians to the House of Representatives Standing Committee on Education and Training Combining School and Work: Supporting successful youth transitions*. Melbourne: Foundation for Young Australians. Retrieved from http://www.fya.org.au/wp-content/uploads/2009/05/fyasubmissionhor160109final.pdf

Kennedy, J., Lyons, T., & Quinn, F. (2014). The continuing decline of science and mathematics enrolments in Australian high schools. *Teaching Science, 60*(2), 34–46.

Leach, L., & Zepke, N. (2005). *Student decision-making by prospective tertiary students: A review of existing New Zealand and overseas literature*. Wellington: NZ Ministry of Education.

Lindahl, B. (2007). *A longitudinal study of student's attitudes towards science and choice of career*. Paper presented at the 80th NARST International Conference New Orleans, Louisiana.

Lyons, T. (2006). The puzzle of falling enrolments in physics and chemistry courses: Putting some pieces together. *Research in Science Education, 36*(3), 285–311.

Lyons, T., & Quinn, F. (2010). *Choosing Science: Understanding the declines in senior high school science enrolments*. National Centre of Science, ICT and mathematics education for rural and regional Australia (SiMERR Australia), University of New England. ISBN: 978-1-921597-19-0. Retrieved from http://www.une.edu.au/simerr/pages/projects/131choosingscience.php

Mack, J. & Walsh, B. (2013). *Mathematics and science combinations: NSW HSC 2001–2011 by gender*. University of Sydney. Retrieved from http://www.maths.usyd.edu.au/u/SMS/MWW2013.pdf

Novak, K. (2009, May 25). Extra points for university entry. *Adelaide Advertiser*.

OECD. (2007). *PISA 2006: Science competencies for tomorrow's world. Vol. 1 – Analysis*. Paris: OECD Publishing.
OECD (2010). *PISA 2009 results: What students know and can do – Student performance in reading, mathematics and science (Vol. 1)*. http://dx.doi.org/10.1787/9789264091450-en
Office of the Chief Scientist. (2012a). *The health of Australian science*. Canberra. Retrieved from http://www.chiefscientist.gov.au/2011/10/health-of-australian-science-report
Office of the Chief Scientist. (2012b). *Mathematics, engineering and science in the national interest*. Canberra. Retrieved from http://www.chiefscientist.gov.au/2012/05/mes-report/
Osborne, J., & Collins, S. (2001). Pupils' views of the role and value of the science curriculum: A focus-group study. *International Journal of Science Education, 23*(5), 441–468.
Osborne, J., Simon, S., & Tytler, R. (2009). *Attitudes towards science: An update*. Paper presented at the Annual Meeting of the American Educational Research Association, San Diego.
Phillips, Y. (2009, December 22). Year 12 students dumbing down to get into TAFE. *Perth Now*.
Ramsden, J. M. (1998). Mission impossible? Can anything be done about attitudes to science? *International Journal of Science Education, 20*(2), 125–137.
Rosier, M. (1973). *A statistical picture of Australian science teachers: Some results from the IEA science project*. Hawthorn: ACER.
Skinner, B. F. (1967). *Teaching science in high school: What is wrong?* Paper presented at the meeting of the American Association for the Advancement of Science, 29 December. New York: Rockefeller University.
Smith, D., & Gunstone, R. (2009). Science curriculum in the market liberal society of the 21st century: Re-visioning the idea of science for all. *Research in Science Education, 39*, 1–16.
Tisher, R. (1971). Verbal interaction in science classes. *Journal of Research in Science Teaching, 8*(10), 1–8.
Tobin, K., & Gallagher, J. (1987). What happens in high school science classrooms? *Journal of Curriculum Studies, 19*(6), 549–560.
Victorian Curriculum and Assessment Authority (2012). *Media manual*. Retrieved from http://www.vcaa.vic.edu.au/Documents/media/mediamanual2012.pdf
White, C., & Hoffrage, U. (2009). Testing the tyranny of too much choice against the allure of more choice. *Psychology and Marketing, 26*(3), 280–298.

Chapter 11
Choice Patterns of PhD Students: Why Should I Pursue a PhD?

Tina Hribar and Slavko Dolinšek

Introduction and Aims

The discussion about choice of *undergraduate* STEM studies is the focus of the majority of studies In this book. The present chapter provides a distinctive focus on the choice of *PhD* study. It may be expected that many of the same factors are at play in the choice process; however, there may also be differences. A clear difference is that undergraduate education is still free of costs (at the present moment) in Slovenia, which is not the case for PhD studies. The educational system in Slovenia will be briefly described in a following section.

In the report "Europe needs more scientists" (EU 2004) the need to increase the number of researchers in the field of science and technology has been addressed and it is recommended that increasing the number of women entering science and engineering careers would significantly contribute to the solution of the problem. The issue of recruiting more women to education and careers in STEM is one of quantity as well as quality: of quantity, because women represent the greatest recruitment potential; and of quality, because a higher participation from women in STEM may expand the scope and ways of thinking, prioritizing and working within this area and contribute to gender equity (see section "Introduction"). Considering this prevailing problem of underrepresentation of female STEM scientists, we felt that it was important within the IRIS project (IRIS 2010a, b) to analyse the choice patterns of STEM PhD students, since these students represent the highest academic achievers as well as research workforce in the academia and industry.

The literature review showed that there are very few studies (if any) that deal particularly with our focus, and that these are mostly within the US (Blickenstaff 2005;

T. Hribar (✉) • S. Dolinšek
Institute for Innovation and Development of University of Ljubljana (IRI-UL),
Kongresni trg 12, 1000 Ljubljana, Slovenia
e-mail: tina.hribar@iri.uni-lj.si

E.K. Henriksen et al. (eds.), *Understanding Student Participation and Choice in Science and Technology Education*, DOI 10.1007/978-94-007-7793-4_11

Ceci et al. 2009; Hazari et al. 2007, 2008, 2010). Here we need to make special attention to the structure of educational system, which varies even among the EU countries (and not only from the US), and together with other social factors (i.e. social background, economic situation, enrolment fees) contributes to decision-making process of future (STEM) PhD students and the recruitment/retention issues in STEM study fields. For example The EU project UPGEM shows that female students in eastern and southern European countries are more easily attracted to physics than in the north, and career paths seem to follow different patterns. The project identified the informal ways in which careers are shaped in various cultural contexts, and their relationship to "the four P's" (prestige, payment, pleasure, policy). They suggested four possible reasons for these distinctions: different paths from school into higher education, the distinction between 'class societies' and 'gender societies', different family patterns and workplace cultures (UPGEM 2008).

The Structure of Tertiary Education in Slovenia

To better understand the relevance of different factors that influence the choice for a PhD in Slovenia, a brief explanation of the structure of tertiary education in Slovenia is needed. Tertiary education in Slovenia is divided into traditional higher education and the newly developed higher vocational education sector. Public and private universities, faculties, art academies and professional colleges provide higher education. There are three large public universities (University of Ljubljana being the largest, followed by the University of Maribor and the University of Primorska) with 53 member institutions; 2 private universities with 7 member institutions and 26 single higher education institutions. The overall structure of the Slovenian education system is presented in Eurydice 2010. The Bologna process started gradually in 2005/06 and concluded in 2009/10, from which time students could choose only the new study programs and the old ones were discontinued.

Enrolment and Gender Distribution in PhD-Level STEM Studies in Slovenia and Internationally

In general, at the doctoral level of tertiary education in Slovenia, gender balance is observed; there are 51 % of women and 49 % men studying doctoral programs. The gender representation in the STEM disciplines studied here shows that the gender imbalance is very similar to international statistics: there are 65 % of men and 35 % women studying natural sciences, mathematics and computer science and technology and construction in Slovenia at the PhD level (MVZT 2010a, b).

International data (Global Education Digest 2010) also show that the highest level of gender imbalance exists in the engineering, manufacturing and construction field of study with 36 % of female PhD graduates, although in North America and Western Europe as well as Central and Eastern Europe, women outnumber men significantly among both the Bachelor's and Master's degree graduates in general across other fields of study. Despite improved access to tertiary education, women face considerable barriers as they move up the educational ladder to research careers, which are an important aspect with regard to the PhD enrolment in general (EU 2006, 2008, 2009a, b).

According to UNESCO (2010), the preferences of women and men for specific fields of study in higher education are shaped by their individual histories and everyday realities as well as their material conditions. Factors that may explain the lower number of female researchers, especially in senior positions, include the work-life balance, sex stereotyping, performance measurement and promotion criteria, governance, and the role of researchers in society (UNESCO 2010).

Research Objectives

Based on our IRIS study on undergraduate students, we were interested to find out the priorities and considerations that determine the rather complex and active process of choosing to become a STEM PhD student. Consistent with the Eccles expectancy-value model (Eccles et al. 1983; see Chap. 2), we assumed that the process of socialization through key persons, shapes young people's (a) interests, priorities and values (forming their identity), and (b) their competence-related beliefs concerning the different career priorities and development. These in turn influence their choice of studying STEM and possibly continuing to a PhD (in STEM).

This study had two main research objectives:

1. To identify important factors that influenced student's choice of STEM PhD
2. To identify which priorities male and female PhD STEM students in Slovenia seek in their future careers.

Theoretical Background

The Transition from Undergraduate to Graduate Studies

According to Ceci et al. (2009), evidence indicates that socio-cultural factors and constraints constitute the most powerful explanatory factor behind females' under-representation in STEM. The results obtained by the IRIS study on undergraduate STEM students in Slovenia showed that good teachers and parents were found to be, among all key persons, rated most influential for students' choice of studying STEM. Female STEM students in Slovenia were found to be significantly more influenced by good teachers and their mothers in terms of their choice of studying STEM, compared to male students (Cerinšek et al. 2012). The impact of social origin on educational achievement has been discussed elsewhere, but Mastekaasa (2006) found that the effect of social origins on students' decision to continue or not in the educational system are strongest at the early stages of education (particularly in the transition from the primary to secondary school) but they are weak or non-existent in the late stages (from undergraduate to graduate studies). One explanation might be that, as they grow older, young people are no longer attached and therefore free from the influence of their parents and families in general, and other social influences become more important (peers and teachers, professors and mentors). A second factor might be the decline of economic dependence on parents.

Disenchantment with STEM Studies and Careers

There is a large body of research (Eccles 2007; Blickenstaff 2005; Brickhouse and Potter 2001), which investigates the underlying reasons for females' underrepresentation in STEM. However, very little research has attempted to investigate why some females do find it attractive (Gilbert and Calvert 2003), i.e. there is a scarcity of studies addressing those females who decide to study STEM (the present book being an exception; see particularly Part IV). Until now, very little research has been conducted in Slovenia studying STEM students' choices (and female students in particular) related to their STEM education, especially at the PhD level.

The Great Expectation(s)

Many factors underlie women's considerations when deciding whether to study for a (STEM) PhD. For some women a key factor is identification with the traditional gender role and the conflict between child rearing expectations and career expectations (Grunert and Bodner 2011; Aschbacher et al. 2009). Many females believe that research careers are incompatible with having a family life, which often means they believe they would need to change their lifestyle to be successful in those careers (Grunert and Bodner 2011). This is closely related to the presumed working atmosphere and usually long working hours in the lab or office which takes complete commitment to scientific research and therefore may involve delaying starting a family (Aschbacher et al. 2009). The perception that women cannot lead a successful career if they have a family is related to cultural stereotypes but can often be misleading. There are many women who handle their family and career obligations well, but who work far from the eyes of public and may therefore not be readily available as role models for young women in STEM. An Australian study

(Vrcelj and Krishnan 2008) shows that role models and social support are very important sources of encouragement for young women's career choice.

Example of Good Practice in Slovenia: Young Researchers Program (YRP)

Considering the possible sacrifices in terms of lifestyle in order for women to be successful in their career, it is important to mention the government initiative to encourage students, both male and female, to enroll for a PhD studies: the Young researchers program (YRP). This is one of the most successful activities in the area of education and training for R&D and innovation in Slovenia. Through this program the Slovenian Research Agency finances students who are selected by their mentors during their MA. or PhD studies as potential candidates for research positions in higher education institutions and public research institutes. This gives the young people an additional incentive to consider the possibility of entering the PhD study. Under the YRP program, female students have the right for 1-year maternity leave without losing social and other material benefits. During the 1-year maternity leave, the status of a young researcher is on hold until she fully resumes her research work.

Methodology

IRIS PHD Questionnaire and Interviews

The main data collection was done using the IRIS PhD Questionnaire (called the IRIS PhD Q). The basis was the IRIS questionnaire for undergraduate students (see Appendix). Different constructs from the Eccles expectancy-value model of achievement-related choices (Eccles et al. 1983), and other factors of relevance to educational choice, are implemented in the IRIS Q and IRIS PHD Q, concentrating on the expectation of success and subjective task value.

The IRIS PhD Q was developed and piloted in the following steps: (1) literature review; (2) using the IRIS Q as the basis for developing IRIS PhD Q; (3) discussion with the IRIS partners on the clarity and appropriateness of the questions; (4) translation of the English version back to the Slovenian language; (5) pilot testing of the translated version of the IRIS Q with five PhD students; (6) finalizing the IRIS PhD Q; (7) designing the electronic form of PhD Q. The final IRIS PhD Q contained a total of 15 closed questions with five-point Likert scales and four open questions (see the Appendix).

Respondents were first asked to provide background information concerning their gender, year of birth, university where they are studying, PhD program and previous enrolments as well as status of young researcher. The study did not identify ethnic, religious or social background of respondents. In the next items in the questionnaire, respondents were asked to rate the importance of the following factors:

1. Importance of age period when they developed the interest for PHD study
2. Importance of key persons in choosing the course of study;
3. Importance of different priorities they seek in their future careers
4. Importance and impact of PhD study on personal life

The Sample Group

PhD STEi[1] students from the largest Slovene University (i.e. University of Ljubljana), that represents the majority population of Slovene PhD students, were invited to complete the questionnaire. Five different STE faculties participated in the study: Biotechnical Faculty, Faculty of Electrical Engineering, Faculty of Chemistry and Chemical Technology, Faculty of Computer and Information Science and Faculty of Mechanical engineering. The target population included all PhD students on study programs at University of Ljubljana within six different International Standard Classification of Education (ISCED) codes. The entire eligible population counted 597 students as presented in Table 11.1.

The respondent group consisted of 134 male and 59 female doctoral students, which represents 32.3 % of the whole target population. The administration staff of the faculties agreed to help us with sending the information to their respective PhD students via e-mail. The students who did not participate in the survey were those who declined to give their e-mail addresses at the time of enrolment, and few foreign students due to language barrier. Males were somewhat overrepresented among respondents. Apart from this, we are not aware of any way in which respondents differ systematically from the population in ways that are relevant to the research questions in this chapter.

IRIS PhD Q Data Analysis

Basic descriptive statistics (mean, standard deviation, and minimum, maximum) were used to describe key features of the data. Statistical significance of gender difference was tested using the repeated-measures analysis of variance (ANOVA). We claim that statistically significant gender difference in means occurs at $P \leq 0$,

[1] The Faculty of mathematics and physics declined the invitation to participate and we were not able to include in the survey the students of Mathematics and Physics; thus, the present study is of STE rather than STEM students.

Table 11.1 Number of eligible PhD students at the University of Ljubljana in 2010/11 within six STE disciplines defined using the ISCED coding system; number of valid IRIS PhD Q respondents, and response rates

ISCED codes	Gender	Total target pop (UL)	No of valid respondents	Response rate (%)
421 Biology and biotechnology (BB)	Males	33	11	33
	Females	76	28	37
	Total	109	39	36
442 Chemistry (C)	Males	37	14	38
	Females	31	14	45
	Total	68	28	41
481 Computer science (CS)	Males	73	25	34
	Females	9	4	44
	Total	82	29	35
521 Mechanics and metalwork (MM)	Males	126	37	29
	Females	13	5	38
	Total	139	42	30
523 Electronics and Automation (EA)	Males	147	42	29
	Females	15	5	33
	Total	162	47	29
524 Chemical and process (CP)	Males	23	5	22
	Females	14	3	21
	Total	37	8	22
Total	Males	439	134	31
	Females	158	59	37
	Total	597	193	32

05. Each figure presents mean ratings for males and females separately. To investigate the size of differences of mean scores, effect sizes were calculated in terms of Cohen's d. According to Cohen (1992), effect sizes of 0,2, 0,5, and 0,8 are regarded as small, medium and large, respectively.

A full analysis of responses to the open-ended questions is not presented here, but example quotes from the open-ended questions are given to illustrate quantitative findings.

Results

Age Period and Key Persons for Influence on STEM PhD Choice

In Slovenia the transition to the PhD level of education takes place at least 5 years after entering university studies, and graduates are usually in their mid-twenties or older. As this is the case we might assume that declining social origin effects at higher transition might be relevant in relation to the transition from undergraduate studies to PhD.

We asked the students *to rank three different age periods* on a scale from 1 (not important) to 5 (very important), concerning when they first became interested in the field of their major.

The results showed that the age period from 19 to 27 is the most important period for developing interest in a STEM research career for both males and females $(\overline{X} = 4,24)$; the results do not demonstrate significant gender differences. It seems that the undergraduate studies are the most crucial period regarding the future career plans in the research and academic field for the students, for both male and female.

Students also answered the question "How important were the following persons in choosing your doctoral study program?" on a scale from 1 (not important) to 5 (very important). Figure 11.1 shows that the results do not demonstrate significant gender differences in means in terms of the key persons who most influenced students' choice for PhD. The mentor (see below) had the greatest influence on both males and females $(\overline{X} = 3,77)$, followed by professors at undergraduate study $(\overline{X} = 3,01)$; all others i.e. high school teachers, other family members, friend and partner, colleagues and career advisors seem not to have had major influence.

Here we need to explain the role of 'the mentor', which is crucial for several reasons. A mentor is a professor who assists the student with the diploma at the undergraduate level. In most cases this professor invites, suggests or encourages his/her best undergraduate students for a PhD study and thus becomes 'the mentor'. Moreover the mentor gets the funding for a PhD student from a National Research Agency for the Young Researcher Program (YRP) and therefore chooses or invites the student to apply for the YRP.

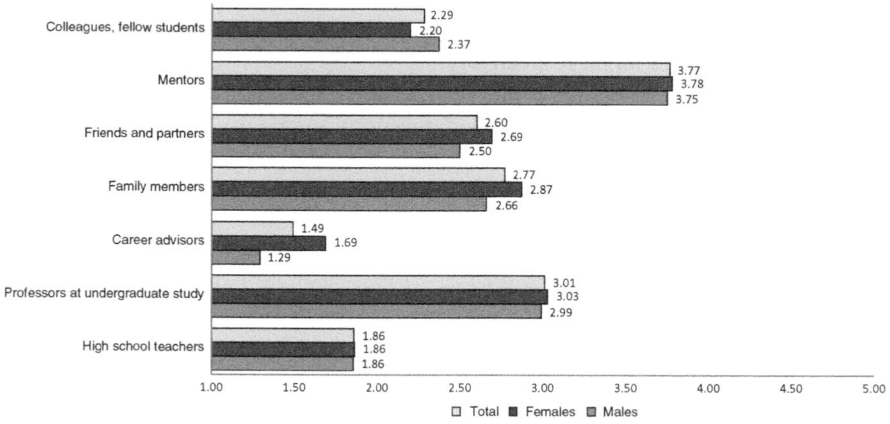

Fig. 11.1 The influence of key persons on female and male students' decision to pursue a STEM PhD, Slovene PhD students' responses on a scale from 1 ("not important") to 5 ("very important") to the question "How important were the following persons in choosing your doctoral study?"

Responses to the open question in the IRIS PhD Q also indicate the two inter-related key factors when deciding for PhD: mentors and the Young researchers program (YRP) described above. The following narratives are representative for PhD candidates answering the question "Tell us how you came to choose your PhD?"

> In the second half of my undergraduate study I **developed stronger interest for my study field** and was thinking about becoming a researcher, but only under the condition that my **material situation will be taken care of as well**, meaning that I would get a job. **The turning point** was when **my mentor** in my senior year mentioned that the following year he will become the 'young researchers mentor' and he finds me as a suitable candidate for a young researcher. So I quickly completed my undergraduate studies and applied for the YRP Call. Luckily I was chosen and that is how I became a young researcher and a doctoral student. (Male, Computer and Information Science student)
>
> The mentor contacted me and asked me if I am interested in a PhD. I immediately agreed, since **I consider myself to be scientific-research type of person**. (Male, Chemistry student)
>
> During my final years I started to work within the research group at the faculty lab and I **was encouraged to pursue a PhD**. They (the professors) suggested both the field of study and the mentor, which suited me perfectly. (Female, Computer and Information Science student)

The results showed that most of those students, who apply for YRP, and are selected, are the top students from the undergraduate programs. They are therefore highly interested in their study, which can be interpreted in terms of interest-enjoyment value of the Eccles et al. model. Also, a STEM researcher identity is expressed in one of the above quotes as facilitating the decision to pursue a PhD.

Priorities for Future Career

Studies have demonstrated that females at undergraduate studies tend to seek more interpersonal values in their future careers, such as helping other people, contributing to the society and protecting the environment, whereas males place more value on extrinsic rewards, such as earning high income (Cerinšek et al. 2012). It is demonstrated through our study that all STEM students at both study levels want to realize their own potential by doing something interesting and fulfilling and by using their talents and abilities in their future careers.

The results from our IRIS PhD study suggest that both male and female students want to "develop their talent and skills", "get a secure job and promotion" as long as they also get the chance to "realize their potential" and "pursue ideals and values" while "improving their professional competences" and "helping others". The students had to answer the questions: "How important are the following characteristics of the career development?" and "How important were the following wishes in choosing your study?" (Fig. 11.2) and rank the answers on the scale from 1 (not important) to 5 (very important). Our results show that there are statistically significant gender differences in the following factors (P-values for statistical significance and Cohen's *d* effect sizes are reported in parentheses):

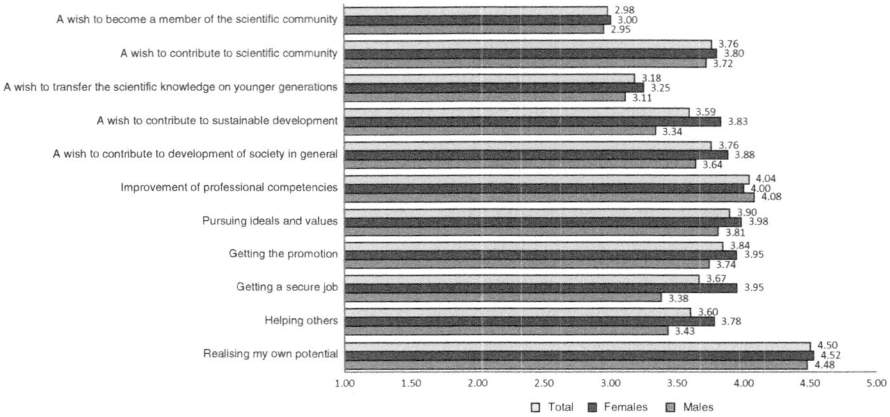

Fig. 11.2 Male and female students' career priorities. Slovene PhD students' responses on a scale from 1 ("not important") to 5 ("very important") to the questions "How important were the following wishes in choosing your study?" (first five items) and "How important are the following characteristics of the career development?" (last six items)

- **Helping others** (P = 0,046, *d* = 0,32): the significant gender difference in means suggest that both male and females want to help others, but women $\left(\overline{X} = 3,78\right)$ show greater interest than men $\left(\overline{X} = 3,43\right)$.
- **Getting a secure job** (P = 0,005, *d* = 0,46): the significant gender difference in means show that getting a secure job is more important for females $\left(\overline{X} = 3,95\right)$ than men $\left(\overline{X} = 3,38\right)$.
- **Contributing to sustainable development and protection of the environment'** (P = 0,008, *d* = 0,42). There is a significant gender difference in means in favour of females $\left(\overline{X} = 3,83\right)$ who rated working on environmental issue higher than males $\left(\overline{X} = 3,34\right)$ and which could be one of the **ideals and values** that females $\left(\overline{X} = 3,98\right)$ are striving more intensely than male students $\left(\overline{X} = 3,81\right)$.

The above results indicate that female PhD STEM students in Slovenia do have different, more inter-personal career priorities and ideals than male PhD STEM students, i.e. they want more than males to work at occupations that enable them to help other people, contribute to the society and protect the environment (Fig. 11.2). The IRIS study on the undergraduate students in Slovenia also similarly demonstrated that females tend to seek more interpersonal values in their future careers, such as helping other people, contributing to the society and protecting the environment, whereas males place more value on extrinsic rewards, such as earning a high income (Cerinšek et al. 2012).

The international comparative ROSE project, which studied affective factors of importance to the learning of science and technology among 15-year old students towards the end of secondary school, also demonstrated that Slovene girls are much more interested to work with people, help other people and protect the environment in their future careers than boys (Dolinšek 2008).

Thus we can assume that PhD students' choice considerations appear to be very similar to those of students choosing an undergraduate education. Those females who do engage in STEM PhD studies perceive jobs and careers in STEM fields as the ones congruent with their priorities and values. Moreover it is evident that both males ($\overline{x} = 3,64$) and females ($\overline{x} = 3,88$) have a desire to contribute to the development of society in general and it is apparent that by their research work they believe they can make this contribution. The Australian study (Vrcelj and Krishnan 2008, p. 45) suggests that a main reason for opting out of an academic career was "the fact that the role of the research as integral and essential enhancement of a society is not emphasized at the undergraduate level".

Influence on Social Life

Students were also asked to answer the question "To what extent does your PhD work weaken the following relationships and areas of life?" on a scale from 1 (no influence) to 5 (strong influence). The results show that there is significant difference among the students on different study programs in all variables (P-values for statistical significance and Cohen's *d* effect sizes are reported in parentheses), but there are no significant gender differences (Fig. 11.3). Hobbies are the area of social life that is rated as being most negatively affected by PhD studies. On the whole, Computer science students, more than students of Biology and Chemistry, express that doctoral study has a (bad) influence on their social life:

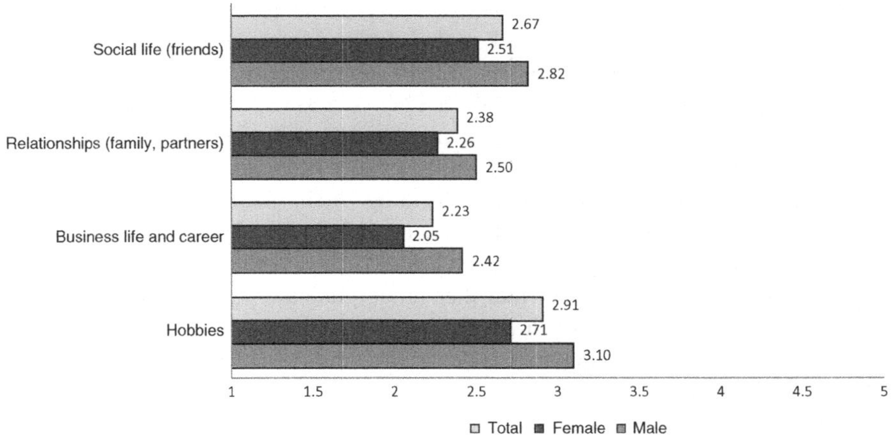

Fig. 11.3 Male and female students' perception of the influence of PhD work on social life: Slovene PhD students' responses on a scale from 1 ("not at all") to 5 ("very much") to the question "To what extent does your PhD work weaken the following relationships and areas of life?"

- **Business and career** (P = 0,000) (*d* = 0,28): Computer science students $(\overline{X} = 3,17)$ in comparison to biology students $(\overline{X} = 1,45)$ feel that their PhD study has a bad impact on their business and career development;
- **Hobbies** (P = 0,002) (*d* = 0,41): Computer science students $(\overline{X} = 3,86)$, biology students $(\overline{X} = 2,45)$;
- **Social life** (P = 0,000) (*d* = 0,45): Computer science students $(\overline{X} = 3,76)$, Chemistry $(\overline{X} = 2,34)$;
- **Relationships/family/partners**: (P = 0,007) (*d* = 0,8): Computer science students $(\overline{X} = 3,11)$, chemical technology students $(\overline{X} = 1,88)$. Moreover, there is no significant gender difference among them.

The results were a bit surprising because we expected students to be more concerned about their relationships and spare time activities when entering a PhD, especially among women (Fig. 11.3). With the exception of the Computer Science students, all other students seem to perceive that their PhD study is in balance with their family and social life. More in-depth studies are needed to explore this further. Moreover we do not know whether these students after completing their PhD study stay in academia or decide to go to the industry sector, and how they are managing there. We might at least say that in the case of Slovenia, the factors of social costs do not appear to influence significantly the experience of being a PhD student.

Conclusion and Implications

This study had two main objectives: (a) to identify which priorities male and female STEM PhD students in Slovenia seek in their future careers, and (2) to identify the important factors (i.e. key persons, career perspectives, social factors) that influenced their choice to study a STEM PhD. We assumed, consistent with the Eccles et al. (1983) expectancy-value model, that the process of socialization through key persons, undergraduate study experiences, shapes young peoples' (a) interests, priorities and values (forming their identity), and (b) their competence-related beliefs. This ultimately influences their choice to study (or not to study) STEM in general (Boe et al. 2011).

Our study demonstrates that, like their undergraduate peers, all Slovene PhD STEM students who took part in our research are strongly in favour of realising their own potential by using their talents and abilities in their future careers. Moreover they showed great interest in their field of study. They also want opportunities to develop themselves and want to get a secure job in their future life, especially females.

Regarding their future careers, females (more than males) were in favour of working in occupations that enable them to contribute to sustainable development, protecting the environment, helping other people and contributing to society. This is an important finding that demonstrates that even females who do engage in STEM studies favour inter-personal career priorities more than male STEM students. It is possible that under-representation of females in STEM courses and careers in Slovenia could be due to females in general perceiving STEM careers as less interesting because they are perceived to offer fewer opportunities for interpersonal involvement (compared to other educational options). Morgan et al. (2001) state that:

> individuals with strong interpersonal work values (who are more likely female) may anticipate experiencing less interest while working on STEM related activities because perceptions of these activities as involving individual achievement and impersonal work environment are incongruent with their interpersonal goals. In contrast, individuals with strong extrinsic reward values (who are more likely male) may anticipate experiencing greater interest when working on STEM activities because perceptions of these activities as affording opportunities for high pay are congruent with their extrinsic reward goals (Morgan et al. 2001).

Gilbert and Calvert (2003) on the other hand found that females who engage in STEM studies were attracted to the rational, unemotional, analytical aspects of science that are often associated with masculinity and men, and that these females pursue careers in STEM because of these aspects, not despite them (see also Part IV in the present volume).

We can elaborate several practical implications from our findings within the Slovenian context. Clarifying how STEM research impacts on societal and environmental issues is important for recruitment (particularly of females) also at the PhD level. We need to be aware that the gender imbalance within STEM starts at the undergraduate level of study. So it is not surprising that the pattern persists at the PhD level. One of the important factors that influenced decisions of both females and males to enrol on STEM PhD programmes were mentors, who can be successful in recruiting their top undergraduate students for the PhD courses. The Australian study (Vrcelj and Krishnan 2008) shows that female PhD students need more networking, role models and retention programs. It also seems that the undergraduate STEM study programs need further development. Young people need to see a range of different possibilities – including those that are people/health/society oriented – in science careers. This appears especially important in order to attract more females to STEM careers. Furthermore, to achieve better gender balance, STEM courses should be more closely focused to real world/contextual and social relevance, e.g. demonstrating how science can contribute to society and environment, since females are attracted to these issues significantly more than their male counterparts, also at the PhD level.

Our study shows that mentors (professors at the undergraduate and postgraduate programs), are the key persons who encourage students to pursue a (PhD) STEM studies and careers. Although family partners appear to be less influential when deciding for PhD, it cannot be said that they do not play important role later on, when supporting their life partner's decision to pursue a PhD and research career.

An extremely important factor that encourages and successfully recruits young graduate students to PhD study is the Young researchers program (YRP). This gives young people the opportunity to continue their studies at the PhD level with high level of economic security. We could argue that the relative costs (social security, time for family duties) seems to be the key factors when making the final decision to enter the PhD. Duration of the doctoral program is 3–5 years. When the students study under the YRP they are employed at the University or at the independent Institutes and receive moderate salary. This is especially relevant for female students who can take a 1-year maternity leave and thereafter resume their research position without losing social security benefits. When this program was started in 1996, Slovenia quickly became one of the EU countries with the highest rates of doctoral students.

Our study of PhD students' priorities and choices is one of the first in Slovenia; therefore one of the challenges in the context of our future work is to conduct interviews with STEM PhD students, both male and female, to get deeper insight into patterns and factors that have a high impact on PhD enrolment and retention.

References

Aschbacher, T. P., Li, E., & Roth, J. E. (2009). Is science me? High school students' identities, participation and aspirations in science, engineering, and medicine. *Journal of Research in Science Teaching, 47*, 564–582.

Blickenstaff, J. C. (2005). Women and science careers: Leaky pipeline or gender filter? *Gender and Education, 17*(4), 369–386.

Boe, M. V., Henriksen, E. K., Lyons, T., & Schreiner, C. (2011). Participation in science and technology: Young people's achievement-related choices in late-modern societies. *Studies in Science Education, 47*(1), 37–72.

Brickhouse, N., & Potter, J. T. (2001). Young women's scientific identity formation in an urban context. *Journal of Research in Science Teaching, 38*(8), 965–980.

Ceci, S. J., Williams, W. M., & Barnett, S. M. (2009). Women's underrepresentation in science: Sociocultural and biological considerations. *Psychological Bulletin, 135*(2), 218–261.

Cerinšek, G., Hribar, T., Glodež, N., & Dolinšek, S. (2012). Which are my future career priorities and what influenced my choice of studying science, technology, engineering or mathematics? Some insights on educational choice – Case of Slovenia. *International Journal of Science Education*. doi:10.1080/09500693.2012.681813.

Cohen, J. (1992). A power primer. *Psychological Bulletin, 112*(1), 155–159.

Dolinšek, S. (2008). *ROSE Slovenija*. Koper: Fakulteta za management.

Eccles, J. S. (2007). Where are all the women? In S. J. Ceci & W. M. Williams (Eds.), *Why aren't more women in science?* (pp. 199–209). Washington, DC: American Psychological Association.

Eccles, J. S., Adler, T. F., Futterman, R., Goff, S. B., Kaczala, C. M., Meece, J. L., & Midgley, C. (1983). Expectancies, values and academic behaviours. In J. T. Spence (Ed.), *Achievement and achievement motives: Psychological and sociological approaches* (pp. 75–146). San Francisco: W.H. Friedman.

EU (2004). *Europe needs more scientists!* Brussels: European Commission, Directorate-General for Research, High Level Group on Human Resources for Science and Technology in Europe. Available from: http://ec.europa.eu/research/press/2004/pr0204en.cfm. Accessed 16 Sept 2011.

EU. (2006). *She figures 2006, women and science – Statistics and indicators*. European Commission, Directorate-General for Research, Information and Communication Unit. Available from: http://ec.europa.eu/research/science-society/pdf/she_figures_2006_en.pdf. Accessed 16 Sept 2011.

EU. (2008) *Education in Europe – Key statistics 2008: Population and social conditions Eurostat data in focus 37/2010, Paolo TURCHETTI, Eric GÉRE*. Available from: http://epp.eurostat.ec.europa.eu/portal/page/portal/product_details/publication?p_product_code=KS-QA-10-037. Accessed 16 Sept 2011.

EU. (2009a). *She figures 2009, women and science – Statistics and indicators*. European Commission, Directorate-General for Research, Information and Communication Unit. Available from: http://europa.eu/rapid/pressReleasesAction.do?reference=MEMO/09/519&format=HTML&aged=0&language=EN&guiLanguage=en. Accessed 13 Sept 2011.

EU. (2009b). *Key data on education in Europe 2009*. European Commission. Available from: http://eacea.ec.europa.eu/education/eurydice/documents/key_data_series/105EN.pdf. Accessed 13 Sept 2011.

Eurydice. (2010). *Gender differences in educational outcomes. Study on the measures taken and the current situation in Europe*. Eurydice. Available at: http://eacea.ec.europa.eu/education/eurydice/documents/thematic_reports/120EN.pdf. Accessed 18 Nov 2011.

Gilbert, J., & Calvert, S. (2003). Challenging accepted wisdom: Looking at the gender and science education question through a different lens. *International Journal of Science Education, 25*(7), 861–878.

Global Education Digest. (2010). *Comparing education statistics across the world*. UNESCO Institute for statistics. Available from: http://www.uis.unesco.org/Library/Documents/GED_2010_EN.pdf. Accessed 15 Mar 2011.

Grunert, M. L., & Bodner, G. M. (2011). Underneath it all: Gender role identification and women chemists' career choices. *Science Education International, 22*(4), 292–301.

Hazari, Z., Tai, R. H., & Sadler, P. M. (2007). Gender differences in introductory university physics performance: The influence of high school physics preparation and affective factors. *Science Education, 91*(6), 847–876.

Hazari, Z., Sadler, P., & Tai, R. (2008). Gender differences in the high school and affective experiences of introductory college physics students. *The Physics Teacher, 46*, 423–427.

Hazari, Z., Sonnert, G., Sadler, P. M., & Shanahan, M. (2010). Connecting high school physics experiences, outcome expectations, physics identity, and physics career choice: A gender study. *Journal of Research in Science Teaching, 47*(8), 978–1003.

IRIS. (2010a). FP7 Collaborative project, Grant agreement no: 230043. Annex 1: Description of work. Available from: http://iris.fp-7.org/data/upload/IRIS_annex1_nov6th_1.pdf. Accessed 13 Sept 2011.

IRIS. (2010b). *Theoretical perspectives for understanding young people's educational choices*. Working document 4.1b, Priorities and educational choices. Available from: http://iris.fp-7.org/reports-publications/. Accessed 13 Sept 2011.

Mastekaasa, A. (2006). Educational transitions at graduate level: Social origins and enrolment in PhD programmes in Norway. *Acta Sociologica, 49*(4), 437–453.

Morgan, C., Isaac, J. D., & Sansone, C. (2001). The role of interest in understanding the career choices of female and male college students. *Sex Roles, 44*(5/6), 295–320.

MVZT. (2010a). *Drzna Slovenija na poti v družbo znanja*. Statistics on higher education in Slovenia. Ministry of Higher Education, Science and Technology of Republic of Slovenia. Available from: http://www.mvzt.gov.si/fileadmin/mvzt.gov.si/pageuploads/pdf/visoko_solstvo/statistika_visokega_solstva_07072010.pdf. Accessed 26 Oct 2011.

MVZT. (2010b). *Resolution on research and innovation strategy of Slovenia 2011–2020*. Ministry of Higher Education, Science and Technology of Republic of Slovenia. Available from: http://www.mvzt.gov.si/fileadmin/mvzt.gov.si/pageuploads/pdf/odnosi_z_javnostmi/01.06.2011_dalje/01.06._RISSdz_ENG.pdf. Accessed 26 Oct 2011.

UNESCO. (2010). *Global education digest 2010; Comparing educational statistics across the world.* http://www.uis.unesco.org/Library/Documents/GED_2010_EN.pdf. Accessed 12 Jan 12.

UPGEM. (2008), About the project. http://www.dpu.dk/upgem/abouttheproject/. Accessed 7 Nov 2012.

Vrcelj, Z., & Krishnan, S. (2008). Gender differences in student attitudes toward engineering and academic careers. *Australasian Journal of Engineering Education, 14*(2), 43–56.

Chapter 12
The Impact of Outreach and Out-of-School Activities on Norwegian Upper Secondary Students' STEM Motivations

Fredrik Jensen

Introduction

In Norway (and in many other countries) a range of STEM recruitment initiatives are launched, organised or maintained every year: Summer camps, physics days, mathematics training, chemistry competitions, web sites, and so on. The adolescents attending these may, or may not, be aware that the initiatives were designed to induce him or her to be more interested in mathematics and science, and potentially in a STEM career.

What is a successful recruitment initiative? One definition could be that it is successful if it leads to a higher number of students entering STEM education. Given the complexity of educational choices, however, such an effect is difficult to measure directly. Moreover, an educational choice does not end on the day the student enters tertiary education. Retention is just as important as recruitment (see Part III in this book), and students may leave their chosen STEM study for many reasons, for example if they have been persuaded to choose STEM on false premises. Thus, one could argue that a successful initiative is one that leads to an increase in the numbers of STEM graduates due to students making more informed choices, and success factors can be defined as any factors in the initiatives that are contributing to a well-informed choice where STEM is a real option.

The expectancy-value model presented in Chap. 2 suggests that choice of a particular education is more likely if the expectation of success, interest and enjoyment, attainment value or utility value attached to it is increased, or the perceived cost is reduced. This means that a recruitment intervention can be successful if it positively influences a choice through these factors. Bøe and colleagues (2011) examined international research on young people's attitudes to

F. Jensen (✉)
Norwegian Centre for Science Education, University of Oslo,
P.O. Box 1106, Blindern, N-0317 Oslo, Norway
e-mail: fredrik.jensen@ils.uio.no

E.K. Henriksen et al. (eds.), *Understanding Student Participation and Choice in Science and Technology Education*, DOI 10.1007/978-94-007-7793-4_12

and participation in STEM in light of the Eccles et al. model (Eccles et al. 1983; Eccles and Wigfield 2002), and concluded that this model is "not only useful for understanding young people's participation in STEM, but also for designing and evaluating initiatives" (p. 63).

In this chapter attention is given to initiatives that are designed with the specific intention of enhancing STEM participation, and which are aimed directly at the target group (school students). This means that the focus is not on museums, popular science, or other outreach efforts aimed at improving young people's attitudes to science more generally. Nor has the focus been on curricular change, teacher training or other initiatives that may affect students' preferences indirectly. Primarily, attention is given to out-of-school activities, although most of the initiatives investigated are in some manner related to the school setting.

I will first give a brief overview of some research literature concerning the effects of recruitment initiatives. Further, I will present studies of three specific recruitment initiatives offered to adolescents in Norway: CERN Masterclass, ENT3R and The Girls' Day. Based on the literature and results from the analyses, I suggest that these initiatives have some factors in common that may contribute to the existing knowledge of what works when it comes to motivating adolescents to engage in science and mathematics. Several of the initiatives described in this chapter are targeted particularly at females, thus recruitment of females will be given special attention.

What Is Known About Effects of Out-of-School Recruitment Initiatives?

To gain insight into outcomes of out-of-school recruitment efforts, a rather extensive search in scholarly databases, journals, and on the internet was performed. Keywords such as "recruitment", "initiatives", "events", "STEM", "attitudes", "interests", "out-of-school", "summer camps" were searched (in particular) in Google Scholar, ERIC and Web of knowledge. Of the articles that appeared, the document and a short summary were saved if both title and abstract indicated that it was relevant. Next, the article's reference list was scanned for related literature, and the "cited by" function in the database was used. In addition to providing more literature, this occasionally gave a few new ideas for search words. These steps were repeated until "saturation" was approached in the sense that few new articles came up. In addition to the search, some articles were found after getting tips from colleagues. Lastly, short summaries of the texts were organised into categories such as "out-of-school", "summer camps" or "hands-on". The following is not meant to be a complete review of the topic, but to show some research that is relevant for the three initiatives investigated in this chapter.

Engaging in out-of-school science activities has been found to be related with science interests and aspirations. Simpkins et al. (2006) found that boys who participated in out-of-school science and mathematics activities in 5th grade compared with their peers held higher self-concept, interest, and perceived these subjects to be more important in 6th and 10th grades. For girls, these correlations were also positive, but not statistically significant. Dabney et al. (2011) found that students who reported that they had engaged in out-of-school science activities a few times a year or more were more likely to be interested in a STEM discipline at a university. Students who participated in science-focused out-of-school time club/competition were 1.5 times more likely, and respondents who had done reading/watching science were 1.3 times more likely to report interest in STEM disciplines at a university. This is in line with the questionnaire study of 298 year 5 and 6 students in four London schools by DeWitt et al. (2011) who found that engagement in science-related activities outside of school were among the strongest predictors for students' aspirations in science. Maltese (2010) found in their sample of chemistry and physics graduate students and scientists that most males reported that self-initiated activities such as tinkering with electronics, conducting home experiments, and reading science or science fiction, triggered their initial interest, while females to a greater extent reported that school-related activities sparked their interest.

Hands-on activities and the personal meeting between participants and university students or professionals are referred to as important factors in several studies on recruitment initiatives. Drawing on a questionnaire study, Woolston et al. (1997) found that participants in Wisconsin-Madison university's day-long campus visit program ranked interaction with students and hands-on activities higher than general introduction to the college and printed material. After investigating a 1-week engineering residential camp for 9th–12th grade young women, Swimmer and Jarratt-Ziemski (2007) wrote that "Participant as well as staff evaluations indicated that the young women learned more from and remained more engaged with the hands-on activities, and small group interactions with engineering professionals" (p. 13). Cantrell and Ewing-Taylor (2009) investigated a program where 130 high school students attended eight weekly sessions where they received presentations from STEM professionals. They found that the social hour after the formal program where participants met presenters and could ask them questions was the most powerful experience contributing to their knowledge about STEM careers. Fry et al. (2008) investigated an annual Girl Scout camp where STEM activities were provided, aiming at motivating the participants to choose science and mathematics later. The researchers stated that the camps influenced the participants' STEM attitudes positively by providing positive role models (the camps are led by female STEM undergraduates), and mastery experiences by letting them master tasks they didn't believe they were capable of doing, in an environment where they felt welcome and could be successful. Survey results showed that approximately 400 (53 %) of the 760 women who attended the "Discover Engineering" summer camp for high school girls at Ryerson Polytechnic University entered engineering studies later on (Zywno et al. 1999).

Providing information about possible STEM careers can be a way to increase participation in these disciplines. Harackiewicz et al. (2012) investigated a three-step recruitment intervention, where parents of high school students received two brochures and were introduced to a website motivating them to talk to their children about the utility value of science and mathematics. Compared to the control group, the students in the experiment group enrolled in nearly one semester more of science and mathematics in the last 2 years of high school. Andree and Hansson (2012), using expectancy-value theory as a framework, investigated a recruitment campaign in Sweden, that mainly involved film format information material. They found that the campaign mainly communicated messages concerning the utility value, attainment value and relative cost of STEM education, rather than its interest-enjoyment value. They found that whereas the campaign did challenge the notion that natural science is for "nerdy people", it did so through displaying personalities who had studied STEM and who had become successful and famous, but NOT because of their STEM engagement. A weakness of the campaign in Andrée and Hansson's view was that the successful people portrayed in the campaign (as role models) "did not claim to love STEM" (ibid., p. 17), and they concluded that the initiative "did not communicate messages that broadened the range of what a 'science self' might look like" (ibid., p. 18).

The duration of an intervention can have implications for how it influences participants. Nugent et al. (2010) investigated two robotics and geospatial technologies interventions. One was a 40-h intensive summer camp and the other a three hour event based on the camp. They found that participants in the longer intervention experienced significantly improved learning compared to students in a control group. The shorter event primarily impacted attitude and motivation.

Furthermore, there can be advantages in evaluating and developing an initiative over an extended period of time. Bischoff et al. (2008) studied a project where upper secondary students participated in week-long science camps at a college in USA. The aim of the camps was to inspire participants and let them experience that they had the skills needed to pursue a tertiary STEM education. During these camps, participants worked in groups solving tasks related to engineering and science. The researchers claimed that the camps were a success. A survey showed that most former participants agreed that the participation made them think seriously about choosing a STEM education. Moreover, even though possible confounding factors exist – like influence from school, peers or other out-of-school activities – the researchers showed that student enrolment at the college had increased after the camps were initiated. The camp was developed over several years and was adjusted after every camp. Initially, the camps were loaded with science activities, but evaluations suggested a more balanced ratio between science and social activities such as cinema visits and sports. Bischoff and colleagues concluded that the many rounds of evaluations and adjustments were critical in making the event successful.

Investigating Three Norwegian Recruitment Initiatives

In the following part of the chapter, studies of three specific Norwegian recruitment initiatives – CERN Masterclass, ENT3R and The Girls' Day – will be presented. For all three initiatives, the aim of the investigation was twofold: First, to investigate to what extent the participants' motivation for pursuing a STEM career was increased as a result of the initiative; second, to identify which factors in the initiative led to such a change (if present).

All three initiatives were investigated using focus group interviews. The interview guides had a general question, typically asking the students how they had experienced the initiative, and then sub-questions related to the Eccles et al. model (Eccles et al. 1983; Eccles and Wigfield 2002. See Chap. 1). In addition, the interview guides had a few questions related to the specific recruitment initiative in question. Audio recordings from the interviews were transcribed verbatim and analysed qualitatively using the NVivo 9 software to code and retrieve quotations. A thematic analysis was performed, following the recommended approach of Braun and Clarke (2006), with the aim of finding repeated patterns of meaning. Naming and interpretation of the codes were also guided by the research questions and the Eccles et al. model. In the investigation of The Girls' Day, questionnaire data were also collected. Responses to the open-ended questions in the survey were analysed using a similar approach as when analysing the focus group data. Data collection for each study is described in the sections treating each specific project investigated.

CERN Masterclass

In the CERN Masterclass initiative, approximately 8,000 secondary school students in 32 countries visit one of about 120 nearby universities or research centres each year for a 1-day encounter with particle physics. Here, lectures from active scientists give insight into topics and methods of basic research on the fundamentals of matter and forces, enabling the students to perform measurements on real data from particle physics experiments. At the end of each day the participants join in a video conference for discussion and combination of their results (IPPOG 2012). Results reported here are from the 2010 Masterclass in Oslo, where the event was carried out for the sixth consecutive year.

Focus group interviews were performed with students who participated in CERN Masterclass at the University of Oslo in spring 2010. The interviews were carried out approximately 10 days after the event. Altogether 11 boys and 5 girls participated in two groups. All the participants were upper secondary school students who had chosen specialisation in physics.

How did the event influence the participants' future career plans? Most of the students claimed that the day was inspiring, but had limited impact on their plans for further education. However, one girl said the following: "I think it was very inspiring that it was a girl who talked first, that it wasn't only boys. Because now there's a clear majority of boys in our physics class and I think no girls have chosen physics next year." This girl said that she had decided not to pursue specialisation in physics in the last year in upper secondary, but after attending the lecture held by a female PhD student she became uncertain whether that choice was right.

The participants were asked about what impression they got of the university students they met at the event. Some responded that the students they met fitted well into a stereotypical view of the geeky physics student. However, the overall feedback on this question was diverse:

> You can't tell that they are STEM students just by looking at them.
>
> . . . I thought they would appear to be geekier.
>
> You can think that those who study there, they actually want something. They are serious.
>
> Enthusiasm is inspiring no matter what. It is fun meeting people who really are enthusiastic about their work.

These quotes indicate that both the group of university physics students and the group of participants are diverse. Moreover, there probably is not a single answer to the question of who may work as good role models, as different persons have different opinions and preferences.

During the event, the participants got a glimpse into what physics can be used for and examples of developments in technology that sometimes follow research projects in physics. One of the boys said: "I thought the Internet was invented in USA by some computer geek or something! I was surprised that it was some physicists that managed to make something like that. So grand and so useful."[1]

Some students brought up that it was inspiring to master the calculations they did on data from CERN. "That made one think that it is actually not impossible to become good at physics and be able to have it as a job. It is not out of our reach at all", one of the boys said.

About the opening lectures in particle physics, one of the girls said: "When you're at school, you only learn stuff others already have discovered before you, and it is like there is not much more to discover. In Masterclass you see how much there actually is to do research on, and how much we don't know. I think perhaps the most interesting part was to discover that." This student's perception of physics is not unique. STEM subjects are traditionally associated with a strong socialisation of students into a well-defined and rigid knowledge base and set of values – which does not harmonize with late modern values of individual freedom of expression, creativity, and self-realisation. In a combined English and American study students reported that they found mathematics rigid, inflexible, and a subject that leaves no room for negotiation of meaning (Boaler et al. 2000). Lectures on unexplored areas in the field can contribute to expanding students' perception of physics.

Above we have seen that the event gave the participants mastery experiences. They got to meet both students and active physicists, and several participants said that the lectures and exercises contributed to an enhanced interest in physics. According to the students, these experiences were the most important in making

[1] "Tim Berners-Lee, a British scientist at CERN, invented the World Wide Web (WWW) in 1989. The web was originally conceived and developed to meet the demand for automatic information-sharing between scientists in universities and institutes around the world." (http://public.web.cern.ch/public/en/about/web-en.html, accessed August 2013).

the event worthwhile attending. However, different students ranked the experiences differently. Some argued that the most important feature was that they had mastery experiences when solving the exercises, others argued that the lectures were the most inspiring feature, while one of them experienced the fact that the first lecturer was a female as particularly inspiring. The expectancy-value model for achievement-related choices (see Chap. 2) highlights the following factors:

- Expectation of success (How well will I do?)
- Interest and enjoyment (Am I interested?)
- Attainment value (How well does this education fit my identity?)
- Utility value (How will this education help me reach other goals I have set?)
- Cost (How much will it cost in terms of time, effort or performance anxiety?)

One positive feature of the event is probably that it provided a variety of experiences. Through these activities most factors in the Eccles et al. model was touched upon. In this way, the project reached a range of different students, with various wishes, interests and preferences.

ENT3R

The ENT3R project is a Norwegian programme initiated to improve recruitment to mathematics and science. In the project, students aged 14–17 go to their local university or university college once a week for mathematics trainings, where they do exercises and STEM activities guided by university mathematics and science students. They also attend monthly theme-nights focusing on STEM career opportunities, with visits from STEM employers. The project is developed and tested over several years at two universities in Norway, and since 2012 it exists at all Norwegian universities and university colleges that offer STEM education.

Four focus groups with ENT3R participants were performed at the University of Oslo during spring 2010. Altogether 11 girls and 14 boys took part in these interviews. We wanted to investigate how participants experienced the project, and chose to recruit only students who had participated in ENT3R for at least one semester. In this way we could be sure that the students had at least participated in a few maths trainings, and thus had some experience with the project. This means that a limitation of the study is that students who only participated once and then quit were not included. The results presented here are reported in more detail by Jensen and Sjaastad (2013).

From analyses of the focus group discussions it was evident that many respondents felt that the trainings provided good mathematics teaching. The mentors are skilled both in mathematics and in teaching and explaining. “The thing is that they teach us stuff in very fun ways. You want to pay attention”, one of the boys said. Two mentors usually lead relatively small groups. Thus, the mentors have plentiful time to help each individual participant with exercises, and this is reported by the participants as an important feature.

If one important factor making the project successful is that the mathematics trainings provide good STEM teaching, does that mean the resources just as well could be invested in enhancing science and mathematics teaching in schools? Our impression is that no matter how good the quality of the teaching in these subjects in school might become, there will still be good arguments for organising out-of-school projects like ENT3R. The trainings take place at universities and university colleges, implying that the participants get a glance at student life at those institutions. The mentors are not involved in assessing the students' performance or setting marks, and the project does not have a fixed curriculum. Moreover, this is an arena where the participants can experience science and mathematics as a social leisure activity. All these aspects of the project contributed to the students' enjoyment, and made them want to spend their spare time at the trainings.

One of the boys said that "I think the best part of ENT3R is that you can learn mathematics in a very social, fun, and informative environment." One of the girls said that "It is easier to work with mathematics when there are several others around you, and you feel that you are not the only one who does not understand the exercise." Based on a synthesis of over 800 meta-analyses relating to the influences on school students' achievement, John Hattie (2009) argues that one of the most important factors facilitating good learning is to create a warm classroom climate where "... errors are not only tolerated but welcomed"(p. 34). One important success factor in ENT3R is probably that the mentors are able to create a social, fun, and accepting learning environment.

Some of the students reported that they appreciated the personal contact they got with their mentors. One of the boys said "My best memory is from when I told my mentor about the mark I got [in school]. He always gets happy." Another one said: "They are a bit like buddies. Just with more knowledge."

By participating in the project, secondary students meet STEM students at the university. "It has changed stereotypes a lot. Previously, I felt that people who do STEM were only skilled in STEM. But now I know that you find different people who do STEM", one of the boys said. This quote provides an example of how the mentors worked as role models. Several students said they had changed their perception of people who are interested in STEM – earlier they assumed that STEM persons were "dull and geeky", but now they perceived them as "ordinary people who like what they do".

So far we have looked at four factors: good teaching, safe and fun learning environment, personal contact, and role models. These factors were important in order to make the participants enjoy the project, wanting to spend their leisure time at mathematics trainings. One question is still unanswered: How and to what extent has the project influenced the participants' career plans?

In the focus groups, many students expressed that ENT3R had given them increased motivation for choosing further STEM education. And the students' responses touched upon all five factors in the Eccles et al. model: Expectation of success, interest and enjoyment, attainment value, utility value and cost. However, different students ranked these factors differently. Some of them said the most important feature of the project was that it had increased their interest in mathematics, some expressed that their expectation of success had increased, and others had learned more about jobs STEM education can lead to, and still others had revised their view of a "typical STEM identity". The project thus affected different priorities related to an educational choice and – similar to CERN Masterclass – one can argue that ENT3R may reach a variety of students this way.

The Girls' Day

The Girls' Day is a 2-day event at the Norwegian University of Science and Technology (NTNU) aimed at girls in the last year of upper secondary school who have chosen specialization in mathematics and physics. At the event, the participants receive information about the STEM study programmes available at the university. In particular, they receive information about study programmes where the proportion of females is low. In addition to attending lectures and stands, the participants get guided tours around the campus, share a dinner and enjoy entertainment. They have to write an application in order to be accepted to the event, and their mathematics marks have to be average or better. The university covers all expenses. In 2011 the event was held early in November, with 251 participants. The Girls' Day has been organised (although with slightly different target groups and aims) annually for more than 10 years, and every event is based on experiences from previous years.

Data were collected through focus groups after the event was concluded on the second day. Three group discussions were arranged, with a total of 17 participants. Approximately 10 days later, an online questionnaire was administered. This received a total of 189 responses, a response rate of 75 %. Six months later, a second survey, with similar questionnaire, was administered, and received 156 responses (62 % response rate). From now on, these two surveys will be referred to as survey #1 and survey #2. The results presented here are reported in more detail by Jensen and Bøe (2013).

How Did the Event Inform Participants' Decision of What and Where to Study in Higher Education?

In survey #1, 64 % responded that they were more certain of *what* to study, and 74 % responded that they were more certain of *where* to study. Thus, many of the participants claimed that the event helped inform their educational choice process.

A common response to an open-ended question in survey #1 was that meeting university STEM students in-between the organised programme, and on the exposition-stands, was one of the most important factors making them more certain about what to study. In a closed-response question in survey #1, 96 % of students responded that meeting university students made them more motivated for choosing a higher education in STEM; 88 % gave the same response in survey #2. One of the benefits of meeting and talking to university students was that they provided information about both the positive and negative aspects of life as a student, thus providing information that participants' judged to be realistic and trustworthy.

> I think it was great that there were so many students around all the time and that we got to hear from them how it really was to be a student in Trondheim [name of city].

In the following sections, participants' responses are related to several of the factors in the Eccles et al. model.

Interest and Enjoyment

A common response in an open-ended question in survey #1 was that the event showed the participants that a STEM higher education can be interesting and enjoyable.

> one learns a lot and the work one does seems really fun and interesting

Eighty-five percent responded in survey #1 that they perceived a STEM education as more interesting after the event; 76 % gave this response in survey #2.

Attainment Value

Participants' responses in the focus groups revealed that university students were perceived as role models. The participants appreciated that the university students were a few years older than them, enjoyed their study, spent time on leisure activities, and that they had a "normal style". Furthermore, university students reported that they too found science difficult and demanding. In such ways, the university students portrayed attainable STEM identities.

> . . . I think she was really great, (. . .) she was like normal in a way (. . .), she was quite good-looking compared to the stereotype of persons, or women that do science and mathematics they are sort of like big glasses and a lot of pimples and stuff like that, but she had a nice style and like, and looked totally ordinary and then she was smart and then it was sort of PhD and she was quite young and, so I think it was a very good role model in a way. . .

Expectation of Success and Cost

Sixty-three percent responded in survey #1 that the event made them think that even more time and work is needed for completing a tertiary STEM education, compared to what they previously had thought. Only 8 % of these reported that their expectation of success was weakened after the event. Fifty-four percent reported their expectation of success to be unchanged, while 38 % answered that it had increased.

The results were very similar in survey #2. Some participants responded that the event made them expect that they would find help and motivation from peers and teachers at the university.

> The Girls' Day showed me that it takes discipline (...), autonomy, a lot of responsibility, motivation and willpower to be a science and mathematics student. It will be demanding too, but with this good a study environment and teachers I got the impression that (...) it will work out wonderfully.

Moreover, at the "motivation lecture" they attended on the last day of the event, they were both told that it demands a lot of hard work to study STEM at a university, and they were provided strategies for making it more enjoyable and manageable.

Utility Value

Participants reported that they did not receive information about many specific jobs within the STEM field, however, they appreciated learning that a STEM education can lead to a wide range of jobs, and that it will be easy to get a job with such background. The information they received about study programmes on the stands was found motivating by 87 % in survey #1, and 69 % in survey #2. 63 % in survey #1, and 51 % in survey #2 answered that the information they received about job opportunities at the stands had increased their motivation for choosing STEM higher education.

Participants' Motivation for Working with School Mathematics and Science

Several students indicated that the event made them more motivated for working hard with school mathematics and science, in order to get the grades needed to be accepted into a study programme at NTNU. However, survey results did show a decline from the first to the second survey. Sixty-nine percent in survey #1 responded that the event made them more motivated for working with school science and mathematics. Forty-four percent gave this response in survey #2.

In the above sections, we have seen that meeting university students was one of the most important factors in The Girls' Day event, as they provided trustworthy information, were role models, and could increase some participants' expectation of success by giving participants an opportunity to identify themselves with persons who are a few year older and have been successful in their study.

Discussion

The outcomes of the recruitment initiatives reported on in this chapter were studied in terms of participants' self-reported intentions of pursuing STEM education and in terms of their self-reports on how their STEM-related values and expectancy of success (as conceptualised through the Eccles et al. model) were influenced by their participation in the recruitment initiatives. Responses in both ENT3R and The Girls' Day indicated that many participants had increased their motivation for choosing a STEM education.

Changes in all five factors in the Eccles et al. model (expectation of success, interest and enjoyment, attainment, utility and cost) can be identified in the participants' responses in all three initiatives. In particular in CERN Masterclass and ENT3R different students reported that they found different parts of the initiatives most important. These responses indicate that by providing a variety of experiences, these initiatives reach a broad group of students. Students who attend recruitment initiatives often have different needs, values and priorities. For instance, a student who already has high expectation of success in mathematics may find an event more motivating if it demonstrates the utility value of the subject and introduces job opportunities, rather than only facilitating mastery experiences.

Although the utility value of STEM appeared to be influenced in all three programmes by the initiatives giving information about job opportunities, ENT3R is perhaps the project that provides most information on specific jobs and businesses through the monthly career nights. In CERN Masterclass the participants mainly learned about the work performed at the CERN organisation, and in The Girls' Day they mainly learned that a STEM degree is highly attractive and opens up for a wide range of well-paid jobs. The responses from the interviews indicate that providing good examples of the utility value of mathematics and science can inspire students to choose STEM, in line with the Harackiewicz et al. (2012) investigation mentioned in the introduction.

The perhaps most important shared factor of the initiatives described in this chapter is the personal meeting between participants and university students, in line with several other studies (e.g. Cantrell and Ewing-Taylor 2009; Woolston et al. 1997). The Girls' Day participants reported that these meetings provided an opportunity to ask questions and get trustworthy information about study programs, social life at the university, and job opportunities. In the ENT3R project university students influenced participants' attainment value, by acting as role models and providing information about STEM choices.

In all three initiatives students reported that they appreciated meeting students who contradicted the stereotypical image of the geeky scientist. Typically they appreciated meeting students who were a few years older, were highly interested in their discipline, enjoyed their student life, who found time to spend on leisure activities and in general had a normal style. The attainment value can also be increased by participants meeting peers and university students with similar interest in science and mathematics as their own. Moreover, students' expectation of success can be raised through meeting university students that participants perceive as similar to themselves, thus inspiring them to reason that "if they can do it, so can we." The women in the study by Zeldin and Pajares (2000) described the importance of peers and the influence of social groups on their developing confidence. This, together with the motivational lecture, was the main factors influencing the participants' expectation of success in The Girls' Day. Although some participants in CERN Masterclass and ENT3R also reported that their expectation of success was raised through meeting university students, these programs were different in the sense that here participants got to do tasks, and mastering these was reported to be the main source for building expectation of success, similar to the summer science camp studied by Bischoff et al. (2008). Particularly in the ENT3R project, an important factor was the mentors' ability to provide a safe and social learning environment, in line with the factors for good learning described by Hattie (2009) and the aims in the Girl Scout camp investigated by Fry et al. (2008).

The largest difference between CERN Masterclass and The Girls' Day on one side and ENT3R on the other side is perhaps the amount of time participants spend in it. Drawing on the participants' responses, one can argue that the length of CERN Masterclass and The Girls' Day imply that these initiatives mainly can strengthen existing interests and contribute in making educational choices more informed. Since the ENT3R project engages participants weekly over an extended period of time, it is in a far better position to support STEM interests also in students who originally participated only because they needed help in order not to fail in school mathematics. This is in line with the results and discussion of Nugent et al. (2010).

Several of the initiatives described as successful in the research literature are developed and tested over an extended period of time. One example is the annual science camps investigated by Bischoff et al. (2008). The initiatives investigated in this chapter are also developed and tested over several years, and although it is possible to create initiatives that are successful from the beginning, there is an advantage in developing an initiative further over an extended period of time, as it is difficult to foresee how all factors will work when it is put into practice.

Conclusions

Success factors in recruitment initiatives include a *variety* of experiences that enhance the interest-enjoyment value the participants attach to STEM, their expectation of success in STEM, that display the utility value of STEM and the various employment opportunities, that reduce their perception of cost through creating mastery experiences in a safe learning environment, and that raise the attainment value of STEM. The latter is particularly related to *personal meetings* between participants and STEM tertiary students and/or professionals who act as models of STEM identities and who may help participants define their own identities as (potential) STEM students and practitioners. Finally, important success factors in recruitment initiatives include development of the initiative over time and (where possible) prolonged 'exposure' of the participants, as in the case of the ENT3R programme.

References

Andrée, M., & Hansson, L. (2012). Marketing the 'broad line': Invitations to STEM education in a Swedish recruitment campaign. *International Journal of Science Education.* doi:10.1080/09500693.2012.695880.

Bischoff, P. J., Castendyk, D., Gallagher, H., Schaumloffel, J., & Labroo, S. (2008). A science summer camp as an effective way to recruit high school students to major in the physical sciences and science education. *International Journal of Environmental and Science Education, 3*(3), 131–141.

Boaler, J., William, D., & Zevenbergen, R. (2000, March 26–31). The construction of identity in secondary mathematics education. In *International mathematics education and society conference*, Montechoro, Portugal.

Bøe, M. V., Henriksen, E. K., Lyons, T., & Schreiner, C. (2011). Participation in science and technology: Young people's achievement-related choices in late-modern societies. *Studies in Science Education, 47*(1), 37–72.

Braun, V., & Clarke, V. (2006). Using thematic analysis in psychology. *Qualitative Research in Psychology, 3*(2), 77–101.

Cantrell, P., & Ewing-Taylor, J. (2009). Exploring STEM career options through collaborative high school seminars. *Journal of Engineering Education, 98*(3), 295–303.

Dabney, K. P., Tai, R. H., Almarode, J. T., Miller-Friedmann, J. L., Sonnert, G., Sadler, P. M., et al. (2011). Out-of-school time science activities and their association with career interest in STEM. *International Journal of Science Education, Part B, 2*(1), 63–79.

DeWitt, J., Archer, L., Osborne, J., Dillon, J., Willis, B., & Wong, B. (2011). High aspirations but low progression: The science aspirations – Careers paradox amongst minority ethnic students. *International Journal of Science and Mathematics Education, 9*(2), 243–271. doi:10.1007/s10763-010-9245-0.

Eccles, J., & Wigfield, A. (2002). Motivational beliefs, values, and goals. *Annual Review of Psychology, 53*, 109–132.

Eccles, J., Adler, T. F., Futterman, R., Goff, S. B., Kaczala, C. M., Meece, J. L., et al. (1983). Expectancies, values, and academic behaviours. In J. T. Spence (Ed.), *Achievement and achievement motives. Psychological and sociological approaches*. San Francisco: W. H. Friedman & Co.

Fry, C. C., Davis, J., & Shirazi-Fard, Y. (2008, October 22–25). Recruitment and retention of females in the STEM disciplines: The annual girl scout day camp at Baylor University. In *2008 38th annual frontiers in education conference*, Saratoga Springs, NY.

Harackiewicz, J. M., Rozek, C. S., Hulleman, C. S., & Hyde, J. S. (2012). Helping parents motivate adolescents in mathematics and science: An experimental test of a utility-value intervention. *Psychological Science, 23*(8), 899–906.

Hattie, J. (2009). *Visible learning: A synthesis of over 800 meta-analyses relating to achievement.* London: Routledge.

IPPOG (2012). *International masterclasses.* http://www.physicsmasterclasses.org/neu/. Accessed 11 June 2012.

Jensen, F., & Bøe, M. V. (2013). The influence of a two-day recruitment event on female upper secondary students' motivation for science and technology higher education. *International Journal of Gender, Science and Technology, 5*(3), 317–337.

Jensen, F., & Sjaastad, J. (2013). A Norwegian out-of-school mathematics project's influence on secondary students' STEM motivations. *International Journal of Science and Mathematics Education, 11*(6), 1437–1461. doi:10.1007/s10763-013-9401-4.

Maltese, A. V. (2010). Eyeballs in the fridge: Sources of early interest in science. *International Journal of Science Education, 32*(5), 669.

Nugent, G., Barker, B., Grandgenett, N., & Adamchuk, V. I. (2010). Impact of robotics and geospatial technology interventions on youth STEM learning and attitudes. *Journal of Research on Technology in Education, 42*(4), 391–408.

Simpkins, S., Davis-Kean, P., & Eccles, J. (2006). Math and science motivation: A longitudinal examination of the links between choices and beliefs. *Developmental Psychology, 42*(1), 70–83.

Swimmer, F., & Jarratt-Ziemski, K. (2007). Intersections between science & engineering education and recruitment of female and native American students. In *ASEE annual conference & exposition*, Honolulu.

Woolston, D. C., Zaki, E., & Winter, C. J. (1997). The effectiveness of freshman recruiting strategies: An analysis at the University of Wisconsin – Madison. In *Frontiers in Education Conference: Teaching and Learning in an Era of Change*, Pittsburgh.

Zeldin, A., & Pajares, F. (2000). Against the odds: Self-efficacy beliefs of women in mathematical, scientific, and technological careers. *American Educational Research Journal, 37*, 215–246.

Zywno, M. S., Gilbride, K. A., Hiscocks, P. D., Waalen, J. K., & Kennedy, D. C. (1999). Attracting women into engineering-a case study. *IEEE Transactions on Education, 42*(4), 7. doi:10.1109/13.804556.

Part III
Staying in STEM, Leaving STEM?

Chapter 13
Why Do Students in STEM Higher Education Programmes Drop/Opt Out? – Explanations Offered from Research

Lars Ulriksen, Lene Møller Madsen, and Henriette Tolstrup Holmegaard

Introduction

According to statistics from the Organisation for Economic Co-operation and Development (OECD), almost one-third of all higher education students drop out of their studies before they complete their first degree (averaged across all OECD countries and all subjects) (OECD 2010). The term 'drop out' is commonly used to describe those students leaving their course before they pass the final examination. The loss of students from science, technology, engineering and mathematics (STEM) studies to other careers presents a particular reason for concern since: 'in many countries, S&T are among the disciplines where the dropout rates are the highest', with science suffering more than technology (OECD 2008, p. 74). This loss of students has been described as a 'leaky pipeline' in the science education literature (Seymour 2002). However, as pointed out by Hovdhaugen (2009), different designations are used within distinct research settings: 'In the USA, the phenomenon is described as 'dropout' or 'student departure' while British researchers usually use the concept 'non-completion' or 'non-continuing students" (Hovdhaugen 2009, p. 2). These different expressions reflect whether we interpret students leaving an educational programme as a push or a pull effect and for whom it is a problem.

In this chapter we provide an overview of how research has tried to explain and understand the issues related to students leaving higher education programmes with a specific focus on STEM courses. We illustrate three significant trends within the

---

This chapter is based on a paper of Ulriksen, Madsen and Holmegaard (2010): What do we know about explanations for drop out/opt out among young people from STEM higher education programmes? Studies in Science Education, 46(2), 209–244.

L. Ulriksen (✉) • L.M. Madsen • H.T. Holmegaard
Department of Science Education, University of Copenhagen,
Øster Voldgade 3, DK-1350 Copenhagen C, Denmark
e-mail: Ulriksen@ind.ku.dk

E.K. Henriksen et al. (eds.), *Understanding Student Participation and Choice in Science and Technology Education*, DOI 10.1007/978-94-007-7793-4_13

literature; the results from Seymour and Hewitt, Tinto's model and studies combining drop-out and theories of identities. We explore whether research on retention and non-completion has produced findings that can identify a direction forward for HE institutions and courses to take measures to reduce the number of students leaving. We show that most of the research focuses on overcoming deficits in students' prior knowledge, but that a more specific focus on identities as an analytical framework for understanding young people leaving STEM higher education programmes is also emerging. A detailed description of method used in the study can be found in Ulriksen et al. (2010).

Leaving STEM Higher Education: Seymour and Hewitt's Research

> Switching is not defined as a problem when it is believed to be caused, on the one hand, by wrong choices, under preparation, lack of sufficient interest, ability or hard work, or on the other, by the discovery of a passion for another discipline. (Seymour and Hewitt 1997, pp. 391–392)

As Seymour and Hewitt state, it is necessary to establish that a problem exists in order to address it. This is also true for the issue of students leaving STEM courses. As discussed in Seymour (2002), the early days of research within this field were dominated by the view that it was the students who were the problem. However, in their analysis of 335 STEM students at seven different types of institutions in a 4-year ethnographic study in the US, Seymour and Hewitt (1997) found that there was no evidence for those beliefs. On the contrary, their study showed that the most common reasons for students to switch degree courses arose in response to a set of common problems experienced by both switchers and non-switchers. They did not, as Seymour puts it in a later article, 'find switchers and non-switchers to be two different kinds of people: they did not differ by performance, motivation or study-related behaviour to any degree that was sufficient to explain why one group left, and the other group stayed' (2002, p. 82). On the whole, Seymour and Hewitt found more similarities than differences between the switchers and the non-switchers. There is a high level of agreement across the whole student sample about the issues that led to defection by switchers and to dissatisfaction among non-switchers, and there are strong similarities in the importance members of each group ascribe to each set of concerns. Based on their findings, Seymour and Hewitt (1997) stated that the problems which arose from the structure of the educational experience and the culture of the discipline (as reflected in the attitudes and practices of STEM faculty) made a much greater contribution to STEM attrition than the individual inadequacies of students or the appeal of other majors.

There seems to be an agreement between Seymour and Hewitt's study of STEM students and the more general research on retention and non-completion of students in higher education (see the reviews of Pascarella and Terenzini (2005) and Harvey et al. (2006)) in focusing less on the students' prior knowledge or preparedness, and more on the teaching and learning experiences the students are presented with once they have entered the programmes.

This point, as well as the fact that switchers and non-switchers, to a large extent, experience the same kinds of problem, result in Seymour and Hewitt using the metaphor of an 'iceberg' to illustrate their conclusion: 'Those who switch represent only the tip of a much larger problem' (Seymour and Hewitt 1997, p. 31).

In general STEM programmes lose students who possess interest and abilities within the field because the pedagogical approach and the study environment are unattractive. To be clear, the learning experiences of the students lead them to lose interest in science. These poor learning experiences are related to the traditions, teaching priorities and ethos of the disciplines, as is the case with the principle of 'weeding out' low performing student, but also the generally low priority that students experience is given to teaching by science faculty (Seymour and Hewitt 1997).

There have been a large number of studies within the field of drop out/opt out since Seymour and Hewitt's (1997) study. However, many of the studies still focus on retention as a matter of increasing students' skills before or during the first year of study and they aim at identifying factors associated with students' academic success (Ariadurai and Manohanthan 2008; Burnett 2001; Dyer et al. 2002; Mills et al. 2009; Yan 2002). Only a few studies have focused on changing university cultures, including teaching practices (for a full review see Ulriksen et al. 2010).

Leaving Higher Education – Tinto's Model

The work of Vincent Tinto has achieved an almost paradigmatic stature (Braxton et al. 2000) for trying to understand retention within higher education in general. Tinto's model (1975, 1988, 1993, 1998) emphasises that student departure from university should be regarded as a process. Students enter with a set of pre-entry attributes (cf. Chap. 14), and these attributes produce a set of goals and commitments that the students bring with them as they start their course and engage in the social and academic environment at the institution.

Tinto criticises psychological approaches to understanding students leaving college: 'such models invariably see student departure as reflecting some shortcoming and/or weakness in the individual', and thus the result of personal failure (1993, p. 85). Instead, Tinto emphasises a more sociological approach, focusing on the institution. Though previous sociological approaches to the study of retention provide relevant insights, Tinto claims that they tend to leave the actual interaction between students and institutions almost untouched (Tinto 1993, p. 86ff). It is precisely this level – the students' interaction with the institution and how this influences student persistence – that is his primary interest. The student's involvement leads to some degree of social and academic integration that again produces a set of goals and commitments that lead to a decision to depart from or stay at the university. The academic integration primarily refers to those parts of university life that are related to the formal education and to the student learning during the course. This interaction mainly takes place in classrooms, lecture halls and study groups. The social integration refers to the student's interaction in informal parts of university life such as unions, cultural gatherings and informal contact with teachers outside of the classroom (Tinto 1993).

Comparing the 1975 version of the model with the one from 1993, the student's interaction with staff/faculty has moved from the social system to the academic system, acknowledging that academic integration is not simply about performing well, but is also a matter of interacting with teachers. However, the academic and the social system of the institution are regarded as two distinct, but 'invariably interwoven' systems (1993, p. 109). Further, in the 1993 version of the model, the process at university is 'nested in an external environment comprised of external communities with their own set of values and behavioural requirements' (Tinto 1993, p. 115). Thus the university is a social system that works within a set of other social systems, and the students are simultaneously engaged in several systems.

Importantly, Tinto also makes the point that the university consists of more than one culture – that there are subcultures, and that students may become integrated into one of these, but not in to the dominant culture (1993, p. 105). These two points, namely, firstly, that the social and the academic systems are interwoven, and therefore influence each other, and secondly, that universities consist of more than one culture leads Tinto to emphasise that educational communities in the classrooms are an important arena for the integration of students at university. This is certainly important for non-residential students, who live outside campus, and commute every day. For these students in particular the social integration usually has to occur during class or in relation to class activities (1993, p. 206, and Tinto 1997, 1998). In his concluding remarks he states that an institution's capacity to retain students:

> . . .hinges on the establishment of a healthy, caring educational environment which enables all individuals, not just some, to find a niche in one or more of the many social and intellectual communities of the institution. This view of the effect of institutions upon student leaving highlights the intricate web of reciprocal relationships which binds students to the communal life of the institution. Rather than single out any one action or set of actions as being the primary cause of student departure, it argues that student leaving is affected by most institutional actions regardless of their immediate referent. (Tinto 1993, p. 204f).

Tinto's model has several virtues. One is that it regards student leaving as a longitudinal process that involves more than one factor. Another is that it includes both the social and the academic aspect of students' integration.

Evidently, holding an almost paradigmatic position does not mean that Tinto's model of student departure is uncontested or uncriticised. At one level, questions have been asked as to whether Tinto's claims can be substantiated by empirical findings, and on another, it has been argued that his study lacks sensitivity towards ethnic minority students' situation in higher education.

In their review, Pascarella and Terenzini state that they can find 'moderate' support for the 15 claims they identify in Tinto's model (Pascarella and Terenzini 2005, p. 425f and 443f). However, as noted by Pascarella and Terenzini, a review by Braxton et al. (1997) reports only 'partial' support for some, and 'frail' support for the other claims. Based on this finding, the authors do not recommend abandoning the model but suggest revising it instead (Braxton et al. 1997, p. 156). Still, the importance of being integrated into the university community is echoed in other findings in the US review. In particular, Pascarella and Terenzini report that different programmatic interventions such as supplemental instruction and first-year seminars have an impact on student persistence (Pascarella and Terenzini 2005, p. 398ff). However, they point out that the dynamics beneath this success are unclear, for instance as to whether the impact is direct (that is, that the skills developed, etc. increase student persistence) or indirect (for instance, earlier socialisation into the university culture and increased interaction with faculty, staff and peers) (Pascarella and Terenzini 2005, p. 403). Likewise, they find that different experiential and inquiry-based learning approaches increase rates of persistence, not least due to the student-faculty contact and active learning involved (Pascarella and Terenzini 2005, p. 406). Similar findings are reported by Braxton et al. (2000), who found that student-active learning activities have a positive influence on student persistence, and inter alia on social integration, and they make the point that 'faculty classroom behaviours play a role in the student departure process' (p. 581).

On a theoretical level, Tinto has been criticised for making general claims from a model that may only fit some groups of students (Pascarella and Terenzini 2005, p. 56). Hurtado and Carter (1997), studying experiences of Latino students' sense of belonging at university, state that Tinto's model does not take account of the importance of racially tense environments at universities. According to Tierney (1999), Tinto's model implies that minority students, or students who in other ways differ from the dominant majority culture, undergo a process of assimilation. It would follow that, as a consequence, minority students must discard aspects of their cultural background in order to succeed at university. Tierney (1999) argues that this philosophy contradicts experiences from his own research with students of colour, which conversely indicates that precisely the inclusion of the family and the neighbourhood of the minority students has led to an increase in students' sense of belonging at university, and, in that sense, to their social and academic integration.

Hurtado and Carter (1997) similarly found that for Latino students at predominantly white universities deliberate inclusion strategies had a positive impact in terms of them their feeling at home at the university through maintaining interactions both inside and outside campus (1997, p. 338), as did participation in some culturally-related activities such as association with social-community organisations and religious organisations (p. 335). For these students, it is not simply a question of being integrated or not, but rather to preserve a relation to multiple peer-groups and cultural environments.

Undoubtedly, there is a risk that social and academic integration could be interpreted simply as assimilation and that measures taken by the institutions to prevent non-continuation could overemphasise that students should conform to the dominant culture. The research reported by Tierney and by Hurtado and Carter indicates that this approach could be detrimental to the persistence of ethnic minority students. But also other groups of students with a particular gender (cf. Chap. 19) or social background (Chap. 14) can be in danger of leaving due to their minority position. Therefore, it is critically important to be aware of whether support activities and structures at universities acknowledge these differences or not.

However, does Tinto's model imply that social and academic integration should be interpreted simply as assimilation? In our view, the answer partly depends on whether the model is read as a normative or an analytical statement. In the 1993 version of the model, Tinto identifies some limitations in using the analogies of the initiation rites and of egotistical suicide on entering university (1993, p. 104ff). Likewise, he emphasises that 'the great majority of colleges are made up of several, if not many, communities or "subcultures," each with its own characteristic set of values and norms' (p. 105) and that for some students 'events external to the college play an important role in community membership' (p. 105). More importantly, what permeates the model is the notion that attending university is a process of socialisation, and as such it is to be regarded as an interactional process between what the students bring with them and the culture they meet. Furthermore, this socialisation does not limit itself to academic features, but affects the tastes and practices of students in a broader context (Huber 1991). Similar observations are made by Becher (1989, cf. Becher and Trowler 2001) who – even though his study concerned research communities and not specifically student communities – points to the different cultures (or tribes as he calls them) that exists within academia, to which students need to gain access (cf. Gerholm 1990). For students at bachelor's level, Hasse (2002), in her study of first-year physics students at a research-intensive university, highlighted that becoming a physics student is more than merely learning the content knowledge; it is a matter of acquiring the right poise, or 'sprezzatura' as she calls it with reference to Italian courts. Conceiving studying as a process of socialisation also partly explains the importance of interaction with faculty members outside the classroom. Such an interaction has an impact due to the process of socialising the students to values and attitudes in the academy.

The idea that facilitating subcultures at university could provide a sense of belonging for students who do not feel related to the dominant social and academic culture at the institution, or whose academic aspirations do not necessarily concur with the dominant academic orientations and paths, sounds convincing. In that sense, not conforming with the dominant culture is apparently a viable way for non-traditional students to survive at university. However, even if the institutions engage in facilitating religious or cultural organisations and institutions at campus, the stance of the institution would still be ambiguous. In his study of the academic field, Bourdieu (1990) remarks that the habitus of those holding the dominant positions in the field serves to select those who are to be included and exclude others:

> What may appear as a sort of collective defence organized by the professorial body is nothing more than the aggregated result of thousands of independent but orchestrated strategies of reproduction, thousands of acts which contribute effectively to the preservation of that body because they are the product of the sort of social conservation instinct that is the habitus of the members of a dominant group (Bourdieu 1990, p. 150)

The socialisation of new students at bachelor's or PhD level, therefore, is not simply to ensure the academic qualification of the newcomers, but rather to make sure that the new members comply with the existing dominant culture. Therefore, when Tierney states that 'educational organizations must also accommodate for and honour students' cultural differences' (1999, p. 83), this may be true if those organisations have an interest in increasing student completion; but from the perspective of the organisations' struggle for position in the academic field, this is not necessarily the case. The interests of the universities are in these cases – from a Bourdieuan perspective (Bourdieu 1984, 1986) – at least ambiguous.

This point also has significance for some of the measures taken by universities to ease the way for minority students. As indicated by both Tierney (1999) and Hurtado and Carter (1997), studies of minority students suggest that for those groups of students to succeed, it may be a more viable path to establish subcultures that value the social and cultural capital of the minority. However, following the analysis of Bourdieu, this approach may well increase the probability of their completing their studies, but it is likely that it will also have the consequence that they are never fully integrated and accepted into the core of the academic community. This should not be an argument for giving up strategies such as the ones suggested in Tierney's study, or for calling for a total assimilation in the white, dominant culture. On the other hand, it seriously questions the impact of targeted sub-cultural services and offers on students' chances of obtaining equal possibilities within the academy.

In our view, Tinto's model provides an approach to examine the student experience that focuses on student departure as a process involving them coming to terms with both academic and social aspects of university life. Consequently, integration becomes a pivotal concept. Furthermore, both Tinto's remarks on the multiple communities and subcultures at university, the critical comments from, amongst others, Tierney (1999) and Hurtado and Carter (1997), and Tinto's further reflections on, inter alia, this critique (Tinto 2006–2007) emphasise that the process of integration is a complex one in which the differences in students' background, the composition of capital, the universities' level of inclusiveness, and the students possible positions in the academic field all influence the students' expectations of success and educational outcomes.

A Focus on Identity to Understand Retention

Based on the above discussions of Tinto's model, it would appear meaningful to apply an approach to the understanding of drop out/opt out among young people from STEM higher education courses that is informed by a narrative psychological conception of identity. If entering a study programme is regarded as a process of socialisation, then identity is a core concept for understanding how students relate to the experience and to the culture and environment they encounter. Since the integration into the culture of the discipline is, inter alia, brought about through the teaching and learning activities and the feedback from the teachers (Hasse 2002), then the relation between these elements in the courses and the identities of the students is of interest. This position is in line with the emphasis that both Seymour and Hewitt (1997) and Harvey et al. (2006) put on the students' study experiences – not least during the first year.

Identity is a concept which, though originally from the field of psychology, has spread to a range of other disciplines, for example, anthropology, history, sociology, linguistics and feminist theories (Holland et al. 1998; Wetherell 2009). Research focusing on identities has been relatively rare in the field of science education, but in recent years it has become a subfield in the study of young people's participation in STEM education (e.g. Hazari et al. 2010; Schreiner 2006; Archer et al. 2010; Hsu and Roth 2009). Identity has been conceptualised from a number of different theoretical perspectives. These positions constitute a continuum from the idea of the individual as stable and coherent to the notion of identity as being multiple, flexible and continually re-negotiated (see Chaps. 3 and 4).

The research focusing on identity draws upon a range of perspectives most of which share an emphasis on the importance of the interaction between the individual student and the culture of the discipline. Secondly, it highlights the importance of being recognised as a legitimate member of the group of science students or 'science people'; and thirdly, it draws attention to the point that some positions are available to some students rather than to others. Overall, there is an emphasis on the socio-cultural aspects of studying, and the analysis of the underrepresentation of particular groups of students.

Research on identity and student persistence in STEM has to a large extent applied quantitative methods (see for instance White et al. 2006; Schreuders et al. 2009; Wasburn and Miller 2004–2005; Xu 2008). In a review of studies on women in computer-related majors, Singh et al. (2007) found that the quantitative studies are primarily based on descriptive analyses, individualized measures, and implicit theoretical frameworks. The qualitative methods used in the research to understand identity issues vary from life history interviews with a small sample of students (Wood 2002), focusing on already ongoing initiatives (Davis 2001) to methods involving a range of qualitative methods (Carlone and Johnson 2007). This research is primarily set in a US context and is mainly related to minority representation problems, in particular the lack of women or non-white students (or both) in STEM programmes (for a full review of this literature see Ulriksen et al. 2010).

In one of the examples of European research addressing identity within STEM education, Stentoft and Valero (2009) state that:

> The notion of identity represents a way to move beyond the existing debate on whether mathematics learning is in essence individual or social. It can be seen as a notion which may assist researchers providing the missing link for grasping the dialectic relationship between the individual and the social dimensions of learning (Sfard and Prusak 2005 p. 15); and therefore it has been taken as a fruitful concept for providing more sophisticated interpretations of processes of mathematics education practices (Stentoft and Valero 2009, p. 56)

Following Stentoft and Valero, applying a socio-cultural post-structural perspective on identities is a way of building a bridge between looking at students leaving university as being either an individual or an institutional problem. It is a way to move away from a dichotomised perception of the problem to a more dynamic understanding in which identity is considered a fragile and ongoing process embedded in the institutional discourses and practices, closely related to the students' actions and participation. In this post-structuralist perspective, identity is perceived to be a process rather than a stable entity, where the individual produces culture at the same time as being produced by culture. This understanding of identity is not widespread in research in science education, but there are some examples of literature that take it into account.

Based on a study of women of colour working on constructing a science identity, Carlone and Johnson (2007) discuss identity as something closely connected to recognition, using a socio-cultural framework:

> Identity is not just something an individual feels; it is not even what an individual does, although both feelings and actions are components of identity. A science identity is accessible when, as a result of an individual's competence and performance, she is recognized by meaningful others, people whose acceptance of her matters to her, as a science person (Carlone and Johnson 2007, p. 1192).

This position strongly connects identity to cultural settings and to other individuals, meaning that the students are not free to construct an identity on their own. They are dependent on recognition from others, and to obtain it they have to make themselves recognisable as legitimate 'science people'. This recognition has to be obtained in a context that is derived from socio-historical discourses of science and what science is, and from historical meanings and societal images of being a woman in science.

In their study, Tate and Linn (2005) use a multiple identities framework that is grounded in situated cognition theory and pay particular attention to the social relations and communities the students engage in. Rather than talking about 'student identity', Tate and Linn distinguish between three identities: social identity (the view of self in society or through society's eyes), academic identity (activities and success) and intellectual identity (desire to be an engineer and insight in the engineering field). They conclude that:

> The multiple identities framework also reveals the intersections of the identities. Students' social identity may affect their academic identity. For example, a student who feels uncomfortable in an engineering environment may experience difficulty in forming study groups helpful to their academic performance (Tate and Linn 2005, p. 491).

The work of Tate and Linn draws attention to the diverse contexts and communities that students engage in, and consequently, studies of students' experiences at university that only address one of these identities may provide a misleading image of the students' situation. Furthermore, their work emphasises how these multiple identities influence each other.

Other research that takes up a more pronounced post-structuralist perspective emphasises that identities is so closely woven into the social and the cultural that they are inseparable. Hughes (2001), in a study of a group of students consisting of both males and females and of students of different ethnicity in a UK city school and post-16 city college, focuses on how identity is connected to recognition and to which positions are available in the construction of a science identity. She points out that different curricula and teaching methods make different potential identities available to students with gender or ethnicity different from that of the majority of students in STEM. Consequently, she cautions against simply linking particular genders to particular sciences. Instead, she concludes how 'socially relevant and more constructivist science can generate a wide range of scientist subjectivities, increase the possibilities for scientist identities and thus open the way towards a more inclusive science curriculum' (Hughes 2001, p. 288). The way science is presented to students set the scene for their participation in science and produce a wide range of subjectivities the students can relate themselves to in their identity-work.

As it is, applying identity as a theoretical perspective in understanding students' experiences and student persistence is primarily found in studies focusing on minority students, which in an STEM context includes both ethnic minority students and women. However, if attending university, as we argued earlier, is a process of socialisation (cf. Tinto 1998; Becher 1989; Becher and Trowler 2001), then it seems relevant to address the identity issue for majority students as well in trying to comprehend the question of persistence or opting out. This point seems even more relevant considering the finding of Seymour and Hewitt (1997), mentioned earlier, that the most common reasons causing students to switch programmes were rooted in experiences shared by both switchers and non-switchers. However, the fact that these experiences were shared does not mean that they were identical. This underlines the importance, not of research into individual traits or characteristics, but research into the intersection of different characteristics and how they are recognised, interpreted and acted upon by both the individual and by others in the academic culture and community.

Conclusion and Implications for Further Research

In this chapter we have reviewed research on students dropping or opting out of higher education with a special focus on STEM studies (for a full analysis see Ulriksen et al. 2010). The research on retention and non-continuation of students across different disciplines shows that there is no one factor determining student success. Instead, whether students persist or not are influenced by a number of factors, and these different factors interact.

The student's social and economic background and the reasons and processes behind their choice of study have an impact, as does the induction into the study programme. Students' preparation for their studies influence persistence, but their academic level and abilities cannot explain why some persist and others opt out. Conversely, the teaching and learning environment and the teaching methods applied prove to be highly important. The teaching and learning activities students are engaged in, the design of the curriculum and the interaction with faculty and peers are also important.

In a substantial part of the research included in this review, the problem of retention is being framed as located in either the student or located in the institution, respectively. However, another research approach to retention highlights the issue of identity construction and of being recognised as a legitimate member of the group of 'science people'. The inclusivity of the study environment and the disciplinary culture provides possible positions for the students to take, and make some identities more legitimate and recognisable than others. Apparently, the STEM culture is still to a large extent distinguished as being competitive, detached, white and male-dominated. Students who for one reason or another (for instance gender, ethnic origin or the part of the discipline the student takes interest in) differ from what is considered normal within the field will often have more difficulties in being socially and academically integrated, and in developing an identity as one belonging to the discipline.

Suggestions of how to increase retention tend to focus on adjusting the students and leaving the institutional or disciplinary side stable and untouched. A few papers move in the direction of organisational change, where the courses and the teaching and learning activities are adjusted according to students' background and experiences. One example is Wistedt (2001) who found how Swedish university technology programmes which with success attacks and keeps female students, were characterized by cooperation based, problem oriented methods, rich interactions with students and staff. She argues how institution must focus on radical changing the study programmes rather than paying attention to recruitment campaigns to retain and attract female students. But these kinds of suggestions risk being rejected because they are considered to be detrimental to the quality of the course, as described in Seymour (2002). This perception of the disciplines as stable and also objective entities with a fixed curriculum (Angell et al. 2004; Hughes 2001) leads any suggestion of changing the curriculum to be regarded as a setback for the science discipline and student achievement. If the discipline is not regarded as an object of negotiation, the point of departure for reducing drop out must be the students. This perspective makes it very difficult to introduce any measures that challenge the identity issue.

From this chapter we draw three important results. Firstly, this perception of STEM within the STEM provides an explanation as to why so few studies have followed the research ideas set out by Seymour and Hewitt (1997). In their work, they rejected the idea that the problem should be located in the student and instead framed it in relation to the match between the institutions and the students. We find that this is one of the prime reasons why it is so difficult to really address the problem of retention in STEM. STEM educators often demand a retention check list that can be imposed without changing the existing framework for teaching and the faculties' relation to the students. Evidently, these are factors that, according to research focusing on identity and the relation between students and institutions, need to be addressed. Further, it is likely that this is the reason why some research addresses this highly complex problem of retention by focusing on the straightforward variables of students' behaviour and capabilities.

Secondly, it makes it even more urgent to further develop research into the culture(s) of STEM disciplines and courses, in the formation of identity during the study, and to expand the scope of this research to all groups of students – not just the minority groups, but also the dominant white male culture. This approach further suggests that the problem of retention should be rephrased from focusing on how to adjust the students so that they can meet the requirements of the existing science programme to a broader perspective on students' experiences with studying science, where not least the question of how STEM programmes can become part of students' identity formation. Will it be possible for STEM programmes to convince future and present students that being integrated in a STEM discipline is an attractive perspective for a young individual trying to find out who she or he is, and what direction her or his life should take?

Thirdly, there is a need to combine research addressing identity issues with pedagogical research approaches that address, for instance, the purpose and objectives of science studies, what content is included and what is excluded in science programmes and the teaching and assessment approaches. Future research as well as future initiatives in higher education addressing the opting out or dropping out of students therefore needs to adopt a broad perspective in terms of both the teaching and learning activities, and on the possible identities made available to students. However, what from our perspective stands out as perhaps the most important finding in this review is that a substantial part of the findings of what could increase student retention within STEM, are at odds with the self-conception, culture and tradition of STEM disciplines and environments. Consequently, if STEM programmes and institutions genuinely wish to increase the number of students completing the STEM degree they enter, their courses need to turn their focus from the students alone on to themselves and the culture and values that are revered there, and consider whether they are perhaps a part of the problem. In our view, this is indeed most likely the case.

References

Angell, C., Guttersrud, Ø., Henriksen, E. K., & Isnes, A. (2004). Physics: Frightful, but fun. Pupils' and teachers' views of physics and physics teaching. *Science Education, 88*(5), 683–706.

Archer, L., DeWitt, J., Osborne, J., Dillon, J., Willis, B., & Wong, B. (2010). 'Doing' science versus 'Being' a scientist: Examining 10/11 year old school children's constructions of science through the lens of identity. *Science Education*, *94*(4): 617–639. Published Online: 21 Apr 2010. doi: 10.1002/sce.20399.

Ariadurai, S. A., & Manohanthan, R. (2008). Reasons for student discontinuation in engineering degree courses offered at a distance. *Turkish Online Journal of Distance Education, 9*(3), 74–86.

Becher, T. (1989). *Academic tribes and territories. Intellectual enquiry and the cultures of disciplines*. Buckingham: The Society for Research into Higher Education & Open University Press.

Becher, T., & Trowler, P. R. (2001). *Academic tribes and territories* (2nd ed.). Maidenhead: The Society for Research into Higher Education & Open University Press.

Bourdieu, P. (1984). *Distinction. A social critique of the judgement of taste*. London/Melbourne/Henley: Routledge & Kegan Paul.

Bourdieu, P. (1986). The forms of capital. In J. G. Richardson (Ed.), *Handbook of theory and research for the sociology of education* (pp. 241–258). New York/Westport/London: Greenwood Press.

Bourdieu, P. (1990). *Homo academicus*. Cambridge: Polity Press.

Braxton, J., Sullivan, A. V. S., & Johnson, R. M., Jr. (1997). Appraising Tinto's theory of college student departure. In J. C. Smart (Ed.), *Higher education: Handbook of theory and research* (Vol. XII, pp. 107–164). New York: Agathon Press.

Braxton, J., Milem, J. F., & Sullivan, A. S. (2000). The influence of active learning on the college student departure process. Towards a revision of Tinto's theory. *Journal of Higher Education, 71*, 569–590.

Burnett, K. (2001). *Interaction and student retention. Success and satisfaction in web-based learning. Libraries and librarians: Making a difference in the knowledge age*. Council and general conference: Conference programme and proceedings, 67th, Boston, MA.

Carlone, H. B., & Johnson, A. (2007). Understanding the science experiences of successful women of color: Science identity as an analytic lens. *Journal of Research in Science Teaching, 44*(8), 1187–1218.

Davis, K. S. (2001). "Peripheral and subversive": Women making connections and challenging the boundaries of the science community. *Science Education, 85*(4), 368–409.

Dyer, J. E., Breja, L. M., & Wittler, P. S. H. (2002). Predictors of student retention in colleges of agriculture. In *Proceedings of the 27th annual national agricultural education research conference*.

Gerholm, T. (1990). On tacit knowledge in academia. *European Journal of Education, 25*(3), 263–271.

Harvey, L., & Drew, S., with Smith, M. (2006). *The first-year experience: A review of literature for the Higher Education Academy*. Higher Education Academy. Retrieved from http://www.heacademy.ac.uk

Hasse, C. (2002). *Kultur i bevægelse – fra deltagerobservation til kulturanalyse – i det fysiske rum* [Culture on the move – From participant observation to cultural analysis – In the physical space]. Frederiksberg: Samfundslitteratur.

Hazari, Z., Sonnert, G., Sadler, P. M., & Shanahan, M. C. (2010). Connecting high school physics experiences, outcome expectations, physics identity, and physics career choice: A gender study. *Journal of Research in Science Teaching, 47*(8), 978–1003.

Holland, D., Lachicotte, W., Jr., Skinner, D., & Cain, C. (1998). *Identity and agency in cultural worlds*. Cambridge, MA/London: Harvard University Press.

Hovdhaugen, E. (2009). Transfer and dropout: Different forms of student departure in Norway. *Studies in Higher Education, 34*(1), 1–17.

Hsu, P., & Roth, M. (2009). To be or not to be? Discursive resources for (dis-)identifying with science-related careers. *Journal of Research in Science Teaching, 46*, 1114–1136.

Huber, L. (1991). Sozialisation in der Hochschule. [Socialisation in higher education]. In K. Hurrelmann & D. Ulich (Eds.), *Neues Handbuch der Sozialisationsforschung* (4th ed., pp. 417–441). Weinheim/Basel: Beltz Verlag.

Hughes, G. (2001). Exploring the availability of student scientist identities within curriculum discourse: an anti-essentialist approach to gender-inclusive science. *Gender and Education, 13*(3), 275–290.

Hurtado, S., & Carter, D. F. (1997). Effects of college transition and perceptions of the campus racial climate on Latino college students' sense of belonging. *Sociology of Education, 70* (October), 324–345.

Mills, C., Heyworth, J., Rosenwax, L., Carr, S., & Rosenberg, M. (2009). Factors associated with the academic success of first year health science students. *Advances in Health Sciences Education, 14*, 205–217.

Organisation for Economic Co-operation and Development. (2008). *Encouraging student interest in science and technology studies*. Global Science Forum. Paris: OECD.
Organisation for Economic Co-operation and Development. (2010). *Education at a glance 2010*. OECD indicators. Retrieved from http://www.oecd.org/dataoecd/45/39/45926093.pdf
Pascarella, E. T., & Terenzini, P. T. (2005). *How college affects students. Vol. 2: A third decade of research*. San Francisco: Jossey-Bass.
Schreiner, C. (2006). *Exploring a ROSE-garden. Norwegian youth's orientations towards science – Seen as signs of late modern identities*. Doctoral thesis, University of Oslo, Norway. Series of dissertations submitted to the Faculty of Education, University of Oslo, no. 58. Oslo.
Schreuders, P. D., Mannon, S. E., & Rutherford, B. (2009). Pipeline or personal preference: Women in engineering. *European Journal of Engineering Education, 34*(1), 97–112.
Seymour, E. (2002). Tracking the processes of change in US undergraduate education in science, mathematics, engineering, and technology. *Science Education, 86*(1), 79–105.
Seymour, E., & Hewitt, N. M. (1997). *Talking about leaving: Why undergraduates leave the sciences*. Boulder: Westview Press.
Sfard, A., & Prusak, A. (2005). Telling identities: In search of an analytic tool for investigating learning as a culturally shaped activity. *Educational Researcher, 34*(4), 14–22. doi:10.3102/0013189x034004014.
Singh, K., Allen, K. R., Scheckler, R., & Darlington, L. (2007). Women in computer-related majors: A critical synthesis of research and theory from 1994 to 2005. *Review of Educational Research, 77*(4), 500–533.
Stentoft, D., & Valero, P. (2009). Identities-in-action. *Nordic Studies in Mathematics Education, 14*(3), 55–77.
Tate, E. D., & Linn, M. C. (2005). How does identity shape the experiences of women of color engineering students? *Journal of Science Education and Technology, 14*(5/6), 483–493.
Tierney, W. G. (1999). Models of minority college-going and retention: Cultural integrity versus cultural suicide. *Journal of Negro Education, 68*(1), 80–91.
Tinto, V. (1975). Dropout from higher education: A theoretical synthesis of recent research. *Review of Educational Research, 45*(1), 89–125.
Tinto, V. (1988). Stages of student departure. Reflections on the longitudinal character of student leaving. *Journal of Higher Education, 59*(4), 438–455.
Tinto, V. (1993). *Leaving college. Rethinking the causes and cures of student attrition* (2nd ed.). Chicago/London: The University of Chicago Press.
Tinto, V. (1997). Classrooms as communities: Exploring the educational character of student persistence. *Journal of Higher Education, 68*(6), 599–623.
Tinto, V. (1998). Colleges as communities: Taking research on student persistence seriously. *Review of Higher Education, 21*(2), 167–177.
Tinto, V. (2006–2007). Research and practice of student retention: What next? *Journal of College Student Retention, 8*(1), 1–19.
Ulriksen, L., Madsen, L. M., & Holmegaard, H. T. (2010). What do we know about explanations for drop out/opt out among young people from STEM higher education programmes? *Studies in Science Education, 46*(2), 209–244.
Wasburn, M. H., & Miller, S. G. (2004–2005). Retaining undergraduate women in science, engineering, and technology: A survey of a student organization. *Journal of College Student Retention, 6*(2), 155–168.
Wetherell, M. (Ed.). (2009). *Theorizing identities and social action*. Basingstoke: Palgrave Macmillan.
White, J. L., Altschuld, J. W., & Lee, Y. (2006). Evaluating minority retention programs: Problems encountered and lessons learned from the Ohio science and engineering alliance. *Evaluation and Program Planning, 31*, 277–283.
Wistedt, I. (2001). *Five gender-inclusive projects revisited*. Sweden: National Agency for Higher Education.

Wood, S. L. (2002, April 1–5). *Becoming a woman engineer in the community of practice: Validity and value in engineering-education research.* Paper presented at the Annual Meeting of the American Educational Research Association, New Orleans.
Xu, Y. J. (2008). Gender disparity in STEM disciplines: A study of faculty attrition and turnover intentions. *Research in Higher Education, 49*(7), 607–624.
Yan, W. (2002). *Postsecondary enrolment and persistence of students from rural Pennsylvania.* Harrisburg: The Centre for Rural Pennsylvania.

Chapter 14
What Makes Them Leave and Where Do They Go? Non-completion and Institutional Departures in STEM

Lars Ulriksen, Lene Møller Madsen, and Henriette Tolstrup Holmegaard

Introduction

Other chapters in this book point to the importance of identity in order to understand the complexity of the process of students' choices and experiences. However, it is also relevant to consider aspects of recruitment and retention on a larger scale. Therefore, this chapter adopts a macroscopic approach. It does so with three purposes. The first purpose is to examine whether there have been changes in the patterns of recruitment and retention of male and female STEM students during the past 10–15 years. This is significant considering the number of research and policy documents published in that period of time (see Chap. 1 in this book for references on policy documents and Chaps. 2, 3, and 5 for research). The second purpose is to explore whether different factors affect the behaviour of men and women differently in relation to entering and staying in higher-education STEM programmes. The third purpose is to explore what happens to the students who decide to leave their STEM higher-education programme without completing. The focus of the chapter will be on overall trends and on gender differences.

Using national data from Denmark, the chapter addresses three questions:

1. *Changing patterns in students' choices*: Have the patterns of men's and women's choice of STEM changed over time?
2. *Risks of non-completion*: Are there differences in the risk of not completing the STEM programme the students have entered according to gender, and do different factors affect the risk of non-completion for male and female students?
3. *Departure trajectories*: Where do students go if the leave their STEM programme without completing it?

L. Ulriksen (✉) • L.M. Madsen • H.T. Holmegaard
Department of Science Education, University of Copenhagen, Øster Voldgade 3, DK-1350 Copenhagen C, Denmark
e-mail: Ulriksen@ind.ku.dk

E.K. Henriksen et al. (eds.), *Understanding Student Participation and Choice in Science and Technology Education*, DOI 10.1007/978-94-007-7793-4_14

The study is longitudinal in two respects. First, it studies the students' trajectories by not only looking at the students' first choice of study after finishing high school, but also whether they complete the study or not, and, in case they leave the study without completing it, what choice of study (or not) they make thereafter. Second, it studies the patterns of choice over time to see whether the patterns have changed during the period 1995–2009. Hence, the study is longitudinal on both an individual level and concerning the patterns that are found.

The Sample

The method of analysis will be presented as we address the three questions. In this section we will present the data the analysis builds upon.

The data used is retrieved from Statistics Denmark, the official statistical bureau of Denmark that receives data from, among others, all Danish educational institutions. Due to the Danish registration policy it is possible to track individuals over time, to combine data from different sources concerning the same individual, and to link individuals to their parents and to data concerning the parents.

After 9 years of compulsory schooling, young people need to decide whether they wish to pursue vocational training for a trade (for instance, carpenter, secretary, or car mechanic) or a non-vocational upper-secondary school preparing for higher education (in Danish: gymnasium). The analysis in this chapter covers all students who in the period 1995–2009 graduated from the non-vocational upper-secondary school in Denmark. The 15 year group were clustered in three cohorts: 1995–1999, 2000–2004 and 2005–2009. Note that a particular year refers to the year the student completed upper-secondary school, *not* the year of entrance at the higher-education programme. Therefore, numbers regarding, for instance, distribution of students at different programmes may differ from other statistics on STEM entrance that usually use the year of entry as reference.

The sample of all students completing gymnasium in the 15 year period consists of 464,607 students (58 % women) (Table 14.1). Not all of these students continued to higher education or to a STEM study programme.

In addition to information concerning students' enrolments on study programmes at Danish institutions, a number of background variables were included in the analysis:

- Sex[1]
- Origin: Danish; Immigrant; Descendant (i.e. a person born in Denmark whose parents are either immigrants or descendants with foreign citizenship).
- Highest acquired education of the parents when the student was 13 years old.

[1] The data uses the biological sex as a means of distinction. In our analysis we use the phrase 'gender' because the behaviour behind the data is, inter alia, a result of gendered practices (see Chap. 4 about the distinction between sex and gender).

Table 14.1 Number of students in the three cohorts and the gender distribution

	Year of completing upper-secondary school					
	1995–1999		2000–2004		2005–2009	
Sex	Number	%	Number	%	Number	%
Men	67,142	41.6	59,627	41.8	68,427	42.6
Women	94,140	58.4	83,057	58.2	92,214	57.4
Total	161,282	100.0	142,684	100.0	160,641	100.0

- The specialisation (or line) of gymnasium (general (stx), commercial (hhx), technical (htx) or higher preparatory (hf))
- Grade-point average (GPA) for upper-secondary school.
- Number of gap years in total.
- Number of years spent at another higher education programme before entering the first STEM programme.

In the present analysis, we focus on gender while other variables are only included to a limited extent.

Groups of Educational Programmes

We divided the higher education programmes that students could enter into STEM and non-STEM programmes. Health programmes (including medicine and the pharmaceutical sciences) were in this analysis considered non-STEM studies because they do not suffer from the same recruitment or retention difficulties as do STEM programmes. Neither do the veterinary sciences, but for institutional reasons (that is, where the programme is offered in Denmark) we decided to include them in the group of STEM programmes.

Next, we distinguished between STEM programmes aiming at Masters level at universities (usually aiming at 5-year programmes), and professional bachelor programmes at university colleges (completed after 3 or 4 years). At the university level we divided the programmes into two: science and university engineering. In this chapter, when we present results for STEM as a whole, all STEM studies are included.

The university sciences (18 sub-disciplines; 26,072 students) include programmes such as biology, chemistry, physics and mathematics. The university engineering programmes (nine sub-disciplines; 13,701 students) include sub-disciplines such as construction, environmental engineering, chemical engineering and nano-technology. Professional engineering programmes (nine sub-disciplines; 15,452 students) include sub-disciplines in structural engineering, environmental engineering and construction.

The number of students entering each of the remaining STEM programmes is too small for robust statistical analysis (even though between 18 % and 15 % of

STEM students attend these programmes) and the programmes are too diverse to allow for grouping.

The extraction and analysis of the data was performed by UNI-C Statistics & Analysis, an agency of the Danish Ministry of Education.[2]

Results (i): Changing Patterns in Students' Educational Choice

The first analysis was done as a simple distribution of the students between different types of studies. The number of students who entered a STEM or non-STEM programme was calculated as a percentage of the total number of students completing upper-secondary school. Note that the number refer to the first time a student entered a STEM programme. A student who first entered a non-STEM programme and then a STEM programme will appear twice, whereas a student who first entered one and then another STEM programme appears only once. The results therefore show the share of students in each set of cohorts who at some time entered a STEM programme and the number who at some time entered a non-STEM programme (Table 14.2).

The decrease in the number of students from the third cohort (2005–2009) who have entered any higher-education programme at all reflects the large group of students who take one or two gap years between upper-secondary school and higher education. According to the Danish Ministry of Science, Innovation and Higher

Table 14.2 Students entering STEM or non-STEM higher education programmes after upper-secondary school. The percentage refers to the total number of students completing gymnasium. That is of the 61,142 men completing upper-secondary school 1995–1999, 24.5 % (16,433) entered a STEM programme

		1995–1999		2000–2004		2005–2009	
		Number	Pct.	Number	Pct.	Number	Pct.
STEM	Men	16,433	24.5 %	14,940	25.1 %	11,659	17.0 %
	Women	8,533	9.1 %	8,053	9.7 %	6,684	7.2 %
	Total	24,966	15.5 %	22,993	16.1 %	18,343	11.4 %
Non-STEM	Men	32,402	48.3 %	29,137	48.9 %	21,619	31.6 %
	Women	61,454	65.3 %	56,560	68.1 %	42,599	46.2 %
	Total	93,856	58.2 %	85,697	60.1 %	64,218	40.0 %
Total entrants	Men	48,835	72.7 %	44,077	73.9 %	33,278	48.6 %
	Women	69,987	74.3 %	64,613	77.8 %	49,283	53.4 %
	Total	118,822	73.7 %	108,690	76.2 %	82,561	51.4 %

[2] We are grateful for the help and assistance in the design, execution, and interpretation of the analysis we have received from Claus Jensen and Tine Høtbjerg Henriksen (UNI-C) and Professor Svend Kreiner (University of Copenhagen).

Education only 24 % of the students entering universities in 2012 came straight from upper-secondary school, and 20 % had taken three or more gap years before entering (Ministeriet for Forskning Innovation og Videregående Uddannelser 2012). A significant group of students in this cohort therefore have not yet entered the programme they wish to pursue, but may do so later.

Overall, there appears to be no, or a very small, increase in the share of students entering a STEM higher education programme (from 15.5 in the first to 16.1 % in the second cohort). The relative popularity of STEM hence is either unchanged, or slightly larger, compared to non-STEM programmes. Likewise, the gender balance within STEM does not appear to change. The decrease in the number of women in the 2005–2009 cohort may be caused by the tendency that STEM programmes with many female students recruit students who have more gap years as compared to, for instance, the technical programmes that are dominated by males.

We did not find any clear changes in the distribution of students between the different types of STEM programmes. About 40 % entered Science, about 20 % university engineering and 30 % professional engineering. In the third cohort, the proportion entering university engineering compared to professional engineering rose, but due to the gap year practice of Danish students the data concerning entering higher education in this cohort should be interpreted with caution.

While no discernible changes were found in the distribution of students between STEM and non-STEM programmes, there may be a slight change in the distribution of men and women between the different kinds of STEM programmes. Over the three cohorts, the share of female students tends to increase in Science and University engineering, while it seems to be decreasing in professional engineering, but the changes are small. The share of female students for the STEM programmes as a whole has been around 35 % through the whole period. For Science is has been between 38 and 42 %, for University engineering between 23 and 29 % (the largest increase), and for professional engineering between 21 and 25 %.

Finally, a multiple Cox analysis was conducted. This included interactional effects of gender on other variables. This analysis therefore examines whether some variables have a stronger effect for one gender than for the other, i.e. any indirect effect of gender differences. We found that the effects of the variables were stronger on women than on men. Concerning the 'risk' of a student choosing any STEM programme after gymnasium, this was increased for both genders with increasing educational level among parents, with higher GPA of the students, and for students who had attended 'htx', the technical branch of the gymnasium. However, if women, for instance, gained higher GPA than the reference person the risk of choosing a STEM course would increase more than would the same risk for a male student. In other words, female students appeared to be more affected by the achievements in gymnasium and by their parents' educational background than their male peers. For students with high GPA and with highly educated parents the differences between the chances of men and women entering a STEM programme became smaller. This was not the case for university engineering where only attending the technical strand htx affected the chances of women entering more than it did men. The gender differences related to choosing university engineering are apparently more consistent than those related to university science or professional engineering.

Risks of Non-completion

The focus in this section is on the factors that affect the risk of students not completing their STEM studies, and whether these factors affect men and women in different ways and to different extents.

According to previous research on student retention and non-completion there is no single, primary explanation for students not completing their programmes, and the findings concerning the effect of particular factors were inconclusive. Furthermore, there has been an increasing focus on issues of identity in relation to student transition and non-completion (see Chap. 13, Ulriksen et al. 2010; Holmegaard et al. 2014). In his model of student departure, Vincent Tinto includes factors related to experiences at the higher-education institution and factors related to what he calls 'pre-entry attributes' such as family background and prior schooling (Tinto 1993).

In this chapter we perform a quantitative analysis focusing on the pre-entry attributes. Even though this kind of analysis cannot uncover the reasons why certain factors appear to predict non-completion, it may draw attention to themes that from a macro level point of view appear to be significant.

Hence, this section seeks to answer the question of whether particular pre-entry attributes increase or decrease students' risk of leaving their STEM programme without completing it and if there have been changes in the influence of these factors over time. We particularly focus on factors related to gender, that is, if factors affect the study course of men and women differently.

Methods

The analysis in this section considers 61,531 students who completed upper-secondary school between 1995 and 2007 and who went on to study at a STEM programme in higher education. Only students who left the programmes within the first 3 years after entry were included. Furthermore, we have only considered entry and departure from the first STEM programme. This means that students who have entered one STEM programme, left it, and then entered another are only included in the analysis in relation to the first STEM programme. Had the student entered a non-STEM programme, left it, and entered a STEM programme the student would be included in the analysis in relation to this STEM programme. Since the focus is on the students who leave their STEM study programme. We have omitted the group of students who completed upper-secondary school in 2008 and 2009 from the analysis, because a substantial proportion of these students will not have entered higher education before 2009, the last year for which we have data.

The gender distribution in the three cohorts is shown in Table 14.3.

A Kaplan-Meier estimate was calculated for the survival time of the students. The Kaplan-Meier estimate can be plotted as a curve to show the survival

Table 14.3 The number of students in the three cohorts and the gender distribution. All STEM programmes are included. The percentages refer to the distribution between men and women within STEM as a whole for each cohort

	Year of completing upper-secondary school					
	1995–1999		2000–2004		2005–2007	
Gender	N	%	N	%	N	%
Men	16,433	65.8	14,940	65.0	8,608	63.4
Women	8,533	34.2	8,053	35.0	4,964	36.6
Total	24,966	100.0	22,993	100.0	13,572	100.0

probabilities for the population. Every time someone from the population leaves the programme (i.e. no longer 'survives') a new survival estimate for the next student is calculated indicated by a small step at the curve. The estimate shows the cumulative risk of non-completion over time, but also the progression in student leaving.

Secondly, Cox analyses were performed. This is a method to calculate the hazard ratio for the survival of persons who have received a particular treatment compared to a reference person who did not receive the treatment. In this case, the treatment is the students' possession of particular prognostic variables and the hazard ratio calculated thus expresses the relative risk of a student possessing one of those particular variables leaving the programme, compared to students without them. The reference person was defined as:

- Female
- Danish origin
- Parents with lower secondary school as the highest educational level (in other words, only compulsory schooling)
- Completed an stx (general upper-secondary school)
- Grade point average for the upper-secondary school exam below 6[3]
- No gap year
- No time spent at a previous higher-education programme
- Living in a city

The analysis therefore expresses the change in relative risk for surviving (the hazard ratio) for, for example, a man compared to the reference person being a woman, or a student who completed an htx exam compared to the reference person completing an stx. The Cox analysis is a regression analysis calculating the relationship between a dependant variable (non-completion) and an independent variable (e.g., gender). The regression analysis was also carried out as a multiple

[3] According to the scale for grading, called the '7-step scale', students can be graded −3; 0; 2; 4; 7; 10; 12. To pass an exam students should have a GPA of 2 or higher. The grade 7 is considered the average level. This scale replaced another scale in 2007, but the GPAs of students graduating before 2007 have been converted to the 7-step scale. Achieving a GPA of 6 is therefore a bit below average, but not much. Between 21 % and 26 % of the students in the three cohorts belong to the group of GPA below 6.

regression where more than one variable were taken into consideration in calculating the hazard ratio. Finally, a multiple model was constructed to estimate the effect of gender on other variables. Does the GPA, for instance, affect choice and behaviour of women more or less than that of men or are there no differences?

In order to achieve a sufficient number of students in the different groups, the analyses were conducted for STEM programmes at sub-group level, including university engineering (a programme aiming at Masters level), professional engineering (a bachelor programme), university science (mainly aiming at Masters level), veterinary sciences (Masters level), architecture (Masters level) and a few others. However, the population in most of these programmes was too small to achieve statistically significant results. We therefore only present results for the STEM field in total and for the subgroups university science, university engineering and professional engineering.

Results (ii): Retention and Non-completion

What Is the Risk of Non-completion?

The proportion of students who left the first STEM higher education programme they entered within 3 years of upper-secondary schooling has been decreasing, especially for the 2005–2007 cohort. Whereas 30 % left within the first 3 years for the group of students leaving upper-secondary school in 1995–1999 only 23 % of the 2005–2007 cohort did the same (Table 14.4). This result for the most recent cohort should be regarded with some caution because the group of students included in this number are mainly those who enter immediately after finishing upper-secondary school. However, even if the completion rate of the most recent cohort is an overestimate, the decreasing non-completion rate is in accordance with statistics from the Danish Ministry of Science, Innovation and Higher Education (Universitets- og Bygningsstyrelsen 2011).

Table 14.4 shows that while STEM programmes in total suffer from higher non-completion rates than non-STEM programmes, there are also variations within each field. Within STEM, university science programmes have higher non-completion rates than the engineering programmes. The most successful programmes in retaining students are university health and veterinary science programmes at universities.

Time from Access to Departure from the Programmes

Kaplan-Meier curves were estimated showing the cumulative percentage of students having left the programme in first 36 months. About half the students who departed

Table 14.4 The percentage of students who within 3 years of entering a university programme leave it without completing, for STEM and non-STEM programmes in total, and for selected programmes within STEM and non-STEM (percentage of the cohorts)

	First opt-out within 3 years (%)									
	STEM programmes						Non-STEM programmes			
	STEM (total)	University science	University engineering	Professional engineering	Architecture	Veterinary science	Non-STEM (total)	University health	University social sciences	University humanities
1995–1999	29.8	37.9	25.0	29.9	18.7	10.2	25.6	14.4	27.9	38.4
2000–2004	28.7	33.5	24.9	29.8	18.0	9.9	26.2	14.3	27.9	35.3
2005–2007	22.6	25.7	22.0	20.4	15.8	4.7	20.3	10.0	20.9	24.5

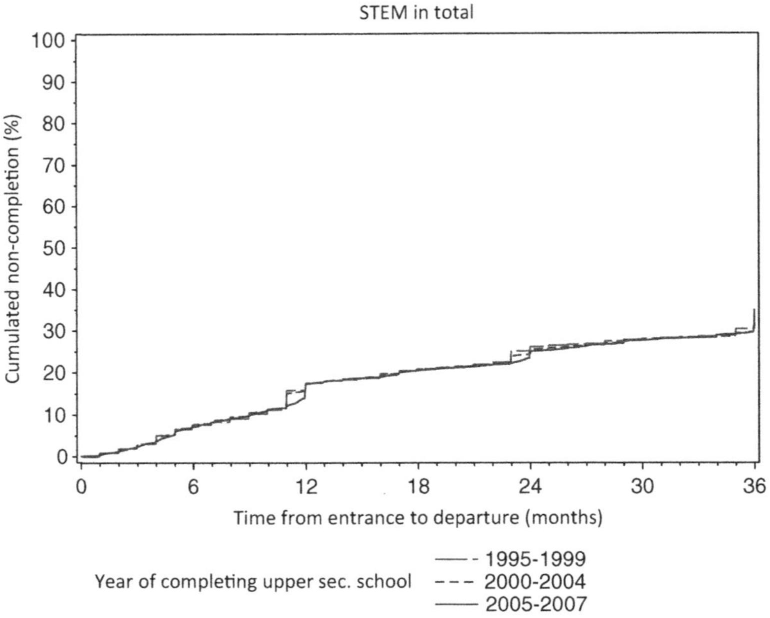

Fig. 14.1 Kaplan-Meier curve for STEM total showing the cumulated departure (%) and the time from entrance to departure (months) for the three cohorts

during the first 3 years did so within the first year. Another 25–30 % of the students not completing left during the second year and between 20 % and 30 % during the third. Student departure occurred throughout the entire 3-year period with the transition between years as noticeable peaks (Fig. 14.1).

These results suggest that even if the first year of study is an important period in the retention of students it is by no means the only time during the study where the programmes are in danger of losing their students. When comparing the Kaplan-Meier curves for the three cohorts we found no clear trend of change in the pattern concerning when students left the programmes.

Factors Influencing the Risk of Not Completing a STEM Programme

The results of the simple and the multiple Cox analyses are presented in Table 14.5. The Cox analyses show that the grade-point average (GPA) from upper-secondary school and the parents' educational background were most strongly related to an increased retention on STEM programmes. Higher GPA reduced the risk of non-completion. This was the case for STEM in total and for the three individual groups of studies (science, university engineering and professional engineering).

Table 14.5 Selected results of the simple and multiple Cox analysis

Estimate of relative risk of non-completion						
	Simple			Multiple		
Gender	1995–1999	2000–2004	2005–2009	1995–1999	2000–2004	2005–2009
STEM	ns	0.902	0.843	0.898	0.804	0.81
Science	1.089	ns	ns	ns	0.846	0.81
University engineering	ns	ns	ns	0.866	0.858	ns
Professional engineering	ns	ns	ns	0.886	0.884	ns
Parental education – masters level (LVU)						
STEM	0.584	0.524	0.625	0.741	0.705	ns
Science	0.555	0.485	0.668	0.752	0.714	ns
University engineering	0.591	0.589	0.677	0.763	0.761	ns
Professional engineering	0.802	0.712	ns	ns	0.784	ns
GPA – upper-secondary school higher than 8						
STEM	0.45	0.416	0.389	0.421	0.406	0.543
Science	0.448	0.379	0.399	0.429	0.388	0.55
University engineering	0.394	0.4	0.36	0.404	0.376	ns
Professional engineering	0.514	0.468	0.459	0.481	0.437	0.479
One gap year						
STEM	0.883	0.888	0.899	0.909	0.881	0.810
Science	ns	0.812	ns	ns	0.860	ns
University engineering	ns	ns	ns	ns	0.834	0.804
Professional engineering	ns	ns	ns	0.865	0.865	ns

Selected variables. ns = no significant difference in hazard ratio compared to reference person. More results can be found in Table 14.9. Numbers refer to the relative risk of not completing a STEM education for a student possessing particular attributes compared to the reference person. Hazard ratios below 1 signify that the a student possessing the particular attribute indicated in the left column is MORE likely than the reference person to complete the education; hazard ratios above 1 mean that the student is LESS likely than the reference person to complete

The effect of the parents' educational background was also clear and substantial, but mainly if the parents had completed a higher education. For some groups, particularly in engineering, there was no statistically significant difference in hazard ratio between students whose parents had no education expect for lower-secondary school (the reference person) and students whose parents had vocational training or (for civil engineering) shorter period higher education of up to 2 years duration. In other words, the decisive feature appears to be whether the parents have attended a university or professional bachelors programme or not.

A second variable with some effect on the hazard ratio was whether there was a time gap between the completion of upper-secondary school and entering the STEM programme taken as a whole. In most cases, having a break reduced the risk of leaving the programme, and the longer the break, the stronger the effect. For some cohorts the effect was only significant for some time gaps and not for others (e.g., for university engineering students in the 2000–2004 cohort). For the 2005–2007 cohort the effect was less pronounced, but still present for the group of STEM students as a whole. However, we found a converse effect of having a 2 or 5 year gap for university engineering students in the 1995–1999 cohort. For those students that particular gap increased the risk of leaving.

The Effects of Gender on the Hazard Ratio

Concerning gender, the simple Cox analysis found an increased risk of non-completion for women in STEM as a whole for the cohorts 2000–2004 (HR = 0.90) and 2005–2007 (HR = 0.84). There was no significant difference for the 1995–1999 cohort. For science programmes we found that for the 1995–1999 cohort men showed an increased risk of non-completion, with no significant gender-related differences for the two following cohorts. In the case of both types of engineering programmes there were no significant differences in the hazard ratios for men and women in any of the three cohorts. Hence, the simple Cox analysis suggests that the there is no significant difference in the risk of non-completion related to gender.

This pattern changed in the multiple Cox analysis. For the first cohort (1995–1999) men had a significantly lower risk of leaving the programmes than women, except in science where there was no significant difference. In the second cohort (2000–2004) there was a significant difference for both STEM and the three individual programmes. In the most recent cohort (2005–2007) the effect of gender was significant for STEM as a whole and for science, but not for the two kinds of engineering programmes.

It therefore appears as if gender indeed has a significant effect in itself since the difference was more pronounced in the multiple analyses where other variables were taken into account. These other components (e.g., the average GPA of men and women) could obscure the gender effect in the simple analyses, but the multiple analyses indicate that there are significant differences in the risk of non-completion for men and women. However, the hazard ratios of between 0.8 and 0.9 are smaller than those related to, for example, parents' educational level. Gender, it appears, is one factor affecting persistence, but not the most significant one.

In the analysis of whether particular variables affected men and women differently, we found very limited differences. Attendance at the technical gymnasium (htx) did affect the hazard ratio of men and women differently. Male students who had attended htx reduced their risk of non-completion more than did women.

However, the effect of htx in itself varied between programmes and across the different cohorts. Furthermore, the varied effect of htx over time may be due to changes in the structure and the position of this particular kind of gymnasium in the 15 years in question. These results are therefore less interesting because it is difficult to interpret any clear direction in the results.

The second variable that affected men and women differently was the GPA for STEM as a whole. This was only significant for the 2000–2004 cohort and not in the analyses at the level of university science or the engineering programmes. It was found that the risk of not completing was reduced more for female than for male students who had a GPA of between 6 and 7 compared to the reference person who had a GPA of 6 or less. However, for students with a GPA between 7 and 8, and of 8 or higher, it had a stronger impact on the male than on female students. In other words, there does not seem to be a consistent difference in the impact of GPA on men and women.

Overall, there do not appear to be any noteworthy significant differences as to how variables affect the non-completion risk of men and women.

Results (iii): Departure Trajectories

The final analysis carried out on the data was to see where students went after having left their higher-education study programme. Tinto (1993) distinguishes between programme departure and institutional departure. Institutional departure refers to students who leave college without entering another programme, whereas programme departure refers to students who leave one programme or college to enter another. From a societal perspective the institutional departure therefore calls for more concern than does the programme departure, because the students who merely change programme still complete a degree.

In an IRIS context, we have a more specific interest in whether students who leave a STEM programme enter another STEM programme or not. We therefore talk of STEM departures as opposed to changes between STEM programmes (programme departures) or students not entering another programme (institutional departure). That is the focus of this section.

The first part of the analysis includes descriptive data where we have tracked the path taken by students who opted out of a programme. Kaplan-Meier curves were made estimating the time from departure to entrance to a new programme within the first 3 years after leaving the STEM programme. Finally, both simple and multiple Cox analyses were carried out calculating the hazard ratio for entering a STEM programme after leaving one. In order to have sufficient number of students to be able to make the analyses, the Cox analyses consider transition to a STEM programme (including veterinary sciences) or a non-STEM programme.

Where Do Students Go After Leaving STEM?

The results concern in total 18,209 students who in the period 1995–2007 within the first 3 years of study left the first STEM programme they enrolled on. Tables 14.6 and 14.7 show where the students went after leaving their first STEM programme (Table 14.6) or non-STEM programme (Table 14.7). Table 14.6 shows that approximately one fifth of the STEM students experienced an institutional departure. The number of institutional departures is increasing slightly, but the students in the 2005–2007 cohort have had a shorter time to enter another programme, and the change for this cohort should therefore be interpreted with caution. Another group entered either vocational training or a 2-year higher education programme (KVU). This was the case for one out of six in 1995–1999, but this group has diminished over the three cohorts, down to one in eight.

Table 14.6 Transition pattern of STEM students after leaving their first STEM programme

	Year of completing upper-secondary school					
	1995–1999		2000–2004		2005–2007	
Transition to	N	%	N	%	N	%
None	1,615	19.8	1,457	20.7	668	22.3
Science, math	1,055	12.9	746	10.6	379	12.6
Veterinary sc	260	3.2	162	2.3	78	2.6
Tech, engineering	1,279	15.7	1,239	17.6	516	17.2
Non-STEM	2,628	32.2	2,500	35.5	975	32.5
Vocational training or 2-year higher education	1,317	16.1	829	11.8	372	12.4
PhD	15	0.2	109	1.5	10	0.3
Total	8,169	100.0	7,042	100.0	2,998	100.0

Table 14.7 Transition pattern of non-STEM students after leaving their non-STEM programmes

	Year of completing upper-secondary school					
	1995–1999		2000–2004		2005–2007	
Transition to	N	%	N	%	N	%
None	7,984	29.5	7,245	29.6	3,540	35.2
Science, math	721	2.7	655	2.7	268	2.7
Veterinary sc	102	0.4	88	0.4	58	0.6
Tech, engineering	597	2.2	576	2.4	285	2.8
Non-STEM	15,117	55.8	13,472	55.0	4,922	48.9
Vocational training or 2-year higher education	2,528	9.3	2,397	9.8	985	9.8
PhD	27	0.1	63	0.3	0	0
Total	27,076	100.0	24,496	100.0	10,058	100.0

Furthermore, one third of the students who left a STEM higher education programme re-entered a non-STEM higher education bachelors or Masters programme – virtually the same share changed from one STEM programme to another. In 1995–1999, 31.8 % re-entered a STEM programme while 32.2 % went to a non-STEM higher-education programme. In 2000–2004, it was 30.5 % to STEM and 35.5 % for non-STEM programmes. In the 2005–2007 cohort, the STEM and non-STEM re-entry was the same: 32.5 %. In summary, approximately one third of the STEM students leaving their programme were institutional departures, but two thirds were STEM departures.

The trajectories of students leaving non-STEM programmes display a different picture. About half the students (a little fewer in the 2005–2007 cohort) continue on to another non-STEM bachelors or Masters programme. Between 40 and 45 % of the students move on to something else – the largest group of about 30 % leaves the educational system. Finally, only 5 or 6 % enter a STEM programme. Most of these come from university bachelor programmes in social sciences or humanities, but the group of non-STEM programmes with the highest percentage of leavers moving to STEM is the university programmes in health (medicine, pharmacy, etc.).

From the point of view of there being too few graduates within STEM these are disturbing results. They reveal a net loss of students from STEM to non-STEM programmes (Fig. 14.2). Even though the non-STEM label in this context refers to a very broad group of programmes including health, it still means that when STEM programmes lose students two-third of them are STEM departures. Therefore it appears to be highly important for the STEM programmes to put even more focus on holding on to the students who decided to enter a STEM higher-education programme in the first place.

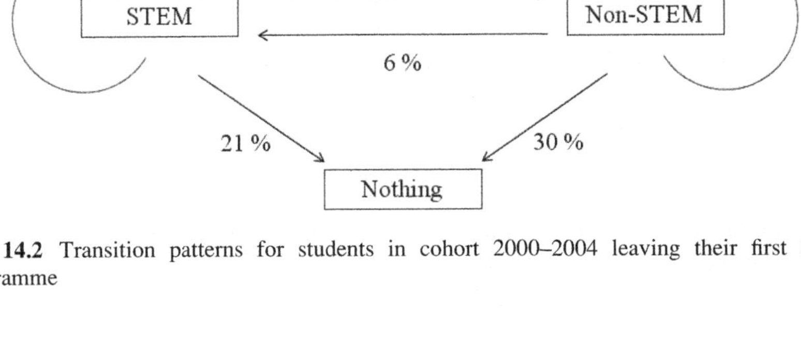

Fig. 14.2 Transition patterns for students in cohort 2000–2004 leaving their first STEM programme

Time Gap Between Leaving One Programme and Entering Another

Kaplan-Meier curves for the time passing between leaving one STEM programme and entering another STEM programme show that for STEM as a whole just over half of the students begin at another STEM programme immediately after leaving (that is, within 1 or 2 months). The remaining half of those re-entering a STEM programme within 3 years enrol at another programme at a constant rate. When looking at the individual programmes this picture changes a little. It appears that the change for the two larger groups (science and university engineering) has a steady flow to other programmes, but after 2 years this fades and the curve becomes flatter with fewer entering another programme. For professional engineering this flattening appears after 1 year. Overall, it seems that most of those students who enter another STEM programme after leaving the first do so within the first 2 years.

Factors Affecting the Choice of STEM After Leaving a STEM Programme

We now examine the multiple Cox analyses concerning which factors affect the inclination of the students who leave the first STEM programme they enter to move on to a STEM programme or a non-STEM programme. Once again we find that GPA from upper-secondary school consistently has a significant effect, but the effect is present both for entering STEM and non-STEM. This means that the GPA from upper-secondary school affects the inclination to enter any programme after leaving, irrespective of the second programme. However, students with a GPA above 8 were notably more likely to re-enter a non-STEM programme compared with re-entering a STEM programme (hazard ratio 2.05 for non-STEM compared to 1.65 for STEM). This could be due to the Danish system of admittance where the selection of students is based on GPA. Most of the STEM programmes have fewer applicants than the number they can admit meaning that there is no GPA-based selection on these programmes. Conversely, many programmes within health and the social sciences have more applicants than they can admit. Consequently, students with a GPA below a given threshold (this varies every year according to the number of students applying and the GPA of the last student admitted that particular year) are rejected. The higher hazard ratio of students with a GPA above 8 may therefore not reflect the fact that students with higher GPA are more inclined to apply for non-STEM programmes than students with lower GPA, but simply that the latter are not accepted at the non-STEM programmes and therefore 'choose' to re-enter a STEM programme.

As for parents' highest education this has an impact on the hazard ratio for both STEM and non-STEM programmes in the 1995–1999 cohort. In the 2000–2004 cohort parental educational level only has a significant effect for choosing STEM

Table 14.8 The relative risk of men as compared to that of women for choosing a STEM programme after upper-secondary school and after leaving a STEM programme. The last cohort covers different time span (5 vs 3 years)

	Hazard ratio for men	
	Choosing STEM after upper-secondary school	Choosing STEM after leaving a STEM programme
1995–1999	2.66	1.09
2000–2004	2.51	1.30
2005–2009/2007	2.21	1.19

programmes, but not for non-STEM programmes. Students whose parents have completed a higher education programme are more likely to re-enter a STEM programme than students with parents without higher education.

Gender is significant for all cohorts. Being male significantly increases the risk of choosing STEM and decreases the risk of choosing non-STEM. However, the effect is stronger in relation to non-STEM programmes. That is, the difference in the relative risk of men and women is larger for non-STEM programmes than for STEM programmes. This means, that even among students who initially entered a STEM programme we found gender differences in the inclination to re-enter a STEM programme. Compared to the differences between the risk of men and women for choosing STEM in the first place, the difference is substantially smaller, but still significant (Table 14.8).

Discussion

This chapter addressed three questions. All of these were scrutinised in a Danish context with Danish data.

Recruitment

The first question concerned whether, alongside the substantial focus in research, policy documents, and recruitment initiatives on gender imbalances in STEM programmes, there were any changes in the patterns of choice of men and women when deciding which higher-education programme to enter. This does not appear to be the case. The share of students entering STEM programmes appears to have remained stable from 1995 to 2009, as has the gender balance. Apparently, the focus on increasing the intake of students in general and of female students in particular has had little effect.

Retention

The second question focussed on retention and non-completion. Firstly, we found a substantial increase in retention within STEM in total and within the different types of programmes. This is encouraging. Still, one in four students who enters a university science programme leaves it within 3 years without completion. For STEM as a whole this is more than one in five. However, these high non-completion rates are not unique to STEM and may be a generic challenge for the higher-education system as a whole rather than specific to individual programmes.

On the other hand, the differences between the programmes suggest that there are reasons to delve further into the decisions of STEM students leaving their programme before completion. The present analysis allows only for a limited scrutiny of the reasons for the students' non-completion, and only for those factors that relate to what is labelled 'pre-entry qualifications' in Tinto's model. The analysis showed that prior school attainment had the strongest effect on persistence with increasing GPA related to increasing persistence. This could explain some of the variations between the non-completion rates within the STEM programmes.

The highest persistence rates were found in the veterinary sciences and architecture. These programmes are highly selective requiring students to have obtained an overall GPA of 9 or more to gain access. With GPA as the most important predictors of persistence it is not surprising that these competitive programmes succeed in holding on to more students. On the other hand, the variance in the proportion of students leaving these programmes (ranging from 5 to 16 %) suggests that there are other elements affecting retention than merely the students' prior academic achievements.

The effect of the parental educational level that we found has been firmly established in a number of other studies (cf. Pascarella and Terenzini 2005). However, we also found that the effect of parental education was not in all cases significant below bachelors level. Furthermore, we noted that the effect size was similar for the first and the second cohorts, but for the third cohort the effect diminished or disappeared. It is unlikely that these results can be interpreted as if equal opportunities now exist for all students, regardless of social background. On the other hand, the results may indicate a development that warrants further investigation.

Finally, there is some evidence suggesting that the students' social background has a larger impact on the non-completion rates in the sciences than in engineering. Particularly, it seems that social background has less importance for students in professional engineering than in the other fields. It is not possible, using the data in this study, to explain why.

Turning to the effect of gender on persistence, we found that while there were few significant effects of gender on the risk of non-completion in the simple Cox analysis, the picture changed in the multiple Cox analysis where a model has been constructed including more than one variable. In the multiple Cox analysis, we found increased risks of non-completion for women in all cohorts, where there were none in the simple Cox analysis. This suggests that even if it appears as if women are not more at risk for dropping out it turns out that they are when other variables are taken into account. This could indicate that some of the conditions students are meeting within the programmes are more hazardous to women than to men (Tables 14.5 and 14.9).

Table 14.9 Selected results from the multiple Cox analysis

	Cohorts – year of completing upper-secondary school											
	1995–1999				2000–2004				2005–2007			
	STEM	Science	Un. eng.	Pr. eng	STEM	Science	Un. eng.	Pr. eng	STEM	Science	Un. eng.	Pr. eng
Male	0.898**	n.s.	0.866*	0.886*	0.804***	0.846***	0.858*	0.884*	0.810***	0.810**	n.s.	n.s.
Parents 2 years HE	0.841*	n.s.	n.s.	n.s.	n.s.	n.s.	n.s.	n.s.	n.s.	n.s.	n.s.	n.s.
Parents 3 years HE	0.808***	0.875*	0.779*	n.s.	0.768***	0.785**	n.s.	0.795*	n.s.	n.s.	n.s.	n.s.
Parents 5 years HE	0.741***	0.752**	0.763*	n.s.	0.705***	0.714***	0.761*	0.784*	n.s.	n.s.	n.s.	n.s.
GPA 6–7	0.680***	0.736***	n.s.	0.623***	0.665***	0.694***	0.792*	0.625***	0.768**	0.758**	n.s.	0.567***
GPA 7–8	0.565***	0.655***	0.644***	0.532***	0.567***	0.561***	0.614***	0.570***	0.697***	0.747**	n.s.	0.531**
GPA 8+	0.421***	0.429***	0.404***	0.481***	0.406***	0.388***	0.376***	0.437***	0.543***	0.550***	n.s.	0.479**
1 gap year	0.909*	n.s.	n.s.	0.865*	0.881**	0.860**	0.834*	0.865*	0.810***	n.s.	0.804*	n.s.
2 gap years	0.913*	n.s.	1.212*	0.806**	0.776***	0.767***	n.s.	0.831*	0.789**	n.s.	n.s.	n.s.
3 gap years	0.755***	0.768**	n.s.	0.777**	0.747***	0.704***	n.s.	0.724**	0.600**	n.s.	n.s.	0.543*
4 gap years	0.678**	0.706**	n.s.	0.692**	0.673***	0.682**	0.649*	0.560***	–	–	–	–
5 gap years	0.561***	0.602***	1.715**	0.563***	0.575***	0.610**	n.s.	0.620**	–	–	–	–
1 year prior HE	n.s.	n.s.	n.s.	n.s.	n.s.	0.746**	n.s.	n.s.	n.s.	n.s.	n.s.	n.s.
2 years prior HE	1.241**	n.s.	0.485**	n.s.	1.287**	n.s.	n.s.	n.s.	n.s.	n.s.	n.s.	n.s.
3 years prior HE	n.s.	1.434**	0.335***	1.396*	n.s.	n.s.	n.s.	n.s.	–	–	–	–
4 years prior HE	n.s.	n.s.	0.380**	1.897**	n.s.	n.s.	n.s.	n.s.	–	–	–	–
5 years prior HE	n.s.	n.s.	n.s.	2.520***	n.s.	n.s.	n.s.	n.s.	–	–	–	–

Un.Eng. university engineering. *pr.eng*. professional engineering. *Parents 2y HE* parents highest education 2 years HE cycle (=KVU). *Parents 3 years HE* parents high

*p < 0.05; **p < 0.01; ***p < 0.0001

Overall, we found effects of factors related to pre-entry qualifications. In an IRIS context, the increased risk of women of leaving their STEM programme without completing calls for concern, not least because it appears as if there are features related to gender itself that are of importance. It would require a qualitative approach to establish which features these could be. Chapter 19 on gender-biased programmes provides an example of such an approach.

The Kaplan-Meier analysis shows that even though half of the students who left their programmes did so within the first 12 months, another half left later. Attempts to reduce student non-completion therefore need to look further than the first year of study. Some of the students who left the programme beyond the first year may not have passed enough courses to make up 1 year of study as measured by the number of ECTS points passed, but they still survived the first year and moved on to the second and third year. We may therefore fail to approach students with difficulties in completing their university studies if we confine our efforts to the students first year of study. This draws attention to the need to address first and second year experiences (as pointed out in the literature review, Chap. 13).

Departure Trajectories

The results concerning departure trajectories are truly worrying because they reveal a substantial loss of STEM students. While students from non-STEM programmes tend to remain within non-STEM programmes (albeit, in this analysis a very mixed group of programmes) only one third of the STEM students opting out decide to re-enter a STEM programme. This suggests that the endeavour to make students who enter a STEM programme stay is highly important if the goal is to increase the number of graduates within STEM.

The Kaplan-Meier analyses indicated that students do not leave one programme to enter another immediately after, but, on the other hand, most of the students enter a new programme within the first 2 years after leaving the first STEM programme. The time gap between leaving and entering may be an indication that the choice of leaving is not necessarily a decision *for* something different, but just as much a decision *away* from the programme first entered.

The third – and also disturbing – point relates to the gender differences in the students' choice of a new study after having left a STEM programme. When students make their first choice after upper-secondary school there is a clear gender difference in the patterns of choice with men being far more inclined to choose a STEM programme than women. However, it is surprising that there is also a gender difference among the students who chose a STEM programme as their first course of study, but then left the programme and entered another. Even in this group of original STEM choosers men are more inclined to enter another STEM programme than women. The gender difference is smaller by the second choice than by the first, but there is still a significant difference. This means that even after the students had decided to pursue a STEM study path at the first point of decision, the gender difference persisted by the second point of decision. The gender imbalance hence increases in two ways when students leave the first STEM programme they entered: firstly, because women have a higher risk of not completing; secondly, because more men than women re-enter a STEM programme.

Conclusion

The results from the analysis of Danish national data reported in this chapter show that the patterns of choice change slowly and very little, both in relation to STEM as a whole and to increasing the number of women entering STEM. In spite of considerable attention over more than a decade, the numbers of STEM applicants have changed very little. On the other hand, the results show that it has been possible to improve the retention rates over time, even though the number of students not completing the programme they enter is still alarmingly high. Furthermore, we found that the pre-entry factors already known to affect retention, namely prior school attainment and the parents' educational background still have a strong impact on the students' risk of non-completion.

However, what we consider the most worrying conclusion to be drawn is the large STEM departure in the departure trajectories combined with an almost non-existing influx of students departing from non-STEM programmes. This means that efforts to hold on to students who initially enter a STEM higher education programme are very important. Such efforts are, perhaps, even more important than recruitment initiatives. Furthermore, the gender imbalance is increased in the process of leaving and re-entering since women to a larger extent seek to move away from the STEM programmes as compared to men.

References

Holmegaard, H., Madsen, L., & Ulriksen, L. (2014). A journey of negotiation and belonging: understanding students' transitions to science and engineering in higher education. *Cultural Studies of Science Education, 9*(3), 755–786. doi: 10.1007/s11422-013-9542-3.

Ministeriet for Forskning Innovation og Videregående Uddannelser. (2012). *Notat 3 om alder på adgangsgivende eksamen*. Copenhagen. Retrieved from http://fivu.dk/uddannelse-og-institutioner/statistik-og-analyser/sogning-og-optag-pa-videregaende-uddannelser

Pascarella, E. T., & Terenzini, P. T. (2005). *How college affects students: A third decade of research* (2nd ed., Vol. 2). San Francisco: Jossey-Bass.

Tinto, V. (1993). *Leaving college. Rethinking the causes and cures of student attrition* (2nd ed.). Chicago/London: The University of Chicago Press.

Ulriksen, L., Madsen, L. M., & Holmegaard, H. T. (2010). What do we know about explanations for drop out/opt out among young people from STM higher education programmes? *Studies in Science Education, 46*(2), 209–244. doi:10.1080/03057267.2010.504549.

Universitets- og Bygningsstyrelsen. (2011, November 18). *Notat om universiteterns bachelor- og kandidatuddannelser*. Retrieved May 9, 2012, from http://www.ubst.dk/uddannelse-og-forskning/statistik/bachelor-og-kandidatstuderende-1

Chapter 15
The First-Year Experience: Students' Encounter with Science and Engineering Programmes

Lars Ulriksen, Lene Møller Madsen, and Henriette Tolstrup Holmegaard

Background: Understanding Students' First-Year Experiences

Each year, higher education institutions succeed in attracting students to STEM programmes. However, recruitment is only the first step towards the graduation of a STEM student. Making the students stay is equally important. Unfortunately, about one third of the students entering tertiary education do not complete the programme (OECD 2010), and this is not least the case for students attending STEM courses (OECD 2008). Analyses of Danish data (Chap. 14 in this volume) found that when students opt out of a STEM higher education programme, only about one third of them enter another STEM programme. In fact, more students leave STEM to go to non-STEM programmes than do re-enter a programme within STEM. Retention, therefore, is a key concern for increasing the number of STEM graduates.

A principal finding in the review of research on retention and non-completion (Chap. 13 and Ulriksen et al. 2010) was that retention should be considered within a broader perspective of the students' learning experiences during first year rather than as an isolated problem. Further, the review pointed to the importance of applying an identity perspective when studying student departure from STEM, rather than merely regarding the students who drop out as less capable or ill-prepared. Even though prior schooling experiences and performance are related to student persistence (cf. Chap. 14), they cannot sufficiently explain the non-completion patterns. Likewise, it is not consistently the less able students who leave their STEM studies prior to graduation. The leavers are in many respects quite similar to the students who stay on the programmes (Seymour and Hewitt 1997).

L. Ulriksen (✉) • L.M. Madsen • H.T. Holmegaard
Department of Science Education, University of Copenhagen, Øster Voldgade 3, DK-1350 Copenhagen C, Denmark
e-mail: Ulriksen@ind.ku.dk

E.K. Henriksen et al. (eds.), *Understanding Student Participation and Choice in Science and Technology Education*, DOI 10.1007/978-94-007-7793-4_15

That different elements affect the students' decisions to stay or leave are reflected in the widely used model of student departure developed by Vincent Tinto (1975, 1993) (cf. Chap. 13 in this volume and Ulriksen et al. 2010). Tinto's model emphasised that student leaving is occurring over time rather than being a discrete event. Further, he included different factors as influential on whether students were leaving or nor, including pre-entry qualifications and family background. A core element in Tinto's model was 'the concept of integration and the patterns of interaction between the student and other members of the institution especially during the first critical year of college and the stages of transition that marked that year' (Tinto 2006–2007, p. 3). This has also been labelled the social and academic integration of the students. The integration relates to the students' experiences with the institution, their interaction with fellow students and with faculty and other staff in formal and informal settings.

The *academic integration* refers to two dimensions. The first is the students' experiences of congruence between their own abilities and skills and the demands of the programme (for example, whether they pass the exams or what grades they get). This can be considered the institutions evaluation of the student, as Tinto put it in an early version of the model (Tinto 1975). The second is what Tinto in the 1975 paper labels 'intellectual development' which can be considered the students' evaluation of the academic system (Tinto 1975, p. 104). It relates to the students' experiences of congruence between their interests and academic orientation and what they meet at the course. Academic integration, in other words, concerns the students' sense of belonging in the academic environment of their study programme in terms of feeling that they can meet the requirements and that they find it interesting and relevant. This integration process both occurs in formal settings of the different teaching and learning activities and in informal contexts outside class where students meet and interact with the staff.

Social integration refers to the process of students becoming part of a social community of fellow students at the programme and gaining a sense of belonging. This also relates to both an informal and a formal context. Examples of formal settings are student societies and unions while the informal parts of the system are when students are simply hanging out together at campus, going to cafés together etc.

The systems of the academic and the social integration are intrinsically interwoven (Tinto 1993). When, for instance, students sit together at university working on exercises for the chemistry class the following day, they are involved in an activity that relates to the academic integration in their doing the course work and their experiences of that. They are also involved in an incident of social integration because the study group provides a sense of being a part of a studying community and because the group may talk about other stuff than chemistry and even may continue going to a café after having completed the assignments. Tinto (2006–2007) made the point that in colleges where students do not live at campus most of the informal integration has to take place in relation to the teaching and in the classroom, because the conditions for out of class interaction are different than at residential universities. As most universities in many European countries, including Denmark, are non-residential, this is indeed an important point.

The model of Tinto has been criticised for being insensitive to social and cultural differences. It has been claimed that the model required students to commit cultural suicide (Tierney 1999) in order to assimilate to the dominant academic culture. Although it is not necessarily the consequence of the model and the process of integration that students need to conform to one culture, there has been articulated a need to develop the model to achieve a more nuanced understanding of the complexity of the process of integration (Braxton et al. 2000; Tierney 1999; Tinto 2006–2007).

However, the idea of integration as a pivotal component in student persistence appears as a viable way to understand students' experiences when entering university. Integration can be considered a process of socialisation. The choice of study is, to a large extent, linked to the students' thoughts about who they wish to become (Illeris et al. 2002; Schreiner and Sjøberg 2007), and therefore the congruence between these ideas and the students' experience of belonging or not is highly important (Holmegaard et al. 2014b; Bøe et al. 2011). Consequently, the academic and social integration are processes where students' prior knowledge, experiences, expectations, and inclinations towards the study meet with the culture, traditions, and pedagogical forms of the programme. Ulriksen (2009) argues that a study programme holds "an implied student". This means that study programmes presuppose that students attending the programme possess a particular study practice, attitude, interest, and behaviour. The structure of the programme, the sequence of the courses and modules, the teaching and learning activities, etc. all presuppose particular traits, attitudes, or competences. The students need (consciously or not) to detect and adapt to these presuppositions in order for the teaching and learning to succeed. For instance, a programme can imply that students have particular interests in the field, whereas if they do not, the students may fail to see the point of the course. The implied student is conveyed through the structure, the curriculum, etc. A programme may hold more than one implied student, and these may even be incompatible in some extent, but the number is limited.

In the process of socialisation, students may assimilate completely, but students may also be forming subcultures where they, for instance, seek to balance the culture of academia they are entering with the culture they bring with them, rather than abandoning their cultural background (Hurtado and Carter 1997). Likewise, students may engage more in some parts of the programme or aspects of the discipline than in others. In some academic disciplines there are differing ideas about what content is the more relevant (Becher and Trowler 2001) and in that case students may orient themselves towards one part of the discipline rather than the other, for instance, a biology student prioritising macro biology before micro biology or a physics student engaging in theoretical physics rather than experimental physics.

Following narrative psychology (cf. Chap. 3 in this volume), the process of socialisation and of balancing involves the students in constructing and reconstructing narratives concerning their previous experiences, their intentions, anticipations, and perspectives, their experiences at the programme, and so forth. The students construct narratives to make meaning and a sense of coherence to themselves and their surroundings (Bruner 2004; Polkinghorne 1988). What narratives the students will be able to construct are framed and confined by the social and cultural environment (what may be recognised by the surroundings as a sensible and legitimate narrative) and by the students' cultural and social background and history (what repertoire does the student have for constructing a narrative, both concerning knowledge, experience, and story "templates").

In this chapter we will analyse the first-year experiences of STEM students at university with a particular focus on the integration and socialisation process of the students as it happens in the encounter with the STEM study programmes they have entered. Our focus is to understand how students engage in the studies, and how they make sense of their experiences compared to the expectations they had. The objective of the analysis, therefore, is to *expand and refine our understanding of how students cope with their first-year experience and what we may learn from this concerning student completion* at STEM higher-education programmes.

Methods

The empirical basis consists of interviews with 20 first-year students at STEM programmes. The students were selected from a sample of 134 students finishing upper-secondary school in the summer of 2009. Of the 134 students, 38 were interviewed 2 months before the completion of upper-secondary school. Based on the students' study plans, 20 students were selected for interviews after having entered first year at university. Three of these were students who in spite of expressing a strong interest in science still opted for a course within the humanities. Eventually, two of these opted out of the humanities to enter a STEM programme. The remaining 17 entered a STEM programme (including one entering veterinary medicine) and 13 of these were interviewed more than once during their first year, some up to five times. Four students did not show up for the second interview. Six of the students were interviewed after having entered the second year of their programme. Some of the 20 selected entered university straight after upper-secondary school while others had a gap year. Eight were interviewed during their gap year (see Holmegaard et al. (2014a) for details on the method).

The interviews were semi-structured (Kvale 1996) and conducted using a narrative approach (Andrews et al. 2008; Bruner 1990; Hollway and Jefferson 2000). The interviewee was encouraged to tell about what it had been like to begin studying at the particular programme. The interviewer's questions mainly aimed at inviting the interviewee to elaborate or expand the narrative.

In the analysis, the interviews were coded using the Atlas TI software. Rather than generating the themes from the text, the coding used codes that were constructed on the basis of Tinto's concepts of academic and social integration and the concept of the implied student. This is what Kvale (1996) calls a theoretical understanding of the interview.

Results

We will begin the presentation of the results by telling the brief versions of the transition of two female students in the sample. The purpose is to offer two more detailed accounts of entering university before we present the results in a more thematic structure.

Emily and Elisabeth[1]

In upper-secondary school Emily became convinced that chemical engineering was the right choice for her and she had visited the technical university to make sure it was. Her experience with the programme, however, was frustrating. She found the quality of the teaching inadequate, she experienced the workload overwhelming and the content difficult. The first semester, she told, was supposed to be the hardest and the saying was that 'everybody fails this course'. At the same time, the content of the teaching did not reflect her interests. In the second interview a couple of months after she entered, she explained that what she liked about science was that 'you describe reality, you can calculate on reality and find out how things work and what we can do to make things better'. So far, she had not experienced much of that but 'I think you have to begin at a basic level, so I think it will come'. During the first months, the teaching was predominantly theoretical with very limited relation to reality or to their later profession. Emily told she missed of seeing things (for instance, in experiments) rather than dealing with them in a table.

In the interviews during her first 6 months at university, Emily tried to find explanations for the difficulties she experienced. She both questioned the quality of what the university offered and of her own study efforts. Hence, the interviews reflected a continuous negotiation of her interests, her sense of her own skills and efforts, and her experience of the teaching and learning environment of the programme. When entering the second semester, she decided to leave. She had been failing exams after the first semester, but this was not the sole reason for her decision. After having opted out, she explained in an interview:

> I felt myself being stupid in all the courses and I couldn't figure things out. I was not motivated to study and it became too tough and I did not feel that I could keep my self-confidence and self-respect when I got the feeling of being stupid every day. Then I thought I needed to make a plan about what to do.

She felt that staying at the programme might undermine her sense of self and be detrimental to her identity. Emily's encounter with university combined a surprise by the difficulty of the subject with an experience that the teaching offered little help for understanding. Furthermore, the content of the teaching had few links to her initial interests and she did not succeed in establishing a social environment at the programme that could support her, just as this was not facilitated by the

[1] All student names are pseudonyms.

programme. Her study group was working poorly both in terms of work discipline and the way the group members talked to each other, and eventually it broke up. As a result, Emily lost contact with two female students she had been seeing quite a lot during the first months. This troubled her.

Emily's integration was neither successful at the academic or at the social level. The social integration suffered from a combination of a poorly functioning study group with no support from the programme and of her feeling that the informal activities were difficult to attend on top of the long hours the students already spent at university. The academic integration was under strain for several reasons: the content, the pace, the mode and quality of the teaching, the academic requirements. The unsuccessful integrations eventually made her leave the programme and the institution, and later enter another non-STEM programme.

A quite different experience was that of Elisabeth who had no doubts, either, about what to study after upper-secondary school. She entered an engineering programme in land management. When interviewed a few months into the first year, her general impression was that 'each time we have been introduced to something new, it has added something interesting to the discipline'. During the first week of introduction, the new students had visited potential future workplaces, offering an impression of what she could do after graduation. Having this was important for the studying to make sense to her. The social life at the programme was positive, not just during class, but the students also went swimming or bowling together in the evenings. The small number of students at the programme (less than 20) combined with project work in groups being the salient mode of teaching at the programme provided a frame for the students to get to know each other. They did not have much contact with the lecturers in mathematics, but each project group had a supervisor assigned that they met with in the group and whom they could call in for meetings. Elisabeth described her interests and motivation as growing; but as she said: 'It doesn't take much before I'm saying: This is fun, this is interesting'.

Elisabeth's narrative presented a successful process of integration, both socially and academically. She got along well with her fellow students, she was involved in social activities, and she found the academic content stimulating and interesting. The future after graduation appeared promising as well.

Compared to the experiences of the 18 other students, Elisabeth and Emily represent each their end of a continuum of successful integration, but more students had experiences similar to those of Emily's than to those of Elisabeth.

Academic Integration

The Academic Content

A great deal of the students told about being surprised that the content they met during the first year was different from what they had applied for. This was not least the case when the students commented on the modules in mathematics that was part of the first year at most of the programmes. An engineering student told that he had asked a professor why they should have mathematics, but the professor had stated that he did not know that either. A student in biochemistry supposed that they should have the module in mathematics of social reasons because they had been told that they would not be using any of the mathematics taught in the module whereas the mathematics they were to use would be taught in a later module. A computer student had expected the study to contain some mathematics in addition to the coding, but it had turned out that it was the other way around, at least in the beginning.

Other students told they had been surprised by other aspects of the programmes: that the programmes were less practical than expected (this was the experience for some of the students at the biology-oriented programmes) or that they were less theoretical (which both a student at sport science and at professional engineering told). The experience of the content of the programme and the balance between theoretical and practical elements were related to the students' sense of identity. An example of this was the female student, Frida, who in an interview during her gap year told that she would apply for admission at biochemistry. She was fascinated by understanding the chemical aspects of the body, that 'it's not just biology all of it' and that she would like to work with medico-chemistry. When entering the programme, she found that all the auxiliary modules were placed at the beginning

> ... which in one way is pretty smart because you need the basic knowledge. But they kind of forget that they need to catch people at the programme, saying "this is what we are going to do" (Frida, biochemistry, second interview)

Even though she acknowledged a need for auxiliary subjects she also regretted that the sequencing of the courses during first year meant that the biochemistry students had to wait until the end of first year before they met courses in biochemistry and maybe 'people will not get caught': 'It's too bad for those who have already dropped out. There are some who have dropped out because they simply didn't find it interesting enough', she said in the second interview at first year.

Frida herself was quite positive about the study and felt like she belonged there. This was partly related to the social environment of the programme, partly to the academic part. The academic dimension not least had to do with the laboratory work. When she was wearing the lab coat, she felt how she

> turned into a professional ... becoming entirely different, straightening the back, becoming proud. [...] And I see myself from the outside and I say: This is actually quite alright (Frida, biochemistry, first interview)

The experience of meeting something different from expected was endemic in the interviews. However, even though some students seemed to have been less careful in their search for information the experience also students who had looked up information about the programmes were facing a different content than expected. This experience seemed to be related to the lack of a meaningful link between the different modules – either because the programmes failed to convey the meaning of the modules or because the meaning was not there.

The Teaching: Pace, Quality, Form

Most of the students needed some time to adjust to the mode of teaching at university, mainly the lectures with large number of students and more emphasis on the students' own reading and doing exercises. The students' sentiments towards lectures were diverse. Some experienced the lecturers as good at explaining the content when it had been incomprehensible when they read it in the textbook. Some told about the lecturers as open and anxious to tell the students more if they approached them after class, while others described them as more remote compared to the teachers the students knew in upper-secondary school.

Obviously, the quality of the teaching could be quite diverse, too. Some lectures could be uninspiring, difficult to follow, and with teachers who judged by the students' descriptions could do with some pedagogical supervision. An engineering student described a teacher in chemistry who was just 'babbling away, writing random chemical equations on the blackboard' and the difficult part was to figure out what was relevant. In other situations, the teaching could ignite the students' interests and fascination. In some of the narratives, the difference between the inspiring and the less inspiring teaching appeared to be related to the pace and to the possibility of engaging more deeply and becoming absorbed in the content. High pace prevented the students from delving into the subject matter.

Frida, quoted previously, experienced that the teaching made the students adopt an approach to studying where they learned how to solve problems at exercises without necessarily learning the theory behind them. She remarked that she supposed it was the theoretical understanding they would need later on, but what they were tested on at the exam was solving problems. Birgitte, another female student, had two modules at the same time: one in mathematics and one in biotechnology. She described the difference between the two. A math day was 'Read. Listen. Understand. Do exercises'. A biotechnology day was 'Think. Rephrase. Explain. Things like that to make you understand it yourself'. Along with Elisabeth, presented in the introduction, Birgitte represents two of the few examples where project work took up a substantial part of the teaching. Most of the students attended programmes where the teaching was organised in lectures, exercise classes or tutorials, and lab exercises.

The student narratives suggested that project work succeeded in conveying a fascination and academic satisfaction to the students. Elisabeth experienced the programmes as fascinating and that they were introduced to new and interesting things. Birgitte worked in a group on a topic within biotechnology they had chosen themselves and investigated that. The project had presented her with an idea of where biotechnology could take her and a sense of 'having come to the right place', as she said. Interestingly, Emily who opted out of engineering, had one of her positive experiences with the programme shortly before she left when they were doing a 3-week project. She liked the teaching being organised as project work even though it was hard having an exam after just 3 weeks, but she appreciated the opportunity to go deeper into something.

The Academic Requirements

Some of the students found the academic requirements in the teaching and the assignments challenging. However, they did link this to the academic level in the sense of how complicated, abstract, or 'difficult' the content was, but rather to the teaching or the kind of learning required. To some extent, this may have something to do with the students trying to construct a narrative where they appeared as competent even though they were struggling to meet the requirements. However, it could also reflect that whether the students meet the standards of the course requirements or not is not solely rooted within the individual student as a particular ability or trait. Rather, the achievement of a particular student depends on the relation between on the one hand the requirements from and the opportunities provided by the learning environment and, on the other, how the student interprets the environment. This interpretation builds on the student's background and prior knowledge and experiences. This means that a particular learning environment may impede some students in expressing their competences, but facilitating the participation of other students.

One example was an engineering student who during the first interview stated that 'I don't think the content is difficult, but there is just so much of it at one time that you are soon falling behind' (Djemal). In the second interview a couple of months later he told that he struggled with how 'to put the formulas together. I always struggled with that – I should be using these formulas and not the other'.

Djemal did not consider himself one of the 'clever heads' at the programme and he experienced having difficulties with the content – even if he thought the level was okay; in fact, he had expected it more difficult. Still, both the pace of the teaching and the textbooks being in English, made his work on meeting the requirements more difficult. Other students told about trying to find study techniques to learn content by heart (for instance, chemical bonds) and a student at computer science realised that he needed to change his way of studying when he failed some exams.

Overall, when the students found the content of the programmes difficult and challenging this was not simply related to the courses presenting them with new and more demanding content. Apparently, the students had difficulties finding a way of coping with the teaching and with how to organise their studying when they were presented with large amount of textbook materials, frequently in English (a second language to the students), and with the expectation that they should be able to both understand and absorb extensive material by heart (Ulriksen 2013).

A particular challenge occurred when the teaching presupposed that the students had particular prior knowledge and experiences that were not explicitly required at entrance. This could be that the teachers assumed that the students had learned some specific disciplinary content during upper-secondary school (which they had not), or that the students had a particular level of knowledge in one of the disciplines at the programme (e.g., chemistry in a biology programme), but where the students eventually had taken the subjects at different levels in upper-secondary school and therefore entered the teaching with different prior knowledge.

A male student at computer science told how the teaching presupposed particular skills:

> JAVA is a language that quite a lot of enthusiasts have used for coding before, and therefore it feels like they [the teachers] expect that most of us already have experiences with programming in JAVA, and then they expect that we almost all of us are able to use if for coding. And I haven't coded before, and it's a bit like offsiding new programmers, because I'm sitting there thinking 'great', and then they are standing there just talking and talking and talking, and you are thinking you're not really learning anything from it. (Belal, second interview).

The narratives of this student indicated that the teaching at computer science assumed the students to be computer enthusiasts who had been playing with their computers as a hobby (there is hardly any formal computer science teaching in Danish primary or secondary school), learning coding on their own, experimenting with writing small programmes. This was one example of courses expecting particular knowledge or approaches from the students. Another was the engineering student, Filip (reported in Chap. 3 in this volume), who after having entered the programme and having met with a professor serving as a mentor changed his perspective of studying engineering from aiming at working with management to focusing on the engineering. In Filip's case, he adjusted his perspective to one more in accordance with the usual and legitimate one.

Overall, the academic integration was by many of the students experienced as troublesome. The content of the courses was different from what they expected; the teaching and learning activities were difficult to get used to and did not always appear to facilitate learning; and the academic requirements were not only challenging because of being at a more advanced level, but also because the students were unprepared for some of the study methods necessary for handling the amount of material and the pace of the teaching.

Still, the students were generally patient and accepting the choices of the programmes they attended, trying to find ways to cope with the sense of insecurity, concerning their academic competence and whether the course was actually the right choice.

Social Integration

One way of coping was to prioritise the social integration. Once again, Frida can provide an example. In the third interview conducted during spring she expressed that the social network established at the programme had been crucial to her persistence:

> I don't think I would have gotten this far. I think the social has been really important for me – both having somebody to study with, but also having a social life in here. [. . .] Those girls [in the study group] have really helped me a lot with understanding some of the theory behind the assignments. (Frida, third interview)

An important point is that the social network both has academic and social implications. When asked what advice she would give to students entering the programme, Frida answered: 'You need to establish a social network. It shouldn't be all work'.

Most interviewees stressed the importance of a social network. Some students, like Frida, considered the social integration as the key to retention. A male student of mathematics who was even more interested in the course content than he had expected to be, had difficulties becoming socially integrated, and said: 'I use quite a lot of effort on that. It's almost more important than doing well at the course. Because, if I don't feel comfortable then I don't think I can make it through' (Bastian). Another student told how she had given priority to becoming socially integrated during the first year, both because she, like Bastian, considered it of paramount importance, and because she expected the first year of study to be somewhat boring due to the auxiliary courses they were to take.

Hence, social integration is important for feeling comfortable and having a sense of relation to the place. The social integration is also important because the social network offers resources for coping with academic content. The students' informal interactions outside class provides access to help and support beyond the study group. Further, the sense of not being the only one struggling was mentioned by some of the interviewees as important in their decisions to persist.

Based on the interviews, it appeared that for most of the students the social integration was more successful and smooth than the academic integration. The experience of Bastian, feeling the other way around, was unusual. There were, however, examples of students who experienced a sense of isolation. One reason could be the geographical distance between the university and the student's home. Another was expressed by a student with an ethnic minority background, who experienced it difficult that most of the social activities involved consumption of alcohol. The apparently successful social integration process of the majority students is a fortunate situation, but it also calls for even more attention to the minority that for different reasons (personal, geographical, religious, etc.) have a harder time finding a social space at the programmes.

The Expectancy-Experience Gap

Virtually all of the 20 students experienced a gap between what they had expected and what they experienced at the courses (for a more detailed discussion of this gap, see Holmegaard et al. 2014c). For most of the students, the gap related to different aspects of the academic integration. The size of the gap differed between the students, but it was experienced by all the students. Consequently, the institutions should expect the students to be faced with a need to adjust their expectations in relation to becoming a higher-education student and that this adjustment process may require some effort. This is an important point, however trivial it appears at first sight.

The students had to find ways of coping with this expectancy-experience gap. For some students the renegotiation of their ideas about studying was a continuous and sometimes arduous process throughout the first year. In that process, students would try to create a sense of coherence between their expectations and their experiences by reconstructing their narrative about what they would meet at the programme. Some students tried to adopt the logic of the programme, as it was, inter alia, expressed through the sequencing or the teaching methods even when this same logic was challenging for their making sense of studying (for instance, that toolbox courses should precede the more interesting courses). Other negotiations could concern whether the choice of programme was in fact the right choice (cf. Chap. 3) and whether the programme suited the student or whether the student was fit for the programme. This negotiation would affect the students' sense of who they were, their construction and reconstruction of their identity. Most of the students included in this study succeeded in this renegotiation process in the sense that they stayed at the programme. Others, like Emily whose story was presented in the beginning of this chapter, opted out because the study experiences were incompatible with her maintaining a sense of self.

Some students managed to reconstruct their expectations in a shorter and faster process bringing them at terms with their experiences. For some, like Elisabeth, this was due to a fairly small gap between the expectancies and the experiences. For others, it was because the students quickly found a way of coping, either by transforming their interests or by submitting themselves to the programme and not expecting much. In some cases there was also an element of the students adjusting and developing their study techniques and strategies, for instance, inventing memory games, involving themselves in study groups working at the university rather than individually at home, adjusting their ways of studying. During this process the importance of the social integration became visible because fellow students, and sometimes older students, could pass on tips and ideas about what to do.

Some students engaged themselves in extra-curricular activities to find resources or experiences that could help them in coping with their programme. A computer science student had his motivation revived at a meeting organised by the trade union presenting possible career paths after graduation. Another student in computer engineering involved himself in working with computers with his friends rather than attending classes at the programme. He did enough to keep track of the courses, but his main focus was on working with computers and systems at the dormitory because it matched his interests better.

Discussion: Integration, Negotiations, and the Implied Student

The experiences from attending first year at a STEM university programme all included the challenge of bridging the gap between the expectations the students had about what the studies would be like and what they eventually experienced after having entered the programmes. As a part of this, students needed to renegotiate their images of studying and of themselves, that is, it required them to carry out identity work in order to establish a sense of meaning and coherence between what they expected and what they experienced.

This effort of the students to bridge the gap could be considered as a process of integration following the model of Vincent Tinto (1993). The students' narratives also showed that both the academic and the social aspects of the first-year experience are of importance in this process. Further, the academic integration consists of meeting the courses' requirements, of being able to handle the teaching and learning activities of the programmes (including the pace of the teaching), and of coming to terms with the content of the courses and modules the students attend during the first year.

Our analysis suggested that the social and the academic integration are related, not least that the social integration may help the students to endure the strain put upon them by the academic life, keeping of the high pace and attending courses dominated by auxiliary disciplines rather than the topics the students had opted for. The social integration can both provide a sense of belonging that can balance the doubts generated by the academic integration process and it can offer resources the students can draw upon in their endeavours to meet academic requirements and endure the long road some students need to travel before getting to the interesting modules.

The different elements involved in this process further emphasise the point that academic and social integration is a complex process. For some students it is a process of assimilation because the programme fits the interests and intentions of the student. This was the case for one, perhaps two, of the students in our sample. For others, it is an assimilation process where the students accept that they have to endure a period of boredom and lack of meaning in waiting for the interesting content at later modules and through this they conform to the way they are expected to study. Others, still, accommodate the study experience in a way that both allows them to become sufficiently integrated to pass the exams and being recognised by the programme as a legitimate student and to engage in extra-curricular activities that meet some of the interests that are not catered for by the programme.

This means that whether students have to commit cultural suicide or not (Tierney 1999) is to a wide extent dependant on the way the students succeed in bridging the expectation-experience gap. While some need to renegotiate their ideas and perceptions in a way that submit their sense of self and of meaning to the logic of the programmes, other students manage to maintain their initial interests and perspectives. However, some of the latter do so by compartmentalising their study experiences so that their subjective sense of relevance is nurtured parallel to rather than integrated with the progression and content of the study programme. Consequently, both assimilation and accommodation can take on different shapes and they represent variations of integration. Hence, Tinto's model draws attention to the pivotal role of integration, but it does not provide an understanding of how this process occurs.

The present analysis suggests that the curriculum is a focal point in the integration and retention of students. The integration of students is related to the content of the courses, to the sequencing and mutual relation of the different modules, to the pace of the teaching, to the kinds of teaching and learning activities students are involved in, and to the kind and extent of student involvement in the courses. This involvement of the students includes to what extent the curriculum leaves room for the students to recognise the content they found interesting when they applied for the course and what kinds of engagement with the content the teaching and learning activities allow.

However, the social context of the studying and the students' opportunities for establishing social networks are also important. Even though establishing social networks could be considered the responsibility of the students themselves, the programme provides a framework that could both facilitate the students' social interaction or hamper it. This framework consists of elements related to the curriculum and teaching practices, but also to the physical options at the campus and to whether the institution consider it a part of its role to facilitate students' social integration. As the social integration is strongly interwoven with the students' academic integration there seem to be a potential for institutions to develop and use this knowledge to support students in their first year of study.

The analysis also drew the attention to how the pre-entry qualifications are part of the integration process. As we noted previously, the students' background (be it social, ethnic, or gender), their prior academic achievements, and the circumstances concerning their choice of programme all feed in to the students' construction of narratives when they meet the university courses. Therefore, these pre-entry qualifications, as they are labelled in Tinto's model, should not be considered as having an impact prior to the negotiation of the students' narratives. They are an integrated part of the continuous negotiation process and therefore the students are differently equipped for the integration process.

In this process, the implied student of the programmes will play a role. The degree of similitude between the implied student and each individual (empirical) student will influence the process of integration because it presents the students with different requirements in their renegotiation. Students with a background and approaches to the studies that differ from those of the implied student will have to perform a more extensive renegotiation than students whose background and approaches are similar to those implied. As a part of this, they will also have to balance how much they are willing or able to adjust in order to meet the courses' requirements.

The integration of the students, however, was not only related to the students' sense of belonging or interests. It also affected their approaches to studying and hence supposedly the quality of their learning. This not least had to do with the students' experience of high pace that persuaded some of the students to adopt a study approach where the course content was 'taken in' rather than understood. Both the narratives and previous research (Entwistle 2009) indicate that the pace and amount of content in the teaching reduces the quality of the learning. These accounts were contrasted by less frequent descriptions of learning situations such as project work and to some extent in laboratory work, that is, teaching formats characterised by more active participation by the students, something that has been seen to improve student retention (cf. Crosling et al. 2008).

Conclusion

The present study has found that the social and academic integration of first-year STEM students involves a process of negotiation and reconstruction where the students balance different elements in order to bridge the gap between their expectations and their experiences. At the core of this process is the identity work of the students, and the outcome may both be that of assimilation and of accommodation. We also found that the social and academic integration are closely related, not least that the social integration provides access to resources among fellow students that can be of help in the academic integration.

In the act of balancing, students' social background and their prior knowledge and experiences encounter the conditions and requirements of the study programme and institution. Therefore, student background and pre-entry qualifications are the resources the students can draw upon when they engage with the course curriculum and when they reconstruct and negotiate their narratives concerning studying. These resources do not just concern academic preparation or careful information seeking prior to entering the programme. They include which patterns of interpretation, understanding, and narratives are available to the students in the balancing and identity work. The integration and retention of the students, therefore, are related to the extent to which students can bridge the expectancy-experience gap, and this partly has to do with the distance between the implied student of the programme and the background and orientation of the student.

The analysis found that the curriculum is crucial in this process, even though the facilitation of the students' social integration is influential as well. The students' narratives of their experiences with the first-year STEM curriculum suggest that the academic integration is hampered by a sequencing that delay the time when the students meet what they applied for, and a pace that is so high that students rely on recollection rather than understanding. A principal challenge of first-year students is how to cope with these experiences in a way that give them a sense meaning and coherence in the study programme.

This means that measures to increase retention should not mainly focus on the preparation of the students, but rather on the curriculum of the first year, both concerning content, sequencing, pace, and the teaching and learning activities students are involved in. This could not just increase retention. It might also improve the quality of learning for the group of students as a whole in accordance with the point made by Harvey et al. (2006) that retention should be addressed as an issue concerning first-year experiences rather than as a separate issue.

It is of paramount importance that the first year at university allows the students to get a sense of where they are going and how the different courses contribute to the overall goal. They should be able to establish a link between who they have been, who they are, who they wish to become, and the course they are attending. Since the students are different and enter with different perspectives this calls for a curriculum that is sufficiently flexible in form and content that it allows for different interests and interpretations of what studying the particular discipline means and where it might take them.

References

Andrews, M., Squire, C., & Tamboukou, M. (2008). *Doing narrative research*. London: Sage Publications.

Becher, T., & Trowler, P. R. (2001). *Academic tribes and territories. Intellectual enquiry and the culture of disciplines*. Maidenhead: The Society for Research into Higher Education & Open University Press.

Bøe, M. V., Henriksen, E. K., Lyons, T., & Schreiner, C. (2011). Participation in science and technology: Young people's achievement-related choices in late-modern societies. *Studies in Science Education, 47*(1), 37–72. doi:10.1080/03057267.2011.549621.

Braxton, J. M., Milem, J. F., & Sullivan, A. S. (2000). The influence of active learning on the college student departure process – toward a revision of Tinto's theory. *Journal of Higher Education, 71*(5), 569–590. doi:10.2307/2649260.

Bruner, J. (1990). *Acts of meaning*. Cambridge, MA: Harvard University Press.

Bruner, J. (2004). Life as narrative. *Social Research, 71*(3), 691–710.

Crosling, G., Thomas, L., & Heagney, M. (Eds.). (2008). *Improving student retention in higher education. The role of teaching and learning*. Milton Park: Routledge.

Entwistle, N. (2009). *Teaching for understanding at University. Deep approaches and distinctive ways of thinking*. Basingstoke: Palgrave Macmillan.

Harvey, L., Drew, S., & Smith, M. (2006). *The first-year experience: A review of literature for the Higher Education Academy*. York: The Higher Education Academy.

Hollway, W., & Jefferson, T. (2000). *Doing qualitative research differently: Free association, narrative and the interview method*. London: Sage Publications.

Holmegaard, H. T., Madsen, L. M., & Ulriksen, L. (2014a). To choose or not to choose science: Constructions of desirable identities among young people considering a STEM higher education programme. *International Journal of Science Education, 36*(2), 186–215.

Holmegaard, H. T., Ulriksen, L. M., & Madsen, L. M. (2014b). The process of choosing what to study: A longitudinal study of upper secondary students' identity work when choosing higher education. *Scandinavian Journal of Educational Research, 58*(1), 21–40.

Holmegaard, H., Madsen, L., & Ulriksen, L. (2014c). A journey of negotiation and belonging: understanding students' transitions to science and engineering in higher education. *Cultural Studies of Science Education, 9*(3), 755–786. doi: 10.1007/s11422-013-9542-3.

Hurtado, S., & Carter, D. F. (1997). Effects of college transition and perceptions of the campus racial climate on Latino college students' sense of belonging. *Sociology of Education, 70*(4), 324–345. doi:10.2307/2673270.

Illeris, K., Katznelson, N., Simonsen, B., & Ulriksen, L. (2002). *Ungdom, identitet og uddannelse*. Frederiksberg: Roskilde Universitetsforlag.

Kvale, S. (1996). *Interviews. An introduction to qualitative research interviewing*. Thousand Oaks/London/New Delhi: Sage Publications.

OECD. (2008). *Encouraging student interest in science and technology studies*. Paris: OECD (Organisation for Economic Co-operation and Development).

OECD. (2010). *Education at a glance 2010*. OECD indicators. Retrieved from http://www.oecd.org/dataoecd/45/39/45926093.pdf

Polkinghorne, D. E. (1988). *Narrative knowing and the human sciences*. Albany: State University of New York.

Schreiner, C., & Sjøberg, S. (2007). Science education and youth's identity construction – Two incompatible projects? In D. Corrigan, J. Dillon, & R. Gunstone (Eds.), *The re-emergence of values in science education* (pp. 231–247). Rotterdam/Taipei: Sense Publishers.

Seymour, E., & Hewitt, N. M. (1997). *Talking about leaving. Why undergraduates leave the sciences*. Boulder: Westview Press.

Tierney, W. G. (1999). Models of minority college-going and retention: Cultural integrity versus cultural suicide. *Journal of Negro Education, 68*(1), 80–91. doi:10.2307/2668211.

Tinto, V. (1975). Dropout from higher education – Theoretical synthesis of recent research. *Review of Educational Research, 45*(1), 89–125. doi:10.3102/00346543045001089.

Tinto, V. (1993). *Leaving college. Rethinking the causes and cures of student attrition* (2nd ed.). Chicago/London: The University of Chicago Press.

Tinto, V. (2006–2007). Research and practice of student retention: What next? *Journal of College Student Retention: Research, Theory and Practice, 8*(1), 1–19. doi:10.2190/4YNU-4TMB-22DJ-AN4W.

Ulriksen, L. (2009). The implied student. *Studies in Higher Education, 34*(5), 517–532. doi:10.1080/03075070802597135.

Ulriksen, L. (2013). Naturvidenskabelige studiekompetencer. Hvad er svært for nye universitetsstuderende? [Study competences in science. What is difficult for the students?]. *Mona, 2013* (2013-3), 44–63.

Ulriksen, L., Madsen, L. M., & Holmegaard, H. T. (2010). What do we know about explanations for drop out/opt out among young people from STM higher education programmes? *Studies in Science Education, 46*(2), 209–244. doi:10.1080/03057267.2010.504549.

Chapter 16
Keeping Pace: Educational Choice Motivations and First-Year Experiences in the Words of Italian Students

Giuseppe Pellegrini and Chiara Segafredo

Introduction

Western societies are characterized by continuing developments in technological and scientific knowledge. This phenomenon entails a constant demand for labour in research and innovation fields (Bucchi 2006). It might therefore be hypothesized that there is a similarly significant growth rate in the number of young students wishing to pursue a career in research in scientific institutions. However, since the second half of the 1990s there has been a relative reduction in the number of university enrolments in STEM studies, especially in Physics, Chemistry and Mathematics, in Europe, Japan, the United States, and the more industrialized countries in general (see section "Introduction" to this book).

According to the European survey *Young People and Science*, while four young people out of ten expressed an inclination to enrol in Social Sciences (39 %) or Economics (36 %), less than a third of the respondents showed interest in scientific courses such as Biology or Medicine (31 %), Engineering (28 %), Natural Sciences (25 %), or Mathematics (21 %). In response to a specific question about intention to study Mathematics at the University, the number of students declaring that they "definitely considered studying" Mathematics was 9 %, while in the case of Social Sciences, the percentage rose to 16 % (Flash Eurobarometer 2008). Along with the reduction in enrolments in science faculties, the phenomenon of withdrawal from studies (also termed 'dropping-out' or 'opting-out'; see Chap. 13), also warrants attention.

The drop-out/opt-out issue has been studied and addressed by several projects and initiatives to promote scientific training. In Italy, for example, the *Progetto Lauree Scientifiche* (the Scientific Degree Project) has been developed since 2004 with the support of the Ministry of Education, Universities and Research, the

G. Pellegrini (✉) • C. Segafredo
Observa Science in Society, Viale A. Fusinieri, 65, 36100 Vicenza, Italy
e-mail: pellegrini@observanet.it

E.K. Henriksen et al. (eds.), *Understanding Student Participation and Choice in Science and Technology Education*, DOI 10.1007/978-94-007-7793-4_16

Federation of Industrialists, and the National Conference of Deans of Science and Technology Faculties. The project has promoted the strengthening of pre-university counselling, the development of more attractive academic programmes, the use of laboratories to make students protagonists in their learning processes, and the development of internships.

But why is it that young people are enrolling in scientific faculties to undertake research activities in ever-decreasing numbers compared with the number enrolling for other non-STEM subjects? There are a number of factors at work: the type of scientific education furnished in schools, socio-cultural and family conditions, and the social representations of science and technology, which influence the motivations and expectations of students, both male and female (Osborne and Dillon 2008, and several chapters in this book).

Students' Participation in Science, Emerging Issues

A large body of research shows that in Europe, the United States and Japan, the underlying problems in science education are shared and widespread (Bizzo et al. 2002; Bøe et al. 2011; Dove 2010; Sjøberg and Schreiner 2005). There are several reasons for the relative crisis of scientific vocations, including: cultural and social factors; the challenges of teaching science at compulsory school; the difficulty of developing and communicating the social utility and social implications of science; the persistence of gender stereotypes; the problem/resource of the guidance for students; and definition of the student's identity and role.

Primary and secondary scientific education encounters serious difficulties in communicating values, social implications, and considerations which extend beyond an explanation of "how science works" (Osborne and Dillon 2008). This is an obstacle for young people, in particular for female students with an interest in science. The perceived utility value of science plays a central role in young peoples' choices, as demonstrated by the data collected during the IRIS survey. 82 % of Italian male and female students responded that they agreed or strongly agreed on the importance of science as a subject at school. As regards their priorities for the future, they gave a high value – 3.8 on a scale of 1 to 5 – to social commitment: working for something which is important for society as a whole, helping others, and contributing towards sustainable development and environmental protection.

Counselling activities, and those which involve mentoring for students through the process of making choices from secondary school to university and during degree courses, are poorly-developed, and yet cannot be abandoned (Fasanella and Tanucci 2006). They are infrequent in Southern European countries, while it has been found that mentoring – which has been developed above all at English and North American universities – is a driver of learning over the long term, promoting the communication of experiences and supporting choices for the future (Felice et al. 2005).

Uncertainties over future careers should not be ignored in certain disciplines such as physics and mathematics. In Italy, for example, there are few job opportunities for graduates in these two subjects, especially if they have taken extremely specialized degree courses with the aim of working in high-quality research laboratories. In 2010, with regard to the Italian employment situation of graduates in the area of mathematical, physical and natural sciences 3 years after graduation, 20.8 % were not in employment and were looking for jobs, while the figure at 1 year after graduation rose to 30.4 % of jobless graduates looking for a job (Almalaurea 2011). This condition is different from the situation in other European countries, where candidates from the physical and mathematical sciences are in high demand (see Introduction to this book).

Students' Expectations and Initial Experiences: Quantitative Analysis

The IRIS survey allows us to study certain aspects of the situation experienced by Italian male and female students towards the end of the first year of scientific studies. In this section we focus on one research question in particular: what are the main factors involved in drop-out decisions? We studied a representative sample of 2,667 students enrolled in Physics, Chemistry, Mathematics, Statistics, Biology, Biotechnology, Computer Science and Computer Engineering, Mechanical Engineering, Chemical Engineering and Electronic Engineering. One section of the questionnaire asked them to give an evaluation of the university environment (relationships, organization of studies and teaching, and opportunities for learning). Their assessments enable us to identify the factors which may influence the processes of withdrawal from studies. The data in Fig. 16.1 show the degree of agreement with statements regarding satisfaction with university studies; the area

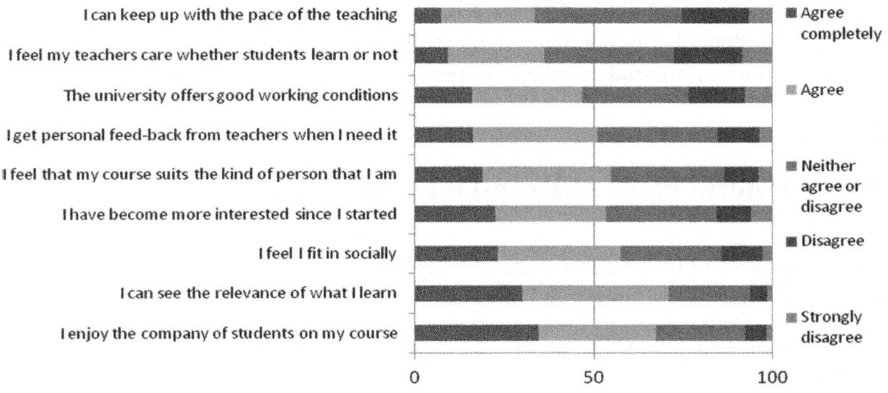

Fig. 16.1 Degree of agreement with statements regarding satisfaction with university studies (n: 2,667) (Source: Young people and scientific pathways; Italian results of the IRIS European survey, March 2011)

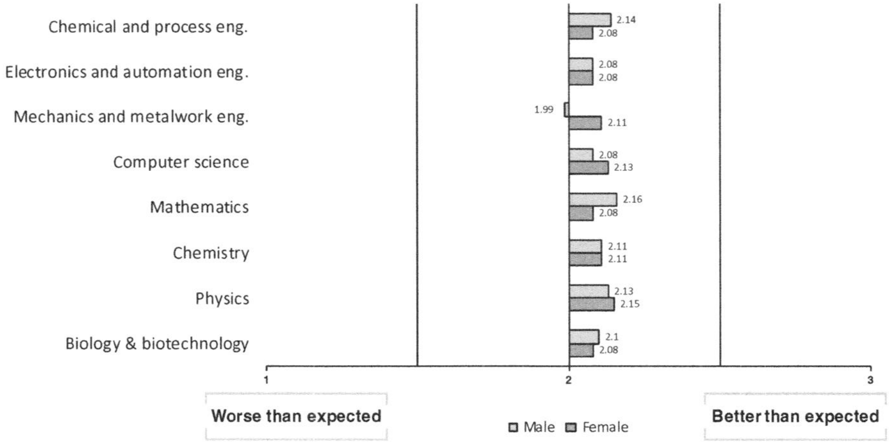

Fig. 16.2 Male-female comparison between expectations and students' experience (n: 2,667) (Source: Young people and scientific pathways; Italian results of the IRIS European survey, March 2011)

where the level of satisfaction was highest is that of social relations. The main challenges, on the other hand, concerned "keeping pace with the teaching" and "getting personal feed-back from lecturers and teachers". This indicates problems related to the learning of content, and the experience that a high level of effort is being made.

The gap between expectations and students' experience is an important factor in measuring the intention to continue with a course or to abandon it (see also Chap. 15) and is also addressed in the IRIS questionnaire. With regard to comparison between initial expectations and experience gained during the first year of the course, one notes a significant level of satisfaction, except for male students on Mechanics and Engineering courses. Figure 16.2 shows an index of five different items; male students of chemical engineering and mathematics show the best level of satisfaction with female students in physics.[1]

Three Determinants of Drop-Out, Opt-Out Intention

The wide range of information gathered on student conditions may be reduced with a factorial analysis to three main dimensions:

- relational well-being, the facility of creating positive relationships with fellow students;

[1] IRIS questionnaire (see Appendix) included five items on students' everyday life: the overall experience of being a student, the social relationship with fellow students, the overall quality of teaching, the interest in the content of the course and the effort to spend on studying.

- results achieved, the student's perception of the level they have reached;
- support from the structure, the well-being obtained from the availability of good structures and support, and attention from the teaching staff.

With regard to relational well-being, one of the most relevant factors was a facility for socializing and for appreciating the company of fellow students as shown in Table 16.1 through a principal component analysis (PCA). This aspect was of great importance for all respondents, especially for those enrolled in Physics, Mathematics, and Chemical Engineering, and it was of greater interest to female students. It was these female students who declared a high level of satisfaction, and who acknowledged that they had developed important relationships. This evaluation is in line with the priorities stated in Fig. 16.1: students, and females in particular, rated social aspects more favourably than other aspects of the study situation.

Concerning the feedback received on learning and the results achieved, males declared a greater level of confidence in their ability to learn, in particular those attending courses in Biology, Physics and Mathematics. These are three disciplines which, compared with the other STEM courses, have a greater female presence, so that we can infer that only the most confident males choose a female-dominated STEM discipline (Table 16.2).

Table 16.1 Principal component analysis of satisfaction with university studies (factor loading greater than 0.30)

Items	Factor 1	Factor 2	Factor 3
I enjoy the company of students on my course			0.89
I feel I fit in socially			0.87
I get personal feed-back from teachers when I need it		0.77	
I feel my teachers care whether students learn or not		0.80	
The university offers good working condition		0.70	
I can see the relevance of what I learn	0.71		
I feel that my course suits the kind of person that I am	0.82		
I have become more interested since I started	0.78		

Table 16.2 Index of self-assessment of one's ability to learn by ISCED and gender (average values, scale 1–5)

ISCED cohorts	Males	Females	Total
Biology and biotechnologies	3.4	3.2	3.3
Physics	3.3	2.9	3.2
Chemistry	3.3	3.3	3.3
Mathematics	3.3	2.9	3.1
Computer science	3.2	3.1	3.2
Mechanical Eng.	3.1	3.3	3.1
Electronics Eng.	3.1	3.1	3.1
Chemical Eng.	3.2	3.1	3.1
Total	3.2	3.1	3.2

As regards evaluations of support from the structure, Italian students attributed particular importance to working conditions, including laboratory activities since the secondary school where they carried out experiments and put into practice what they had learnt in their theoretical lessons. Laboratory work and fieldwork were elements crucial for students' motivational development. This relevance emerges both from a qualitative analysis of the open questions and from the quantitative analysis: the importance of these experiences was fundamental especially for those who enrolled in Chemistry and Biology, and for female students.

Motivation, Interest in the Subject and Risk of Abandonment

An analysis of motivation and interest in the subject reveals important aspects of the student experience. Here we assume that there is a strong relationship between the two elements, given particularly the notion of intrinsic motivation described by Ryan and Deci (see Chap. 2). Students mention motivation when talking about the course as a whole, whereas interest in the subject is one of the main elements mentioned by students when describing their satisfaction levels. The replies to the open-ended questions confirm that recognition of the intrinsic value of the subjects being studied enabled a student to continue with his or her studies and to avoid abandoning them, notwithstanding the considerable amount of work required. In this regard, the choice of a study pathway is developed around a reciprocal relationship between interest and self-representation (Rosenberg 1979), understood as the perception of oneself, which can influence an individual, and therefore his or her choices (Beier and Rittmayer 2009; see also Chap. 3).

We can depict the relationship among interest, self-description and realization in a given area in the form of a triangle: success in a discipline (for example, the positive outcome of a mathematics project) influences perception of one's potential in this field ("I'm good at mathematics"), which has a positive effect on interest (Beier and Rittmayer 2009; Guay et al. 2003). An increasing interest in the subject leads to a search for increased achievement in the field.

Lessons on the practical applications of science also heighten interest and reduce drop-outs (see next section). It should be recalled that during adolescence, and in the successive phases during which the choice of an educational pathway matures, it is fundamental for boys and girls to provide an answer to the question "what is the usefulness of my actions?", and therefore to consider the practical effects of scientific studies (Eccles and Wigfield 2002).

For the purposes of an analysis of the drop-out and opt-out mechanisms, it is important to consider the number of students who faced a situation which was worse than had been anticipated (see Fig. 16.2 for an overview): a percentage between 8 % and 13 %. Males encounter more unexpected difficulties in Computer

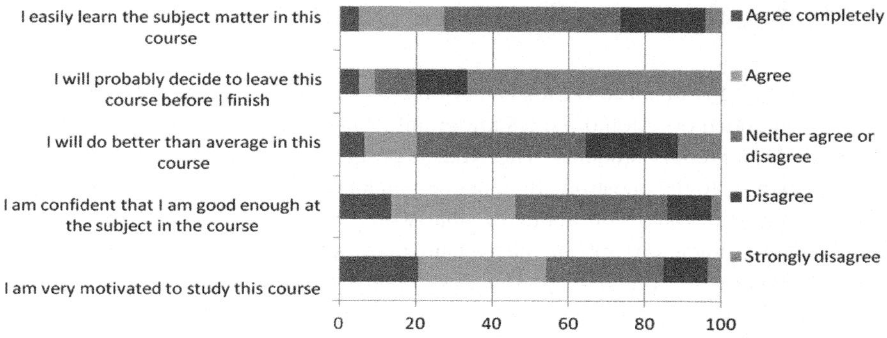

Fig. 16.3 Level of agreement with statements on present and future academic performance (n: 2,667) (Source: Young people and scientific pathways; results of the IRIS European survey, March 2011)

Science and Mechanical Engineering, while it is women who face more problems with Chemical Engineering.

Figure 16.3 shows the level of satisfaction with academic performance and perceived learning level. A minority of respondents (fewer than 1 in 10) saw themselves at risk of abandoning the course, notwithstanding the fact that the course work was regarded as onerous in the vast majority of cases. Only one-fifth of the sample thought that they would obtain better than average results; only a quarter learnt subject matter easily on the course; and less than half were confident that they were good enough in the subject. Furthermore, many respondents placed themselves in the middle of the scale, thus expressing uncertainty about their academic futures. What seems to prevent a greater tendency towards abandonment, despite the considerable effort required in the courses, is the students' motivation: more than half of the respondents stated that they were highly motivated to study the course subjects.

The greatest likelihood of dropping/opting out was indicated among students enrolled in Biology, where the share of students who manifested doubts about completing their studies represented over a quarter of the sample. It should be made clear that many female students had selected Biology as their second choice; their first, at the moment of enrolment, was medicine, which in Italy requires an admission test. These students therefore intended to leave Biology in order to retry the admission examination for Medicine. The two cases where the intention to abandon was instead the lowest were Physics and Mathematics. With regard to Computer Science, it was female students who expressed more uncertainty about the future, while in Mechanical Engineering, the (very few) female students were nearly all sure that they would complete their studies.

Evaluating the Course Experience: Qualitative Analysis

Together with the quantitative analysis of the data, a qualitative analysis of replies to an open-ended question provides important insights into the experiences of male and female university students. The Atlas.ti program was used to study the 2,192 open responses to the question "If someone you know were thinking of enrolling on your course and asked you about it, what would you say to her or him?", using labels which made possible a macro-categorization based on the achievement-related choices interpretative model developed by Eccles (1994). The opinions expressed in responses to this question can be classified into three groups: those who regret choosing their course, those who value it positively, and those who take an intermediate position, suggesting the risks and the potential. From the dissatisfied replies, we can therefore gather the factors which are crucial to course abandonment; while from the positive and enthusiastic responses, we can find the reasons for continuing. The intermediate replies show more balanced positions, but they are often highly uncertain.

In order to study the drop-out and opt-out processes, we used the dimensions of the achievement-related choices model (Eccles et al. 1983; see also Chap. 2): interest value, which is used to gain an understanding of all the elements of a subject involving interest and pleasure; attainment value, which concerns statements referring to achievement values and strong identification with the duties and the role of the scientist, also in "vocational" terms; utility value, the aspects of individual usefulness in relation to a person's development project; and perceived cost, or the weight given to the costs which must be sustained in order to complete a course of study.

We finally analysed the responses with particular regard to the learning environment (student relations and teachers' role) and to external factors (future job opportunities).

Enjoying Science: Interest Value

The quantitative data showed that intrinsic interest in the subject was a strong factor in pursuing a scientific career. Many respondents used emotive adjectives such as "fascinating", "exciting", "stupendous" and "great", and intensely emotive nouns such as "passion", "interest", and "pleasure":

> The course is very interesting, and "enjoyable" in certain ways, because it's always possible to see and feel what you have studied. It's very hard because of the effort and workload (female, Chemistry)
>
> Independence in one's studies is the key element in scientific faculties, and chemistry is no exception. If you're curious and chemistry reflects your passion and the career you would like to have in the future, this is the degree course for you (male, Chemistry)

Intrinsic interest was cited as a priority especially by Physics, Chemistry, Biology, Biotechnology, and Mathematics students, males and females alike. This is probably due to their greater difficulty in finding jobs compared with their colleagues enrolled in Engineering and Computer Science, for whom utility and career move are the absolute priorities

> Interesting, and an excellent choice for the future, since electrical engineers are sought after and well paid because they are in short supply, although it is currently one of the most demanding degree courses (male, Electrical Engineering)
>
> As a course which provides great mental training, which prepares us for the problems of every day in what will be our profession one day male, Mechanical Engineering)
>
> Important! We learn to be the engineers of tomorrow (male, Computer Science)

Fitting the Identity: Attainment Value

Besides the condition of students on a university course, consideration must also be given to the issue of constructing a distinctive identity (Illeris et al. 2002; see Chap. 3), which is especially significant during the phase of choosing a future career. Transmitting the most interesting, pleasant self-image possible is important for young students, and the choice of a course is therefore connected with the perception that the course is interesting as to content, future expectations, and identity.

There are numerous responses with a powerful vocational connotation. The aspect of science as a "life choice" is present, and is cited most often by Chemistry, Physics, Mechanical Engineering and Biology students:

> Incredibly difficult, but if this is what you want to do with your life, I would recommend it, because it's the best (male, Mechanical Engineering)
>
> Very demanding, but important for both educational and personal growth (female, Mechanical Engineering)
>
> A course for those who want to learn the truth about the world and everything that surrounds us (male, Biology)
>
> Useful, demanding: a life choice (male, Mechanical Engineering)

For many students, especially those studying Physics and Mechanical Engineering, the inevitable problems of studies and workload were overcome by strongly recognising the significance of the subject being studied, as regards both the level of interest and personal and cultural growth:

> Very demanding, but important for both educational and personal growth (female, Mechanical Engineering)
>
> Helps us to understand the world around us, and create a personality which is critical, curious and always in search of the truth (male, Physics)
>
> Great, if one recognizes the fact that one is studying the most important thing that there is (male, Physics)

Students who are highly motivated to study in order to fulfil themselves are unlikely to abandon their studies or to opt for a different course because of the cost of success.

Planning the Future: Utility Value

An analysis of the need to reach external goals, which a large majority of the respondents believed to be important and inevitable in terms of efforts (using terms such as "demanding", "tiring", and "difficult") shows that it was considered to be tolerable when there was a high level of recognition of professional utility and passion for the subject:

> It is a difficult course to follow because the issues are very complex, but it offers good training for the near future (male, Mechanical Engineering)
>
> I would certainly describe it as more demanding than it might appear. The amount of time devoted to studying is very considerable. But I think it's a very interesting course, and useful for the future (female, Biology)

With regard to opt-out, a considerable number of students stated that they would like to change course and apply to a different faculty. This was primarily true for Biology, and in particular women rather than men. This is apparent from the replies to the question "Describe how you came to choose this course", to which around 100 male and female students answered that they had selected it as a second choice because they had not passed the Medicine admission test. In some cases, after attending the classes, and appreciating the subject and the organization of the Biology degree course, they might decide to continue along this route:

> An excellent course which prepares you well for the medicine admission exam. It offers stimulating subjects with great attention to the practical aspects (male, Biology)

Most of those who had selected their university course as a second choice were motivated to retry the medicine route; for some, the intervening year might have been experienced with a sense of inadequacy and demotivation, which was often the case of female students:

> It is a course to be chosen if you have a passion for it, and not to opt out of a medicine exam which went badly! Biologists accept themselves; failed doctors don't! (female, Biology)
>
> It's a very interesting course which must not be undervalued. It needs to be selected consciously, and not as a second choice. It is very demanding, and so one needs to have a serious predisposition for it (female, Biology)

The Right Effort: Perceived Costs

Considering the resistances and difficulties in pursuing a course of study, the students especially emphasised the perceived cost. More than a third of the respondents referred to the effort required, another group referred to difficulty, and many spoke of "complex" or "complicated" courses.

Few students passed a clearly negative judgement on their university course, but it is interesting to describe how they expressed this judgement. First of all, they cited factors such as workload (within the perceived cost dimension), using words such as "fatigue", "stress", "tough", and "sacrifice". From an expectancy-value perspective, the burden of taking part in an activity – the cost – is the price to be paid for bringing a task to its conclusion. Completion of the task is closely correlated with a person's reaction to the cost to be sustained (Eccles 2005; see also Chap. 2):

> Too heavy, and mentally destructive (male, Chemistry)
>
> Hard with regard to the sacrifices we make for our studies, due to stress factors and sometimes frustration with others, requires motivation and will-power (male, Mechanical Engineering)

It is apparent that many responses, even the most positive ones, refer to cost, in terms of effort, energy, and workload. Nonetheless, as also found by the quantitative analysis, effort is compensated for by a strong interest in the subject and/or practical motivations, such as job prospects. Many students associated problems with the subject with its fascination:

> Demanding but fascinating, and repays being taken (male, Physics)
>
> The course is very demanding, but if one is highly motivated, or at least interested, it is not hard to keep up with the teachers (female, Biology)
>
> More difficult than I had thought, but very, very fascinating (male, Physics)

The Learning Environment, Student-Teacher Relationship in the Tertiary Education

Students often mentioned the relational context: when considerable results-based competitiveness is created, it is more likely that elements of cognitive and emotional tension will make progress difficult:

> Most of my colleagues disrupt the classes. The teachers are often not up to their jobs (male, Electronic Engineering)

This is a kind of two-faced Janus, however. Competition is seen by others as a motivating factor. Although the workload has a significant influence on students, if it is linked with a strong interest in the subject, it is unlikely to become a drop-out factor. The students' accounts highlighted a number of crucial factors: the teaching methods, or the quality of the teachers, and their ability to support students, both male and female.

Group work was fundamental, especially for female students in Biology, Mathematics and Chemistry. Group work allowed for active involvement, and an open exchange of ideas by using experimental methods. This element was to be desired above all in courses where laboratory use was crucial, such as Biology, Biotechnology, Chemistry, Computer Science, and Physics. Learning how to work in a group was also mentioned as important:

> My course enables students to fully understand the why of what surrounds us, the right approach to a life that is waiting for you. Learn how to be practical and work in a group in an organic way (female, Chemical Engineering)

In courses involving group work, practical research in the laboratory, and fieldwork, involvement was held to be a strong point of a university career:

> My course lets me acquire a deep understanding of what is around us, the right approach to the life which awaits us. It teaches us to be practical and how to work in a group systematically (female, Chemical Engineering)
>
> You are never bored, between practical classes in the laboratory and theoretical classes. You test your practical abilities from the start, and you understand if you have a talent for the subject or not (male, Chemistry)

The role of well-prepared teachers able to guide, motivate and support their students was regarded as decisive, as was the use of involvement methods which stimulated participation:

> It's great. The teachers motivate you. We do laboratory work and go on trips which help us better understand what we are studying in the books (female, Biology)
>
> An excellent course. The teachers are very good, and always available to help their students (female, Electronic Engineering)

We now consider some critical responses on the role of teachers:

> As regards organization and the teaching and explanation methods of certain teachers, I would recommend not enrolling (female, Biology)
>
> Interesting, but unfortunately not much help is given to those who don't have the basics from high school, and this causes demotivation (female, Chemistry)
>
> Disorganized, disorienting, and difficult, with professors who are not very helpful and are too abrupt with the students, but maybe this is one of the few ways to guarantee yourself a future (male, Electronic Engineering)
>
> No good if you don't like chemistry, and somewhat 'improvised' by the teachers. Two of the seven I have come across so far didn't seem to me of university quality (male, Chemical Engineering)

Discouragement due to inadequate teaching methods was cited by a minority, but this aspect is fundamental for analysing the opt-out and drop-out processes. A quantitative analysis of the Italian IRIS questionnaire, in fact, shows that one student in two of those who enrolled in Mathematics, Electronic Engineering and Chemical Engineering attributes a significant role to good high school teachers, and this figure increases for Physics and Chemistry. Significantly, females enrolled on university courses in Biology, Biotechnology, Physics and Chemical Engineering most clearly state that their teachers have a strong influence on their university careers.

The role of teachers therefore appears as a key element in educational choice, both for enrolling and for staying on in STEM higher education, especially for female students. The significance of the teacher/female student relationship takes root at the time of compulsory education, and through secondary school: in this phase, physics, chemistry and mathematics teachers are mostly women and, curiously, a certain disaffection on the part of female students towards science may be attributed to the part played by certain teachers in perpetuating gender stereotypes by offering greater support to males at the expense of females (Jones and Wheatley 1990; Liu 2006).

Future Job Opportunities

One reason for the possible abandonment of studies mentioned almost exclusively by students enrolled in Biology, Physics and Chemistry, and mainly by women, was the lack of post-degree job opportunities, which can be a strongly demotivating factor:

> I would describe it as a useless faculty, because it provides few opportunities for the future (female, Chemistry)
>
> As a course which doesn't offer many job opportunities in Italy, and with very little benefit (male, Biology)
>
> Interesting and fascinating, but few job opportunities (male, Physics)

Considering the complexity and the effort which scientific degree courses demand, it is especially important to be able to rely on satisfactory job opportunities upon graduation. Professional utility is an essential factor for balancing the costs and the sacrifices required. When this sense of utility is lacking, only students with exceptionally strong motivation manage to continue:

> It is a demanding course, but one which does not guarantee high economic rewards. So if you want to enrol you must do it with passion (male, Chemistry)

Conclusions

Having analysed the quantitative data and the qualitative materials, and having earlier summarized the international, European and Italian situation regarding scientific vocations, we now briefly describe the elements which might help combat drop-out and opt-out processes.

Motivating and orienting young people in the choice of university studies is one of the main challenges. Support from guidance services in the pre-enrolment phase is important, as well as within the university. This is because educational choice is a complex process which is ongoing, even after enrolment, and is characterized by constant reflection on the choices made.

Given the beliefs, values, experiences, and the reasons proposed by students, it is very important that teachers should use the most participative teaching methods with effective feedback procedures in order to enhance the students' capacity to deal with scientific content: the provision of clear explanations is a key factor which respondents remarked on frequently.

Teaching staff need to support and dedicate attention to students' diverse needs concerning teaching and learning styles and preferred ways of communicating. This introduces the topic of mentoring or other forms of tutoring, which have been shown to be extremely useful (above all for female students) in keeping up with classes and examinations. One answer among many, from a female mathematics student, clarifies the role of a university open to the needs of individual students: "It's hard, you need to have a lot of passion, you have to like mathematics, like it a lot. You also need to be diligent and constant in your studies, so you can manage to keep up. The environment helps a great deal – the teachers, the teaching assistants, and the tutors are very helpful".

It is important to consider the transition between university and the working world so that professional expectations are made real and credible. Job opportunity problems for graduates in Physics, Chemistry and Mathematic partly discourage students. An interest in, and passion for, the subject motivate enrolment in science faculties, but the choice must be made in line with other factors, such as professional attainments and career aspirations. Competition and cooperation can be seen as two sides of the same coin, and it should not be thought that one is independent from the other. The scientific enterprise is fuelled by elements of conflict and confrontation together with moments of full and constructive exchanges of opinion.

Finally, developing a good relational climate can enhance cooperation among students. It offers the opportunity to share comments and proposals on learning difficulties, considering that teamwork is especially appreciated by female students (see also features of the Tinto model in Chap. 13). This is demonstrated by the fact that Biology and Biotechnology – where group work is common, for example in laboratories – record only a few cases of abandonment (on the other hand, cases of opting out due to a choice preference for medicine are more common, as we have seen). Of course, not even the most positive relational situation among course-mates can eliminate the inevitable difficulties, and it would be superficial to think that all the difficulties involved in learning and adaptation processes can be remedied.

References

Almalaurea. (2011) Condizione Occupazionale dei laureati, marzo 2011 – Indagine del Consorzio Interuniversitario AlmaLaurea, data ultimo aggiornamento: 10/03/2011.

Beier, M. E., & Rittmayer, A. D. (2009). *Predictors of commitment and success in STEM majors.* Paper and poster presented at the ADVANCE research symposium at Rice University. Houston, TX.

Bizzo, N., Kawasaki, C. S., Ferracioli, L., & Rosa, V. L. (Eds.). (2002). *Rethinking science and technology education to meet the demands for future generations in a changing world.* Sao Paulo: International Organization for Science and Technology Education.

Bøe, M. V., Henriksen, E. K., Lyons, T., & Schreiner, C. (2011). Participation in science and technology: Young people's achievement-related choices in late modern societies. *Studies in Science Education, 47*(1), 37–72.

Bucchi, M. (2006). *Scegliere il mondo che vogliamo. Cittadini, politica, tecno scienza.* Bologna: il Mulino.

Dove, A. (2010). *Science education crisis intervention.* New York: The New York Academy of Sciences.

Eccles, J. S. (1994). Understanding women's educational and occupational choices, applying the Eccles et al. model of achievement-related choices. *Psychology of Women Quarterly, 18*(4), 585–609.

Eccles, J. S. (2005). Subjective task value and the Eccles et al. model of achievement-related choices. In A. J. Elliot & C. S. Dweck (Eds.), *Handbook of competence and motivation.* New York: Guilford Press.

Eccles, J. S., & Wigfield, A. (2002). Motivational beliefs, values and goals. *Annual Review of Psychology, 53*, 109–132.

Eccles, J. S., Adler, T. F., Futterman, R., et al. (1983). Expectations, values and academic behaviors. In J. T. Spence (Ed.), *Achievement and achievement motivation.* San Francisco: W.H. Freeman.

Fasanella, A., & Tanucci, G. (2006) Orientamento e carriera universitaria. Ingressi ed abbandoni in cinque Facoltà dell'Università di Roma "La Sapienza" nel nuovo assetto didattico, Franco Angeli.

Felice, A., et al. (2005). *L'accompagnamento per contrastare la dispersione universitaria. Mentoring e tutoring a sostegno degli studenti.* Roma: Isfol.

Flash Eurobarometer. (2008). *Young people and science, series n. 239, analytic Report*, The Gallup Organisation.

Guay, F., Marsh, H. W., & Boivin, M. (2003). Academic self-concept and academic achievement: Developmental perspective on their causal ordering. *Journal of Educational Psychology, 95*, 124–136.

Illeris, K., Katznelson, N., Simonsen, B., & Ulriksen, L. (2002). *Ungdom, identitet og uddannelse.* Frederiksberg: Center for Ungdomsforskning, Roskilde Universitetsforlag.

Jones, M. G., & Wheatley, J. (1990). Gender differences in teacher–student interactions in science classrooms. *Journal of Research in Science Teaching, 27*, 861–874.

Liu, F. (2006). School culture and gender. In C. Skelton, B. Francis, & L. Smulyan (Eds.), *The Sage handbook of gender and education.* London: Sage.

Osborne, J., & Dillon, J. (2008). *Science education in Europe: Critical reflections: A report to the Nuffield Foundation.* London: Nuffield Foundation.

Rosenberg, M. (1979). *Conceiving the self.* New York: Basic.

Sjøberg, S., & Schreiner, C. (2005). How do learners in different cultures relate to science and technology? Results and perspectives from the project ROSE. *Asia Pacific Forum on Science Learning and Teaching, 6*, 1–16.

Part IV
Applying Feminist Perspectives to Understand STEM Participation

Chapter 17
When Research Challenges Gender Stereotypes: Exploring Narratives of Girls' Educational Choices

Marianne Løken

What Kind of a Girl Chooses Science?

Women's roles in Norway have changed greatly over the last few decades and women have made significant gains in many fields. Despite these gains, much attention has been focused on the limited participation of girls and young women in STEM[1] (Science, Technology, Engineering and Mathematics). It is not easy to interpret and make sense of the limited participation of girls[2] in STEM. Feminist analyses have linked socio-cultural[3] ideas of feminine ideals and gender to the historical under-representation of girls in STEM, arguing that the professional qualities most valued in science are not consistent with the acceptable social behaviors prescribed for girls (Schiebinger 1999). However, *how* social-cultural ideals shape educational choices is still a relatively open question.

The research and recommendations given to stakeholders, politicians, media and the public often contribute to and correspond with what I refer to in this chapter as the "meta-narrative" about girls and/in science. A meta-narrative in this context, is a story about stories, or how to tell a story. In this case, the meta-narrative is used as an analytical tool to clarify the dominant public discourses about girls, which in different ways can affect girls' choices through normative practice and gendered

[1] For a historical review of girls and science, see for example Brotman and Moore (2008).

[2] I will mainly use the term "girl" throughout the text. The category ranges from early childhood to young women in higher education and young women generally in society, the latter constituting the context for this chapter.

[3] See discussion of the concept "socio-cultural" in James Wertsch's book, *Voices of the Mind: A sociocultural Approach to Mediated Action* (1991), where he also includes the historical dimension in socio-cultural.

M. Løken (✉)
Norwegian Centre for Science Education, University of Oslo, P.O. Box 1106, Blindern, N-0317 Oslo, Norway
e-mail: loken.marianne@gmail.com

E.K. Henriksen et al. (eds.), *Understanding Student Participation and Choice in Science and Technology Education*, DOI 10.1007/978-94-007-7793-4_17

expectations. Norms exist in the form of tacit knowledge, stereotypes and cultural barriers, in research communities, in the world of academia, in the world of work and in society-at-large. Such discourses can be reproduced without resistance, because we take knowledge for granted. Providing a meta-narrative is an attempt to synthesize the results of multiple studies that support and reproduce dominant discourse of girls and/in science, tacit knowledge and common gendered perceptions,[4] to describe the current understandings of girls and/in science. The concept of a meta-narrative is here positioned as a dominant public discourse, which includes: (1) culturally dominant discourses of *girls in general*, (2) perceptions about girls' attitudes and aspirations to science and a scientific careers – *girls and science*, and (3) discourses and perceptions about girls who have chosen an education in natural sciences, and/or a career within natural science – *girls in science*.

The meta-narrative as a dominant public story, or dominant discourse, is consistent with what Nancy Brickhouse and her colleagues describe as: "... a story that was constructed by the comparison of boys and girls; studies which focused on the differences between the two groups" (Brickhouse et al. 2000, p. 442). In the same article Nancy Brickhouse and colleagues describe this story as follows:

> Girls are alienated by science. Science is masculine, competitive, objective, impersonal – all qualities that are at odds with our images of what girls are. The more masculine the branch of science (e.g., physics), the less likely it is that girls will like it or do well... [...]...Girls are disadvantaged in science before they even get to school because they are encouraged to play with dolls rather than blocks. They rarely accompany their fathers while they fix items around the house. Parents rarely purchase chemistry sets or microscopes for their girls, nor do they take them camping. As adolescents, girls become interested in being attractive to boys, they take on more feminine roles that often exclude science. Girls become women who cannot and do not engage in science. (p. 442)

Brickhouse and colleagues (2000) say that this is the story we tend to hear about girls, and that these claims are well known and supported by research. Girls' underrepresentation in STEM has been a key feature in science education over the last four decades, and research has documented that girls: are more collaborative than boys, are less competitive (Chetcuti 2009), are more concerned about context (Stadler et al. 2000), wish to know why things happen in science rather than what happened (Osborne and Collins 2001), have a more theoretical approach to science (Staberg 1994), are more people oriented (Miller et al. 2006) and think it is important to have a job where they can help others (Holter et al. 2009), have lower self-efficacy in science (Kjærnsli et al. 2007; Boe et al. 2011), have fewer relevant science experiences from their early childhood (Sjøberg 2000b) and are alienated by science (Brickhouse et al. 2000). These stories have been important in terms of creating gender awareness in science classrooms among other things, and often illustrate that "inequity is a social problem that can be fixed" (ibid., p. 442).

[4] Such tacit or implicit knowledge can be seen in relation to what Svein Sjøberg (2000a), calls the "body-language" of science, which is a metaphor to "describe the often hidden and implicit messages about the nature of science (as well as scientists as persons), aspects relating to the perceived values, norms and ideologies of science".

However, these stories also represent a gender-stereotyped and homogenized image of girls, which may limit individual choice. The complexity of gendered meanings and diversity can, therefore, be ignored if this picture is not more nuanced.

To what extent are we in danger of reproducing broad generalizations and, thus, creating or cementing barriers? I will argue that the meta-narrative supported by much research gives a stereotyped and oversimplified picture of *girls in general*, of *girls and science* and of *girls in science*, that do not help us to understand the diversity between girls. As research also shows, there are greater differences within each gender category than between in many areas (Hattie 2009; Fine 2010). Many of these studies also show that gender constitutes an important part of identity, but not the only part and, therefore, highlights the importance of exploring diversity within gender groupings, in order to work against cementing stereotypes and essentialist binary oppositions between boys and girls as unitary groups (Brotman and Moore 2008). This is consistent with post-structural feminist theories and the idea that gender is one of many factors contributing to a person's identity project (Brickhouse 1998, 2001).[5] Work inspired by post-structural and post-constructivist theories acknowledges the diversity that exists among boys and girls, arguing that many girls engage with science in positive ways, but as Brickhouse et al. (2000) argue, we need to "know more than that they are girls," we need to know "what kind of girls they are" (p. 457). I, therefore, also question the consequences of research that distinguishes girls in comparison to boys, without any reference to gender as a cultural discourse, and thus assuming a naturalized conception of difference. It should be clear by now that I in this study, in tune with perspectives inspired by postmodern feminist theories, conceive of gender as socially constructed.[6]

In this chapter, I will argue that communicating broad generalizations based on sex/gender differences, stands the risk of losing important nuances that again might lead to the cementation of gender stereotypes.[7] To illustrate this, I give examples from an empirical analysis of female students' written narratives collected through the "Write your choice" project.

After a brief description of my methodological approach, I will, examine three issues that became apparent during the narrative analysis, namely; (1) negotiating identity and participation, (2) the significance of role-models and (3) questioning feminine appropriate values. Addressing these issues I will finally discuss: whether gender generalizations based on differences might reproduce stereotyped images of girls, and thus act as a self-fulfilling prophesy; and how discussing narratives about

---

[5] See Chap. 4 in this book, for a more detailed conceptualization of gender in third wave feminism.

[6] But I also sympathize with theories which move beyond postmodernism and post-structuralism in the sense that they theorize gender as historical-socio-cultural and / or semiotic-discursively constructed, while also emphasizing agency of materiality (artifacts, technology, body, clothes, time, text, etc.) See for example Lykke (2012) and Barad (2007).

[7] Norms exist in the form stereotypical perceptions about girls' educational choices, cultural barriers at school, in research communities, in the world of academia, in the world of work and in 'society-at-large'. Such discourses can be reproduced without resistance, if we take knowledge for granted.

girls in science can give a more nuanced picture of females in science and challenge and refine dominating discourses – thus reducing barriers and obstacles to allow more girls to see attractive possibilities within the broad spectrum of the sciences.

The Study "Write Your Choice": A Narrative Approach Through a Gender-Critical Lens

To obtain valid data on people's lived experience is not easy. One way of doing this is to engage in narratives (see Chap. 3). A narrative approach emphasises the individual's understanding and perception of her place in relation to her surroundings, and attaches importance to historical, structural and socio-cultural factors and embodiment in relation to the social and physical world (Daiute and Lightfoot 2004). Thus the narrative "is constructed on a background of memories and thoughts about what happened in real life on this journey of transformation" (Horsdal 2012, p. 88). According to Denzin (1989) "stories then, like the lives they tell about, are always open-ended, inclusive and ambiguous, subject to multiple interpretations" (ibid., p. 81). Like my informants' written stories, narratives are always situated interpretations of lived experience. My methodological approach is mainly hermeneutic and inspired by literary narrative analysis.

The female informants in "Write your choice"[8] were recruited among first- and second-year college and university students in Norway who had chosen studies where women are underrepresented; technology, engineering, mathematics or physics. Female students in subjects such as biology, dentistry etc. are not a part of my study. The 17 narratives were collected in the autumn of 2009 and through two follow-up email-interviews conducted in the spring of 2011 and the spring of 2012. The informants were partly recruited from the same population as the Norwegian IRIS respondents, but were at an earlier phase of their studies. The point was to invite girls to share their story with me, as a researcher. The girls themselves chose to write their stories, and, thereby, constitute a group that cannot immediately be said to represent all girls who choose sciences. The sample consists of girls aged 18–22, who became science students in a subject with a low percentage of female students. They are in a minority and some are the only girl in their class.

The invitation to participate was launched through a website in the autumn of 2009 and the texts were immediately entered into the software NVivo, to help me achieve order early in the analytical process. The website provided some guidelines to encourage the authors to focus on experiences of importance for their educational choices. At the same time, the informants were urged to emphasise what they themselves wanted to emphasise. The method was somewhere between the open

[8] For more information about the project, see: naturfagsenteret.no (in English). http://www.naturfagsenteret.no/c1515605/prosjekt/vis.html?tid=1519446

qualitative interview with a semi-structured interview guide and dialogue as the ideal on the one hand, and the biographical reporting that maintains something akin to a storytelling style in a diary or a chronicle on the other. In the method literature, my choice of methods can be recognised as solicited stories (Hammersley and Atkinson 1996). The written narratives are research-generated personal documents that have to be read against the background of the context in which they were written. The approach is also characterised by the belief that people's choice of words can be of great importance; "The 'local vocabulary' may provide useful information about how members of a specific culture organise their opinions of the world, and, thus, participate in the social construction of reality" (Hammersley and Atkinson 1996, p. 210). Of course, this study's limited sample is not statistically representative, but its strength is to give a more in-depth and broader qualitative view, applying a critical gender lens on educational choices.

Analysing educational choice through a critical gender lens means investigating how gender is constructed in relation to complex social institutions through lived experience and through intra-acting with the material world. The focus has shifted over time but much feminist research is committed to forms of "situated knowledge" (Haraway 1989). Especially since the early 1970s after Gayle Rubin's formulation of the distinction between sex and gender (see Chap. 4), the idea that biological differences could not provide a universal basis for social definition emerged as an established orthodoxy (Rubin 1975). Current feminist analyses often focus on the recognition of the specifics of historical and cultural contexts, and most particularly on the intersections between gender and other forms of difference. The consequence of this is the attention given to the diversity of women's experience, the differences within each woman (Braidotti 1994),[9] situations, powers and resources, rather than simply documenting cultural variability. The major outcome of this work was to position gender as an analytic category; not as a fixed category, but as the performance of a set of regulatory practices.

My focus on girls' stories is consistent with Haraway's discussion of the "Inappropriate others", which she has borrowed from the US-Vietnamese theorist Trinh Minh-ha (Haraway 1992). The discussion of women's stories in science can also be traced back to Harding (1986) and her discussion of how "The Woman Question in Science" turned into to "The Science Question in Feminism", which is about feminist inspired transformation of the epistemological basis for all scientific knowledge production (see Chap. 4). The girls in the study are part of this knowledge production, as agents within and outside of science, but also in-between (science and everyday life).

Baker and Leary (1995) argue that quantitative methods seldom reflect females' opinions because they are decontextualized, therefore, we need qualitative methods

---

[9] This is consistent with postmodern philosophy that does not believe there exists an entity that provides a stable inner core. But we are, according to a postmodern mindset popularly said, different people at different times and in different situations, what Gilles Deluze calls a nomadic subject – a term Braidotti (1994) borrows from Deluze.

to provide a deeper understanding of girls' experiences. My point of departure is lived experiences transformed by my informants into written stories in an attempt to achieve a deeper understanding, or meaning-making of choices. The focus is not primarily on the individual person, how she "really" is beyond the surface, but how she tries to make sense of her educational choice through her story-telling. What can the young women's own reports teach us? "Us" refers to researchers in the field, but also to teachers and school administrators, politicians, industry, and the media. To understand more about educational choice as a phenomenon, I have approached the material with an open mind.

A narrative approach attaches importance to how the informants express themselves and what this expression, through interpretations, can say about a phenomenon (see for example Chap. 3, Johansson 2005; Daiute and Lightfoot 2004). The narrative approach has thus been chosen to understand, describe and explain girls' choice of science, on the basis of an understanding that a construction of text is a construction of meaning. By going deeper into the stories of girls making "atypical" choices, I unravel diverse narratives about girls choosing an education in technology, engineering, mathematics and physics. These girls' breach of conventions can shed new light on dominant discourses and gender stereotypes. Stories are created *by* and *in* social life – and they help to create social life. In other words, my own stories and those of my informants are part of life itself. Telling a story can be understood as a desire to project identities and self-images by telling about one's self. I understand a narrative "self" in the same non-essentialist way that social anthropologist Marianne Gullestad describes the concept in her book *Everyday philosophers*: "As a perpetual process to bring together an individual's many experiences and adventures" (1996, p. 25).

There is no shortage in explanations of why education continues to be gender segregated. The explanations range from lack of confidence to lack of aspiration, from lack of guidance and lack of knowledge of opportunities to socialization by gender, peer influence and lack of role models among others. Stereotypic understandings of science and scientists, and ideas about appropriate behavior for men and women have also been put forward as explanations for this phenomenon. Recognition that students are not passively situated in educational discourse, but agents who actively negotiate subject positions within discursive constraints, points towards new ways of understanding the complexity of gender issues in science, which do not rely on universalized gender categories and stereotypes.

"Until now, stories about the diverse roles and paths girls take have not been told. We want to tell those stories" wrote Brickhouse and her colleagues in 2000. Twelve years later I want to tell these stories, illustrated by excerpts from 6 (out of the 17 submitted) stories in the study "Write your choice". The stories belong to Vanja who studies technology, Maria and Tina who are both students in computer science, Kate who studies marine technology, Sandra who is a cybernetic student and Stella who studies chemical processing. They are all underrepresented in their field of studies. In common with all the girls in my data, they have different experiences with STEM, and different expectations and dreams for the future. But they still represent some phenomena, or patterns, that I will focus on in the following analysis. There is no room for further detailed profiles of the six Norwegian girls here, but excerpts from their written narratives were strategically selected to shed light on the question raised in the chapter: whether more diverse stories can challenge the meta-narrative of girls in/and science. I could have chosen others to represent the findings, but found these excerpts/stories well suited to illustrate some of the tendencies I will examine.

I will now go on to show how findings from my study challenge conventional discourse about girls and/in science – an established discourse with which young women are confronted through the processes which lead them to their educational choices.

Destabilization of Dominant Discourses

"I am like most girls. I spend too much time in the shower. I'm a bad driver. I use impractical clothing in winter. I use every opportunity to dress up a little bit extra, even when I am at the lab." This is how Stella begins her story in a gender-stereotypical way. The participants in "Write your choice" all question the stereotypical pictures painted of girls and/in science, while they also write themselves into such an understanding of their own life-world. And they all use different strategies to meet the socio-cultural expectations of them as "girls against the current" – their own and others' expectations. They are "the others" because of their choices, but they still describe themselves as similar to their female peers. This ambiguity appears in different ways in the data. Stella begins her story by describing herself as a "typical woman" with feminine abilities and values, and hereby positions herself as being like "most girls". In the end of the story, after reflecting upon her educational choices, she describes her "science identity" by saying that "science students are characterized as featureless, nerdy, antisocial and bad dancers. So what? We may not be the hippest, most pretentious students. It is often an all-weather jacket and rubber boots that counts. Nerdy, yes we are to a pretty high degree. So what?" This awareness of her situation as someone "within" and someone "outside" at the same time, can also be interpreted as rhetoric one should master to be both an appropriate girl *and* an appropriate science student.

I will now focus on three themes derived from findings in the narrative analysis; negotiating identity and participation, the significance of role models and questioning "feminine appropriate values", before I go on discussing the results in the light of theoretical perspectives and the overall aim of the chapter.

Negotiating Identity and Participation

A lot of research indicates that girls feel alienated by science, and that boys to a greater extent that girls express an intention to study or work in science (see for example Schreiner and Sjøberg 2007), but there is also research which to some extent challenges this view, such as the Australian study "Choosing Science" (Lyons and Quinn 2010).[10]

The majority of my informants express a positive image of science in general, although ambiguities are prevalent in the data. Sandra is one of the girls who appears as sporty, active and ambitious. After she participated at a gründer-camp[11] she decided to study something more "practical" (than mathematics): "Until then, I was convinced I was going to study maths, since I loved maths and because it was very easy for me". With a positive attitude towards science she considered studying nanotechnology: "Me and my friend from the physics class began to play with the idea to study nanotechnology, primarily because of the high entrance requirements and because we thought it was for the elite. And we liked the idea of being in the elite." One could argue that it is not surprising that Sandra and other girls in these fields of science do not find themselves fitting into a stereotyped description of girls, since they themselves have chosen to study science. They have made different choices, but still view themselves as like "most girls", as Stella expresses in her story. What is certain is that girls in general are not a homogeneous group, neither are girls who choose to study a science where they constitute a minority group. Several of my informants state that they wish to be viewed as unique *and* invisible at the same time. Stella writes, "I wanted to take an education that made me unique.", and Maria writes this about being visible as a representative of a minority in her field: "Boys have the advantage that they can easily fit into the surroundings of male science students (. . .) In a way it would be nice to be invisible. I like to distinguish myself through clothing, interests and general behavior, but I hate to stand out negatively. On the other hand, it gives me the opportunity to distinguish myself positively, which I should embrace with open arms." This is one kind of ambiguity that runs through the written narratives. The informants are aware that they are visible by virtue of being in a minority, and that this provides opportunities. At the same time they state that they do not want to be labelled as different or stereotypic, and certainly not as victims or someone who needs special treatment because they are girls. Vanja represents this view by saying that she is looking forward to the day we "can all be individuals and not gender/sex".

Vanja is only one of my informants who expresses that she feels alienated by the descriptions of "typical girls", and thus the meta-narrative. As she so clearly puts it, "Maybe I am a boy-girl, although I am also a girl-girl (. . .) I generally enjoy being with people who are not so keen to categorize the characteristics of people according to their sex (. . .) people are first and foremost individuals. Most people are not stereotypical, but have a mix of typical girl and boy qualities (. . .) it is stupid

[10] See Quinn and Lyons (2011) for a critical look at students' perceptions of school science and science careers, which is most relevant for the discussion in this chapter.

[11] May be translated as "innovation camp" or "entrepreneurship camp".

when people try to explain my behavior based on what sex I am. I get the impression that I react more quickly to being gender-labelled in this way than others might. I'm not quite sure why, but I obviously feel more unfairly treated than others. I can imagine that those who feel offended by being reduced to a "number" have similar feelings as me when I'm reduced to what gender I am. I really do not know if I feel it is useful to react to this, wish I did not have to, really. Maybe I see too few opportunities in being a woman. Perhaps it is simply that I do not identify with any particular stereotype of women." Vanja also refers to the issue of gendered attitudes as resistant and is surprised by readers' reactions to articles in the media which "present a positive attitude to girls", and how people (males in particular) have "quite peculiar views as to what are the right and wrong things for girls to do." Vanja emphasize her own ambivalence to gendered matter by saying that she is "a little confused as to how I relate to my own sex".

The uncertainty Vanja describes in relation to her own gender identity can be related to the discussion in Anna Danielsson's dissertation (2009), about how female physics students balance the norms of femininity and the standards of what a good physicist is. One strategy is to reject the traditional female image, thereby positioning herself into the male-dominated culture of physics (ibid). An excerpt from a physics student from my own data, who previously studied nursing, illustrates a similar position: "As a former nursing student, I know what it is like to be overrepresented as a girl. To be quite honest, it was hard to find like-minded people there. I did not thrive in the big "women's club" as it was. As a physics student, however, it is different (...) I'm glad, that the atmosphere in the class is a little bit more masculine." Her way of dealing with the transition from a nursing identity to a physicist student, is to reject the nurse identity and to portray herself as different from women in the so-called "women's club". (See also Chaps. 18 and 19 in this book for further discussion on available identities to females and males in STEM).

Everyday language usage often reaffirms how well-established stereotypical opinions of gender are. This is why the choice of education is far more than a choice based on interest and subject-related motivation. It is about the cultural discourse in which the subjects are located, about others' expectations and prejudices. This means that educational choices also are about "doing identity". Identity in this context is understood as what is formed when an individual creates stories about herself (Solsvik 2004). According to Shanahan (2008) "identities are constructions, inextricable from both the individuals and their surroundings and relationships" (p. 44). (Gender) identity can be regarded as a process, as something we *do*, rather than something we *are* or *have*, thus similar to the concept of "doing gender", which refers to how gender is constituted and in a constantly renewed negotiation of meaning (Lykke 2012; Butler 1990). This process refers to previous experiences in the girls' lives, seen in the light of new experiences, and understood on the basis of what they think about the future; about imminent education and future jobs – and, thus, how stereotypes act as forceful "organizers" of practices, choices, identities, negotiations and participation.

The Significance of Role Models

The meta-narrative suggests that girls lack significant role models in science. This fact is hard to deny. To achieve a sustainable society based on diversity and equal status, and to make girls less alienated by science, we might focus on a variety of girls that other girls can relate to and identify with. However, good role models alone are not enough to recruit more girls to science. Maria refers to women from the business community who make company presentations and illustrates how role models can represent a kind of ambiguity in being inspiring but also threatening: "The student association often organizes company presentations. Now and then I meet women who hold that very special position that I myself would like to have in the future. They are sociable, professional and clever. They do their jobs well and enjoy the daily challenges they meet. Every time I think to myself that I really have to shape up to have the slightest chance of ever becoming like them. In a way they are my role models, but they are also a kind of ghost. They haunt me when I have time to think, stirring up my fear. Sometimes I want to give up." Such "heroines" can also have a negative effect; they can give young girls the impression that these women are unattainable and alienating.

One of the recommendations in the Norwegian "Lily"[12] report is that one must show that STEM subjects are consistent with a feminine identity and therefore women need feminine role models:

> This means that the conscious use of role models may be a relevant measure to strengthen the recruitment of young people in general and girls in particular. We think then, of course, of social, athletic, talented, competent female role models and representatives of science. (Schreiner et al. 2010, p. 92)

This recommendation implies some assumptions about gendered values, norms and expected or socially appropriate behavior (Sinnes and Løken 2012). Such "normality" can be understood as an expression of social order, or what "at a given time in a given society is perceived as a prudent and good way of life" (Solvang 2006, p. 168). Hazari and colleagues found in their study that female role models such as scientists and guest speakers had no significant effect on girls' identity formation as scientists (Hazari et al. 2010). However, as Quinn and Lyons argue (2011): "These are not arguments against using role models, but underline the need for additional strategies to foster girls' enjoyment of science and science-related self-concepts and identities" (ibid p. 233).

In her story, Kate writes about women from the business community who make company presentations: "From time to time we meet female representatives from different companies, for example giving a business presentation for marine students on campus, I have noticed that they distinguish themselves not only as women but as more committed, more inspiring and simply happier." It is not necessarily any contradiction between being "social, athletic, talented, competent" and being

[12] The Norwegian research project Lily has served as a pilot to the IRIS project. See Sinnes & Løken (2012) for a more detailed analysis of gendered assumptions in the Lily report.

"committed, inspiring and happy", but my informant's description can be interpreted as a less gender stereotypical description, which emphasize personal engagement rather than abilities and appearance. Nevertheless, it shows that there is a need to rethink and to refine the image of what a "good" female role model is, or what "proper" female values are, and to conceptualize gender identity more broadly. How female students form and negotiate their identities as "becoming" scientists is a key issue. These are processes that Aikenhead (1996) refers to as "cultural border crossings".

An excerpt from one of the girls can illustrate how she becomes what culture conventionally supports as a "proper participant in science": "Something happens when you enter the science community. I did not know I was a nerd before I started here [at the University]. But during the first year I had already seen all six Star Wars films for the first time, solved Rubik's cube, and gotten glasses, and had developed a kind of dry humour." This strategy can be understood as a form of assimilation, where she adopts the dominant codes to be a proper physicist while she marks the distance from girls she describes as "chatty", "I just cannot stand assumptions and opinion and empty talk that do not introduce anything." Thus she refers to the meta-narrative and a dichotomous understanding of girls as "chatty" and men as active or "vigorous" – emotion versus action.

Parents are also role models, providing girls and boys with different experiences and messages (Eccles 1994). A majority of my informants describe the influence of their parents. In particular, they emphasize the father's role as a source of inspiration, even in families where the mother or other close relations have a scientific background. Vanja writes that "my father has been an inspiration through his own studies and his master's degree. My mother is also interested in technology and works as an engineer with an ICT company, but she is mainly a self-made woman. In my family, several people, aunts and uncles and a grandfather, have higher education. I strongly feel that my interest in science is a part of my identity. But it is hard to say how much my family has influenced me. People say that the children of parents with higher education also choose higher education. To me this seems right. Nevertheless, I think I have identified more with my father, sharing my interest in science with him. My mother was different, she was the one who told us to work hard and be good at school, but she was not present in the same way as my father was." Fathers as significant others concerning girls' choice of science-related studies is broadly documented by research (Sjaastad 2011; Meece et al. 2006), but why do girls who choose to study STEM refer to their fathers as a source of inspiration – and not their mothers to the same extent? Is it because the majority of the fathers themselves possess a science degree, or because they generally show a greater interest in science and technology? Or is it because it is most natural to refer to fathers because the subjects they have chosen are already associated with masculine values and norms, and therefore, associated with men and fathers more than women and mothers? It also shows a "possible violation" of a general understanding that it is fathers who inspire their daughters to study science. Such assumptions, documented by research or not, should be regularly challenged and not left as taken-for-granted "truths". The picture might be more complex. For example, one study found that the more mothers believed in their children's science and mathematics skills in grade 7, the more likely those students were to pursue science careers at age 24 (Bleeker and Jacobs 2004). Whether young people report that they are influenced by the recognition and support they received from their parents, or not, such social influences might be troubling, because parents (like peers and teachers) often have gender stereotypical views of children's interests, aspirations and abilities in science and mathematics (see for example Eccles et al. 1983).

Questioning "Feminine Appropriate Values"

Statements such as 'girls prefer to work with people' or 'girls are more idealistic', construct social gender categories that are inscribed in the original biological categories (Johnston and Dunne 1996). One way around this is to move towards a deeper understanding of women's motivation for studying STEM, and to question traditional female roles as "caregivers". Stella has no plans to use her education to work with people: "I am interested in health, but I will not work with people. (It may sound selfish or cynical, but I certainly have enough insight to recognize my weaknesses.)". What she does here can be interpreted as an excuse for not being an "appropriate" girl, because she does not want to work with people being well aware that this is a culturally gendered expectation. Vanya also shows such implicit contradictions between subjectivity as a female science student and socio-cultural discourses of femininity. She writes that she would not consider "typical women's work because the term female work is somehow understood as something bad. Working with children, the sick or the elderly is not quite me, anyway." Vanya is not comfortable with the label "men's work" either: "Speaking of women's and men's careers, I am probably not very motivated by the term male work either. I do not like things that are gender labelled". This statement confirms an overall pattern in the analysis, as stated previously; the girls state that they want to be treated as individuals, not according to their gender/sex, despite ambiguities in their stories.

Furthermore, the stories reveal how problematic it can be to connect symbolically gendered constructions with the gendered preferences and identities of human subjects. Is it the case that girls who choose atypical educational routes "are undoing gender"? Choosing traditional masculine educational routes should not imply that women "lose" their femininity. Nevertheless, a large body of research has shown that girls respond to a greater extent than boys that they would like to work with people. Such findings are often understood as an expression of personal interests and aspirations, but the picture might be more nuanced. Girls might respond more positively to questions related to care and empathy, because they reflect the *socio-cultural expectations* of femininity and female behavior.[13]

[13] See also Chap. 18 which describes how traditional male STEM students, more than females, tend to "rely on pre-established roles, which in the case of science and technology are easily available and provide them with reassurance."

This chapter aims to take the gender and science education literature a step further by exploring the available discourses and practices of girls' educational choices, but the excerpts from the narratives demonstrate that the process toward a choice is ambiguous and complex. The girls' stories challenge the stereotypes of girls, and through their "atypical" choices and practice, they also challenge the image of what it means to do science. But they also challenge various attempts of specially designed initiatives, based on so-called female- or feminine values. Stella is one of several girls in the study who indirectly challenge a more "girl-friendly" approach by distancing herself from what she sees as "typically girlie". She is critical of "pink blogs" as a phenomenon, and writes: "In the past few years the so-called pink blogs have become popular, where young girls blog about clothing, makeup, hair and various "female topics". While these blogs are only meant as entertainment, they seem to make girls stupid. I hope young girls now know that there are more important things in the world than conditioning treatments, and that they will do more than be concerned about how they look." Stella is well aware of her opportunities as a scientist and has no intention to use her knowledge in the cosmetic industry: "I read somewhere that a chemist's number one dream employer was the cosmetic brand L'Oreal. I hope it is because of ignorance of other employers. I am certainly not one of those who aim to use my education to develop body lotion." Other girls express a similar attitude. They will use their education in various fields, and not on what "others" might think is appropriate for girls.

The informants in the "Write your choice" – study do share some values, interests and attitudes to STEM, but they also differ in various ways, even though they all have an "untraditional" choice of education in common. They tell different stories about their choices, although there are overlaps. In other words 'girls who choose STEM' are not a homogeneous category. Therefore, they will, in different ways, influence the discourse of female educational choices through diverse participation *in,* and approaches *to*, STEM. Thus, their differing participation in science might influence stereotyped images – both of what it means to be a girl, of what it means to be a female scientist, and of the image of scientists and science in general.

So far I have tried to show how telling local stories can open up a more plural representation of girls in/and science. I will now go on to discuss some of the implications of this.

Discussing Gender Stereotypes

Firstly, I will discuss whether communicating stereotyped images of girls might act as self-fulfilling prophecies, and secondly how challenging narratives about girls in/and science can give a more nuanced picture of the kind of girls actually choose to study science.

The point is not to show whether the meta-narrative, to some extent supported by research, is "false" or not, but to show that reality and everyday experiences are more nuanced than the dominant stories we tend to hear about girls and/in science.

Repeated Citations and Self-Fulfilling Prophecies

Words can lead to action. Words can trigger actions as impacts of the spoken word. This is in line with post-structuralist perspectives and in particular with Judith Butler's theorization of gender as performative (Butler 1990). In line with Butler's theory of performativity, educational choices might be seen as an effect of repeated citations – or actions. By repeating the meta-narrative of girls and/in science, norms about gender, aspirations and interests, behavior, and of educational choices, are "naturalized":

> Gender is the repeated stylization of the body, a set of repeated acts within a highly rigid regulatory frame that congeal over time to produce the appearance of substance, of a natural sort of being. (Butler 1990, p. 33)

This performative practice is consistent with the idea of gender as something we are *doing*; not something we *are* or *have*, but rather is a repetition and a ritual, understood, in part, as culturally sustained over time. A key point is that performativity is not necessarily a willing act of the subject, but rather is the reiteration of various gendered norms that tell us how to act, dress and think in terms of gender identity (Butler 1990). Using performativity as a lens for analysing girls' educational choices shows how such choices are gendered, and a part of the process of "doing" and "becoming" a female scientist. By repeating the meta-narrative without emphasizing the nuances, we can contribute to self-fulfilling prophesies:

> A self-fulfilling prophesy refers to a type of circumstances that take place when social participants believe that something is going to happen, and act accordingly, then it happens. In other words: The prophesy fulfils itself, so to speak, in that those who make the prophesy also act in a way that brings it about. Initially, the prophesy is based on a mistaken assessment of the state of things, but through the course of action elicited by the prophesy, it ultimately works, then the prophesy is fulfilled, as though it had been based on a correct assessment. (Østerberg 2003, p. 62)

Such repetitive practice includes an element of power, because it reproduces a normativity which people relate to and "define" themselves in relation to. Researchers' repeated stories about how girls "are", what kind of aspirations they share as a group, and about the grounds on which they choose, are absorbed into language and cognition as "common sense".

We base our choices on our perceptions of the world, like Tina. She has chosen to study computer engineering, but she feels she does not fit the subject's public "image": "Many have really been surprised when I tell them that I have started studying computer engineering. Are you one of *them*, they ask me? Well, I guess I can be, I reply with a hesitant voice and a half smile, because I am still very uncertain. Am I really the right girl for this?" Tina wonders whether she really wants to be a computer nerd, which underlines the conflict between her academic interests and motivation on the one hand, and society's expectations, myths and prejudices on the other. She is not alienated from the ICT profession, but society's scepticism and gendered perceptions make her feel alienated from the potential opportunity to become a computer programmer. Such paradoxes open the door to several interesting interpretations of the data, which must be situated in a historical and socio-cultural context. The choices my informants make, are a part of themselves. They are what they choose – and if they choose outside the box, they stand out from their peers and outside of the public view on gendered behavior, values and norms. Girls are "free" to choose whatever career they would like in countries such as Norway, but are also limited by the meta-narrative, and therefore have to negotiate their choices and participation in science, because of their sex.

Generalizations might contribute to negative and limiting stereotypes, which in turn shape our perceptions of normality and variance. Research does not occur in a vacuum but is part of a set of mutually constitutive intra-active[14] processes. As studies of gender stereotypes among adults reveal, gender polarization is resistant and does not necessarily wane as people grow older. Perhaps as a result of these tendencies, researchers often expect to find sex differences. Even researchers are not immune to popular essentialist beliefs about gender, and we should therefore address the ethical dimensions of generalizing findings. As the physicist and feminist Karen Barad (2007) argues, politics and ethical issues are always part of scientific work, but specific historical circumstances encourage people to fail to see those connections.

As history inside and outside the field of science shows, dominant discourses of science are open to reformulation, reinterpretation and destabilization, offering possibilities for widening the range of scientist subjectivities available, for those in, and for those often excluded from science. A key question in the process of becoming a scientist may still be "who we think we must be to engage in science" (Barton 1998, p. 379). A relevant follow-up question might be; *why do girls think that way, and what is there to do about it?*

A More Nuanced Picture Towards Diversity and Sensitivity

On the one hand, gender inequity is a problem within technology, engineering, mathematics and physics, and stakeholders want clear answers as to how the problem can be solved. On the other hand, in the process of making broad generalizations based on differences due to sex we stand the risk of losing important nuances that again might lead to the creation and cementation of stereotypes. As Cathrine Hasse puts it: "When we look for gender differences we might overlook differences that are not simply sharply defined and cannot be distributed in two groups defined by the categories male and female" (Hasse 2002, p. 253). I wish to emphasize the importance of more detailed knowledge of the variation between women who choose science, to avoid categorizing women who choose science as a

[14] See Karen Barad (2007) for a discussion of the concept of intra-activity.

homogeneous group. They do share some interests, aspirations and values, but the way I interpret my data, they also describe and justify their choices differently.

Challenging the meta-narrative of girls and/in science implies a consideration and consciousness of the language in use, including what type of language we use as researchers, teachers and parents. The French feminist thinker, Luce Irigaray, says that researchers must innovate and reformulate the age-old socio-cultural values and concepts on which the understanding of gender rests (Irigaray 2004). Is there a need to create a new language with more room for women in the natural sciences? Some would claim that inventing new words to describe "old content" is the same as "the emperor's new clothes"; the content will remain the same. An ambition can be to elevate girls' own descriptions and experiences, which I have tried to do in this chapter, to create more room for more choices available to girls (and boys) enrolling in higher education or embarking on their career.

Theories about normality and non-conformity are related to ideas about what is perceived as "pure" and "impure" in a society; or what the social anthropologist Mary Douglas (1966) theorizes as "matter out of place". Anything impure is "matter out of place", meaning things (or persons) that do not fit in. Is it the case that the absence of girls in the sciences is ascribable, among other things, to girls being perceived (and perceiving themselves) as "impure" in a research tradition and discipline where the norm includes values such as: neutral, objective and value-neutral research? This is, perhaps, not putting too fine a point on it, and it is certainly a provocative assertion; but language is power, and power can be symbolic. Symbolic power is, according to Bourdieu (1996), the power to construct reality. The researcher as a producer of knowledge, and the teacher as a communicator both have the power to define a part of reality through their choice of words. Therefore the researcher has a significant responsibility to avoid one-sided and oversimplified language and a repetitive practice which might help to confirm negative stereotyped perceptions.

According to feminist scholars such as the biologist Donna Haraway (1991) and physicist Karen Barad (2007), scientific research produces realities and worlds that are never without consequences. Therefore, researchers cannot evade moral responsibility for the consequences of their research. The researcher is always located in, and part of, the reality he or she explores (ibid). Research on educational choices is no exception. We are all socialised into gender-specific values and norms that tell participants in a society what counts as valuable research, what being a girl implies and what being a boy implies – and how we are expected to choose educations and professions based on sex.

More nuanced stories about girls and/in science may provide us with a better understanding of the variety of ways girls engage in science, and how this engagement is shaped by their experiences and views of what kind of girls "they" are. Like Haraway, I argue that although not all stories about the world are equally valuable, several stories are better than one (Haraway 1989). My interpretation of this statement – and an implication of it, is that several stories of girls' educational choices are better than one. This means both telling more stories, and refining the stories we tell about girls and/in science.

In the article *Gendered education in a gendered world: Looking beyond cosmetic solutions to the gender gap in science*, Sinnes and Løken (2012) suggest looking at research and recruitment initiatives from a perspective developed by feminist critics of science, to reveal the implicit gendered assumptions that can be found within projects focusing on gender issues in science (see also Brickhouse 2001; Harding 2001; Sinnes 2006, for reviews of the historical development of feminist perspectives on science education). Use of feminist theory to elicit the view of gender reflected in interpretations of IRIS findings can be useful (see also discussion of different feminist theoretical approaches in Chap. 4 in this book).

It is beyond the scope of this chapter to go deeper into the study of girls' diverse lived experience in a more qualitative way. I have argued that there needs to be increased attention to the level of local stories; without it we will not obtain a nuanced picture of what kind of a girl does science. Failing to attend to these issues might, as argued, lead to a cementation of negative stereotypes and thus limit diversity and plurality in science. To focus more on diversity and plurality within the gender groups may be a step towards a more (gender) sensitive (science) education, which could avoid contributing to enhancing stereotypical images that young people of both sexes feel alienated towards.

References

Aikenhead, G. S. (1996). Science education: Boarder crossing into the subculture of science. *Studies in Science Education, 27*, 1–52.

Baker, D., & Leary, R. (1995). Letting girls speak about science. *Journal of Research in Science Teaching, 32*(1), 3–27.

Barad, K. (2007). *Meeting the universe halfway: Quantum physics and the entanglement of matter and meaning*. Durham: Duke University Press.

Barton, A. C. (1998). *Feminist science education*. New York: Teachers College Press, Columbia University.

Bleeker, M., & Jacobs, J. E. (2004). Achievement in math and science: Do mothers' beliefs matter 12 years later? *Journal of Educational Psychology, 96*(1), 97–109.

Boe, M. V., Henriksen, E. K., Lyons, T., & Schreiner. (2011). Participation in science and technology: Young people's achievement-related choices in late-modern societies. *Studies in Science Education, 47*(1), 37–72.

Bourdieu, P. (1996). *Symbolsk makt (Norwegian edition)*. Oslo: Pax Forlag.

Braidotti, R. (1994). *Nomadic subjects. Embodiment and sexual difference in contemporary feminist theory*. New York: Columbia University Press.

Brickhouse, N. W. (1998). Feminism(s) and science education. In K. Tobin & B. Fraser (Eds.), *International handbook of science education* (pp. 1067–1082). New York: Kluwer.

Brickhouse, N. W. (2001). Embodying science: A feminist perspective on learning. *Journal of Research in Science Teaching, 38*, 282–295.

Brickhouse, N. W., Lowery, P., & Schultz, K. (2000). What kind of a girl does science? The construction of school science identities. *Journal of Research in Science Teaching, 37*(5), 441–458.

Brotman, S. J., & Moore, M. F. (2008). Girls and science: A review of four themes in the science education literature. *Journal of Research in Science Teaching, 45*(9), 971–1002.

Butler, J. (1990). *Gender trouble. Feminism and the subversion of identity*. London: Routledge.

Chetcuti, D. (2009). Identifying a gender-inclusive pedagogy from Maltese teachers' personal practical knowledge. *International Journal of Science Education, 31*(1), 81–99.
Daiute, C., & Lightfoot, C. (2004). *Narrative analysis. Studying the development of individuals in society*. Thousand Oaks: Sage.
Danielsson, A. (2009). *Doing physics – Doing gender. An exploration of physics students' identity constitution in the context of laboratory work*. Uppsala: Uppsala University.
Denzin, N. K. (1989). *Interpretive biography: Vol. 17. Qualitative research methods*. Thousand Oaks: Sage.
Douglas, M. (1966). *Purity and danger: An analysis of the concepts of pollution and taboo*. London: Routledge & Kegan Paul.
Eccles, J. (1994). Understanding women's educational and occupational choices. *Psychology of Women Quarterly, 18*, 585–609.
Eccles, J. S., Adler, T. F., Futterman, R., Goff, S. B., Kaczala, C. M., & Meece, J. L. (1983). Expectancies, values and academic behaviors. In J. T. Spence (Ed.), *Achievement and achievement motives* (pp. 75–146). San Francisco: Freeman.
Fine, C. (2010). *Delusions of gender. The real science behind sex differences*. London: Icon Books.
Gullestad, M. (1996). *Hverdagsfilosofer*. Oslo: Universitetsforlaget AS.
Hammersley, M., & Atkinson, P. (1996). *Ethnography: Principles in practice*. London: Tavistock.
Haraway, D. (1989). *Primate visions. Gender, race and nature in the world of modern science*. London: Verso.
Haraway, D. (1991). *Simians, cyborgs, and women: The reinvention of nature*. London: Free Associations Books.
Haraway, D. (1992). The promises of monsters: A regenerative politics for inappropriate/others. In L. Grossberg, C. Nelson, & P. A. Treichler (Eds.), *Cultural studies* (pp. 295–337). New York: Routledge.
Harding, S. (1986). *The science question in feminism*. Milton Keynes: Open University Press.
Harding, S. (2001). Comments and reply comments on Walby's "Against Epistemological Chasms; The Science Question in Feminism Revisited". Can democratic values and interests every play a rationally justifiable role in the evaluation of scientific work? *Signs, 26*(2), 511–525.
Hasse, C. (2002). Gender diversity in play with physics: The problem of premises for participation in activities. *Mind, Culture, and Activity, 9*(4), 250–269.
Hattie, J. A. C. (2009). *Visible learning. A synthesis of over 800 meta-analyses relating to achievement*. New York: Routledge.
Hazari, Z., Sonnert, G., Sadler, P. M., & Shanahan, M.-C. (2010). Connecting high school physics experiences, outcome expectations, physics identity, and physics career choice: A gender study. *Journal of Research in Science Teaching, 47*(8), 978–1003.
Holter, Ø. G., Svare, H., & Egeland, C. (2009). *Gender equality and quality of life. A Norwegian perspective*. Oslo: The Nordic Gender Institute (NIKK).
Horsdal, M. (2012). *Telling lives: Exploring dimensions of narratives*. New York: Routledge.
Irigaray, L. (2004). *Key writings*. London/New York: Continuum.
Johansson, A. (2005). *Narrativ teori og metod*. Lund: Studentlitteratur AB.
Johnston, J., & Dunne, M. (1996). Revealing assumptions: Problematising research on gender and mathematics and science education. *Gender Science and Mathematics, 2*, 53–63.
Kjærnsli, M., Lie, S., Olsen, R. V., & Roe, A. (2007). *Tid for tunge løft. Norske elevers kompetanse i naturfag, lesing og matematikk i PISA 2006. (Time for heavy lifting. Norwegian students' competence in science, reading, and mathematics in PISA 2006)*. Oslo: Universitetsforlaget.
Lykke, N. (2012). *Feminist studies. A guide to intersectional theory, methodology and writing*. New York: Routledge.
Lyons, T., Quinn, F. (2010). *Choosing science: Understanding the declines in senior high school science enrolments*. National Centre of Science, ICT and mathematics education for rural and regional Australia. (SiMERR Australia), University of New England.
Meece, J. L., Bower Glienke, B., & Burg, S. (2006). Gender and motivation. *Journal of School Psychology, 44*, 351–373.

Miller, P., Slavinski-Blessing, J., & Schwartz, S. (2006). Gender differences in high-school students' views about science. *International Journal of Science Education, 28*(4), 367–381.

Osborne, J., & Collins, S. (2001). Pupils' views of the role and value of the science curriculum: A focus-group study. *International Journal of Science Education, 23*(5), 441–467.

Østerberg, D. (2003). *Sosiologiens nøkkelbegreper*. Oslo: Cappelen Akademisk Forlag as.

Quinn, F., & Lyons, T. (2011). High school students' perceptions of school science and science careers: A critical look at a critical issue. *Science Education International, 22*(4), 225–238 (special issue).

Rubin, G. (1975). The traffic in women: Notes on the 'political economy' of sex. In R. R. Reiter (Ed.), *Toward an anthropology of women* (pp. 157–210). New York: Monthly Review Press.

Schiebinger, L. (1999). *Has feminism changed science?* Cambridge, MA: Harvard University Press.

Schreiner, C., & Sjøberg, S. (2007). Science education and youth's identity construction – Two incompatible projects? In D. Corrigan, J. Dillon, & R. Runstone (Eds.), *The re-emergence of values in the science curriculum* (pp. 165–248). Buckingham/Philadelphia: Open University Press.

Schreiner, C., Henriksen, E. K., Sjaastad, J., Jensen, F., & Løken, M. (2010). Vilje-con-valg: Valg og bortvalg av realfag i høyere utdanning. *Kimen* (2/2010): Naturfagsenteret.

Shanahan, M.-C. (2008). Identity in science learning: Exploring the attention given to agency and structure in studies of identity. *Studies in Science Education, 45*(1), 43–64.

Sinnes, A. T. (2006). Three approaches to gender equity in science education. *Nordic Studies in Science Education, 1*, 72–83.

Sinnes, A., & Løken, M. (2012). Gendered education in a gendered world: looking beyond cosmetic solutions to the gender gap in science. *Cultural Studies of Science Education*, 1–22. doi: 10.1007/s11422-012-9433-z.

Sjaastad, J. (2011). Sources of inspiration: The role of significant persons in young people's choice of science in higher education. *International Journal of Science Education, 33*, 1–22.

Sjøberg, S. (2000a). Kjønn og naturvitenskapens "kroppsspråk". *Nordisk Pedagogikk, 2*, 80–89.

Sjøberg, S. (2000b). Interesting all children in "science for all" curriculum. In R. Millar, J. Leach, & J. Osborne (Eds.), *Improving science education* (pp. 165–186). Buckingham: Open University Press.

Solsvik, A. (2004). Likeverd i barneoppdragelsen. In I. Frønes & T. S. Wetlesen (Eds.), *Dialog, selv og samfunn* (pp. 61–86). Oslo: Abstrakt Forlag.

Solvang, P. (2006). Problematisering, utdefinering eller omfavnelse. Om normaliteten. In J.-K. Breivik & E. Thomas Hylland (Eds.), *Normalitet* (pp. s. 167–185). Oslo: Universitetsforlaget.

Staberg, E. M. (1994). Gender and science in the Swedish compulsory school. *Gender and Education, 6*(1), 35–46.

Stadler, H., Duit, R., & Benke, G. (2000). Do boys and girls understand physics differently? *Physics Education, 35*, 417–422.

Wertsch, J. V. (1991). Voices of the mind: A sociocultural approach to mediated action. London/Sydney/Singapore: Harvester Wheatsheaf.

Chapter 18
Italian Female and Male Students' Choices: STEM Studies and Motivations

Alessandra Allegrini, Giuseppe Pellegrini, and Chiara Segafredo

Introduction: Young People and Their Study Paths

Research has pointed to some differences in the priorities of males and females choosing higher education (Blickenstaff 2005; Brotman and Moore 2008). Published studies on this topic offer numerous explanations and adopt various methodological approaches, but the factors taken into consideration are common to all of them: the social, economic, and ethnic context; the family context; gender issues; the influence of the school and the quality of teaching; and interest in and aptitude for the sciences (Scantlebury and Baker 2007; Brotman and Moore 2008).

The literature in this field has highlighted the influence of the family context on the development of educational motivation and student progress, and the importance of positive parent-child relationships for the creation of adaptive capacity in the educational context (Ryan et al. 1995; James 2002; Wildhagen 2009; Munk 2011). Data from the IRIS survey conducted in Italy confirm the influence of the family on orienting students towards scientific studies, and they point out a considerable gender difference: one quarter of females have parents coming from a university background while seven out of ten males are the first in their family to study at university.

The Italian survey also shows that students enrolling in Engineering and Computer Science faculties come from families with a lower level of tertiary education, and they attend courses where there is a large male presence. Biology courses, on the other hand, receive a high level of female enrolment coming from families with high "science capital" (see Chap. 6); it is therefore more often the case that a female student enjoys support for her choice from university-educated parents. One might infer from this that families with a higher "science capital" help female students overcome cultural pressures aimed at orienting them towards traditional gender

A. Allegrini • G. Pellegrini (✉) • C. Segafredo
Observa Science in Society, Viale A. Fusinieri, 65, 36100 Vicenza, Italy
e-mail: allegrini_a@yahoo.it; pellegrini@observanet.it

E.K. Henriksen et al. (eds.), *Understanding Student Participation and Choice in Science and Technology Education*, DOI 10.1007/978-94-007-7793-4_18

roles and behavioural models, and assist them in overcoming the obstacles in their future professions that arise when they are at the university (Blickenstaff 2005).

On the other hand, as we will further describe, a large proportion of male students enrolled in Engineering and Computer Science courses comes from technical institutes which in Italy are male-dominated secondary school contexts, traditionally oriented to work after school rather than academic education. This is a crucial factor since it suggests that parental role models also have a central influence on these male students, but not tied to educational and cultural priorities, but more focussed on work improvement (Allegrini 2004).

An analysis of pre-university study shows that males mainly come from secondary schools which are similar to their current course of study, while female students have more often attended schools of other kinds; the largest share of students continuing with their previous studies are enrolled in Computer Science and Biology. These data confirm what has been shown by various studies on women's preferences for biological, bio-medical and care-giving studies: this is a tendency which explains the greater presence of women in Biology courses (Osborne and Collins 2000). Male students, on the other hand, more often come from technical secondary schools. Normally, the majority of female students attend Biology or Biotechnology courses in higher education, and come from humanities and science-based secondary schools.

Method

Four open questions proposed to Italian students in the Iris survey enabled us to investigate the factors which encourage students to choose science studies, and allowed us to verify gender differences, and differences between the various degree courses, through a qualitative analysis of 2,192 open responses to the following questions:

- Q 9. Describe how you came to choose this course.
- Q 13. If someone you know was thinking about enrolling on your course and asked you about it, what would you say to her or him?
- Q 15. Do you attend a course where one gender is over-represented? If so, why do you think this is the case?
- Q 16. Do you see any reason why the situation described above should be changed – and if so, what do you think could be done to change it?

Using the Atlas.ti programme package for qualitative analysis, the research team used an open-inductive coding process to identify central concepts appearing in the students' responses. Responses were then categorised (Strauss and Corbin 1990) to reduce the complexity of the corpus. This analysis resulted in seven conceptual dimensions: five taken from Eccles' achievement-related choices model (Eccles 1994;

Eccles and Wigfield 2002, see Chap. 2), and two developed inductively from the data material. The seven dimensions used for the analysis were:

- *intrinsic value*: all the elements of interest in, and enjoyment of, a subject;
- *attainment value*: statements about personal attainment, identification with the task and social role of scientists;
- *utility value*: aspects of individual utility in relation to a project for personal growth;
- *perceived cost*: perceptions of the costs required to complete a course of study;
- *expectation of success*: the expectation of success, including self-perception and self-assessment of the student's abilities (Rosenberg 1979, 1986);
- *cultural features*: scholastic and educational factors (the role of the teachers, the educational background, and previous experiences), and also factors such as the role and expectations of family and friends, as well as influences outside the school environment;
- *natural features*: beliefs that a certain aptitude and predisposition for scientific subjects are connected with, and necessary for, academic success.

These last two interpretative dimensions have been evaluated as consistent with, and relevant to, a gender issues analysis, most of all in the light of qualitative outcomes collected through the open questions. Further details can be found in Chap. 4 for general perspectives and in Chap. 20 for traditional and non-traditional aspects that underpin gender and science representations. In this chapter, we focused our attention on priorities and values, as discussed in Chap. 9 for the data sets from Denmark, Norway and England.

Results: The Motivations Which Lead Young People Towards Science

Intrinsic Value

The first open-ended question – Describe how you came to choose this course – allowed students to track the aspects which had encouraged them to undertake a scientific course of study. In their short accounts, we frequently found words such as "interest", "passion" and "pleasure", which are notable for the emotional connotation given to the main motivation expressed by students: an interest in scientific subjects. The intrinsic value dimension showed two clearly distinct response tendencies: one is personal, but still more rational and well-considered, connected with interest in and attraction to the subject:

> Since I was a small child, I have been very interested in mechanics, especially in motor engineering (male, Mechanical Engineering)

The other tendency is more emotional and emotive, an affective response couched in terms such as love, appeal, enthusiasm, magic, passion, and pleasure:

> After careful reflection, I eliminated the subjects which did not excite me. Chemistry was what I loved most, and it is what I still love; I like to know how something happens, and what lies at the bottom of the phenomena which surround me (female, Chemical Engineering)

A high level of correspondence between one's interests and the environment to which one belongs leads to greater satisfaction, better performance in the field, and therefore a greater determination to carry on with one's decisions (Ackerman et al. 2001; Eccles and Wigfield 2002; Schiefele et al. 1992):

> I chose this degree course because of my passion for it, and because it involved scientific subjects, and because I wanted to be an engineer (male, Electrical Engineering)

An academic interest in a particular subject is therefore correlated with success in it (Ackerman 1996), and vice versa. This was evident in the answers provided by many students – as can be inferred in the texts from the number of times that terms communicating intrinsic value were used together with words which referred to a sense of satisfaction, achievement, and success:

> I was very attracted by ecology, and I was doing very well at secondary school (male, Biology)

Aspects of pleasure, interest, and real passion for the subject, or for certain specific features of it, were common to both men and women in the intrinsic and interest value dimension. Certain differences are worth noting, however, especially with regard to different disciplinary areas. Students enrolled on different tracks in Engineering often emphasized a specific interest in the world of machines, both mechanical and electronic, while women's interest was more often described as a generic "interest in or leaning towards scientific subjects". Numerous responses emphasized the former aspect:

> A strong propensity for Physics, Mathematics, and other scientific subjects, as they apply to the real world and to their social and technical function (male, Mechanical Engineering)

> I chose engineering because of my passion for scientific subjects and I hope to gain a passion for the more technical subjects I will be coming across in my course (male, Mechanical Engineering)

Among the replies provided by women, on the other hand, we have:

> Interest in the course subjects (female, Mechanical Engineering)

> The course subjects fascinate me (female, Mechanical Engineering)

In other words, the interest expressed by men attending engineering faculties was especially in technical aspects and technological subjects, whilst women's interest was generically defined in terms of an interest in scientific subjects, science, or the techno-scientific field. Male students' experiences could also be related to their childhood experiences with technology (electronics, vehicles etc.), thus giving males a stronger sense of "ownership" of technical aspects of STEM (Sjøberg 2000; Jones et al. 2000).

This same trend was apparent when we compared the responses of men and women studying Computer Science and Technological and Computer Sciences, where a passion for electronics and computers was a factor strongly orienting males, and less the females, who, except in one case ("Electronics fascinate me" –

female, Computer Science), gave a reason for their choice based on a generic interest in techno-scientific subjects. Among the men's replies were:

> Electronics have always interested me (male, Technological and Electronic Sciences)

> I've always been interested in electronics and computers in general (male, Electronics)

Women's responses included:

> Because I'm interested in techno-scientific subjects (female, Electronics)

> Because of an interest in techno-scientific subjects (female, Electronics)

Among students of Mathematics, Statistics, and Computer Science, a specific interest in, predisposition or inclination – a true passion, in some cases – towards the study of mathematics constituted a major orienting factor from primary school onwards. There were numerous responses similar to the following:

> My passion for mathematics developed as early as middle school, but it was only thanks to the support of my mathematics teacher in my final year of high school that I decided to take this uphill path (female, Mathematics)

> I've been interested in mathematics since primary school. As time passed, this interest was confirmed, and now I've chosen the course I was most interested in (male, Mathematics)

Finally, in all courses, the word "passion" – one of the terms most frequently used by male and female respondents, together with "interest" – was often accompanied by other terms such as "expectations", "aspirations", and "ambitions" in male cases:

> I carefully evaluated my knowledge and abilities in many areas, but above all in those which interested me the most, and concluded that this faculty could offer me what would best reflect my expectations (male, Information Engineering)

In the case of females, the purer and less tangible aspects were emphasized, and the stress was placed on a speculative interest in knowledge not tied directly to concrete results:

> I felt myself particularly attracted by "conceptual" scientific subjects, at the end of High School, (not for calculation purposes)! (female, Biology)

Attainment Value

Both male and female students cited various elements in their responses which explained orientation towards science courses in terms of life-choice. This concerns the personal self-fulfilment which is attained by the idea of becoming a scientist, thereby satisfying particular aspirations: challenging oneself to attain a career goal, and pursue what is perceived to be a socially important career.

We often encountered phrases such as "I've always wanted to do this", "since I was a boy", which are expressions referred to a deep-rooted conviction behind a life choice. It was apparent that the "vocational" aspect, and the desire for attainment, generated and supported the choice of a university course. There were also references to various images: dreams, future aspirations and desires, expressed in their "purest" and most open form, frequently unconnected with the tangible and practical relevance of the course:

> I chose it so that I could fulfil my secret dream (female, Physics)

The vocational element was mainly present when students talked about their choice of beginning a path of personal growth or the definition of their passions and aspirations, and when they underlined the opportunity for self-fulfilment that their educational choices gave them. Those who mentioned the vocational side revealed a sort of 'calling' to perform a certain role. An example of that was the reply given by a female Mathematics student:

> I would like to become a teacher, and I know that society needs teachers (female, Mathematics)

As regards public impact, various responses suggested the desire and responsibility to change and improve society with the aid of scientific research:

> I chose it because it is my wish to help my neighbours, and research seems to me to be the best area to express my passion (female, Biology)

Although these responses were equally significant for men and women, an analysis of the texts from a gender perspective showed some differences, mainly in the kind of language used and the content proposed; and these acquired further – and specific – meaning when they were related to a certain disciplinary area.

Social relevance aspects, for example, were cited by female and male students with different meanings. The former offered reasons which more closely concerned mankind, the environment, and nature, with a specific connotation of "taking care of others", while male students mentioned issues concerning the exploitation of science and technology for the betterment and development of society, which were often described in general and abstract terms. Female students' orientation towards aspects of life tended to shift their university choices to the so-called life sciences (Biology, Biological Sciences, Chemistry, or Biotechnology). The prevalently male orientation towards technical and technological aspects was most evident in technological faculties, especially Engineering.

> I've chosen this course because my desire is to help others, and research seems to me to be the best area for expressing my passion (female, Biology)

> I have always dreamed of being an inventor. Although it may seem stupid, the thing which is likely to get closest to this is mechanical engineering. And finally because I want to be able to build machines which can improve life and also increase the development of society (male, Mechanical Engineering)

References to social relevance, such as helping others, were most frequent among respondents attending faculties mainly chosen by women, such as biological sciences; and they were also common in the case of biotechnology, where female students emphasized the bio-medical aspects of care-giving, despite the technological nature of the degree course. As also emphasised in Chap. 20, the technological aspects typical of the biotechnology faculty, were related to the care dimension enriched by a future-oriented attitude. In fact, "future" is a word which appeared in many responses: "the future of mankind", "the society of the future", "the future of incurable diseases", as suggested by certain female Biotechnology students:

> The world of research fascinated me, the study of diseases and experimental techniques for healing the sick, and I chose the faculty which I thought came closest to my expectations (female, Biotechnology)

> I followed my passion for science. The future will be increasingly based on biotechnology (female, Biotechnology)

Utility Value

A sense of utility and future career prospects were important motivating factors: male and female students showed that they were very attentive to, and aware of, job opportunities when they completed their studies. Students stated that they were studying sciences in order to earn high salaries, and attain their career and job objectives (see also Chap. 9). This concerns career objectives and lifestyles, but also the social role which the students intended to perform:

> I chose this university course because of the enormous range of job opportunities it provides (male, Electronic Engineering)

> Good job prospects (female, Mathematics)

> I tried to combine my passion for scientific subjects with the chance to attain ambitious goals in the workplace (male, Chemical Engineering)

A considerable amount of research into the future emerged from the answers:

> Initially, I was undecided between humanities and chemistry, but I chose the latter because it was more likely to offer a future with greater opportunities (male, Chemistry)

This search was linked with the perception, or certainty, that scientific training offers rapid entry into the labour market and certain career prospects, due to opportunities for the practical use of knowledge acquired at university:

> I believe it is fundamental to do courses which have real utility in life, which are not just 'hot air' but provide good job opportunities at the end of the course. I also wanted to plan and carry out something concrete (male, Computer Science)

Another utility factor was the significance of the study area: students showed a desire to undertake degree courses in order to do something of note:

> I chose this course because the future is increasingly linked to technology, so you need to understand it and have a solid base (male, Computer Science)

> To try to understand the future, and I'm convinced that scientific subjects are the only way to do this, and I also think that they are the only subjects which give you the hope of getting a job (female, Biology)

Job prospects in research provided a powerful motivation, and enabled students to formulate their objectives as far as academic success and self-confidence are concerned:

> I like being active in the field of research. It's an environment which opens doors to many professions (male, Biology)

> I made my choice mainly by thinking about the job I would like to have one day, and so I thought that Physics might open a lot of doors in the research field (female, Physics)

An analysis of the responses from a gender perspective shows that male and female students attach great importance to their future employment and professional opportunities. In the case of men, this significance was more frequently expressed in terms of remuneration and, in some cases, the social role – in the sense of public image – which the course allowed them to occupy. This was especially the case for the "applied" sciences – Biotechnology, and most of all the various Engineering disciplines.

> My cousin was doing Chemical Engineering, but quit, and we decided that we'd attend Biotechnology together. Prospects of becoming important people (male, Biotechnology)

> It seemed to me the course which would give me the best chance to get on in the labour market in the future, and it was the nearest one to my passion (male, Mechanical Engineering)

> I looked at which path would give me good training and an excellent income at the same time. To make an impact on society, be someone, and make money (male, Mechanical Engineering)

To a lesser extent, men – in contrast to women – also cited aspects of utility in the "pure science" faculties (Mathematics, Mathematical Sciences, and Physics), whereas women attached more significance to a passion for knowledge or an interest in the subject. The following are examples of responses given by males enrolled in these faculties:

> I looked at all the choices I would need to make, what I would like to do, and finally the employment opportunities I would have once I would complete my studies (male, Mathematics)

> Looking at course programmes for the various faculties, and choosing the one which gave me the most job opportunities and which I liked the best (male, Mathematics)

In the case of women enrolled in Engineering, this dimension was an important element in making their decisions. More than one female student who enrolled in Mechanical Engineering replied:

> I made my choice on account of the opportunities I would have in the future if I chose this course;
> I chose my university course because of my interest in scientific subjects and the future opportunities it could offer me

Finally, there were some noteworthy differences among the replies regarding a career in research. Women, above all, expressed a specific motivation to engage in scientific research, in particular Biology and Biotechnology students. They considered the social relevance of a research career both in laboratories and in biomedical clinics.

> My choice for this type of course was conditioned by the fact that it allows me to work in scientific research and to have a certain level of responsibility on advancing scientific and technological development (female, Biotechnology)

> Because I like researching, finding vaccines for diseases or finding something better than what already exists (female, Biotechnology)

It should also be noted that, in cases where the importance of income was cited, most of the responses came from students enrolled on Engineering courses, in particular Mechanical Engineering, which is probably the course that provides the most professional type of training.

> I chose this course because I think it guarantees that I'll find a good job and it will be well paid (male, Mechanical Engineering)

> To achieve my goals, which are to have a good job which I like and a high income (female, Mechanical Engineering)

Perceived Cost

Both male and female students attached particular importance to the difficulties which had to be faced on their courses: the workload to be managed, the length of the course, and the difficulty of the degree course.

> I had to choose between veterinary science and biotechnology, but the number of years you have to study for medicine and veterinary science scared me, so I chose biotechnology (male, Biotechnology)

Here, the heavy financial burden of university fees, the ongoing economic commitment, the cost of university services, and the need to adapt to a new city and to a university environment emerge.

> Reasons: personal propensity for studying technical subjects. Best relationship between the quality of the courses and the cost of the university: the most beneficial and effective (male, Mechanical Engineering)

Opinions relating to academic organizations were closely connected with the previous evaluations. Administrative efficiency, effective management of teaching, and lack of overcrowding were other factors appreciated by students who selected the university to attend with special care.

> I chose only on the basis of possible job opportunities and because other "interesting" faculties were overcrowded, so there was less chance of "excelling" (male, Computer Science)

Other students performed a cost-benefit calculation of the available resources, and found it impossible to move to a university outside their home town: this meant excluding faculties a long way from home. Apparent in these cases was the significance attached to a graduate career, regardless of the career itself, in order to gain access to a profession:

> In today's labour market, you are supposed to have a degree (male, Computer Science)

From a gender perspective, there are no marked differences among the overall responses within the 'perceived cost' category. Evaluations of the difficulties to be faced were common to men and women attending different faculties and courses. Nonetheless, it is interesting to note that the choice process of some female engineering students was more problematic than in the case of their male counterparts or female students of other faculties:

> The choice of my university degree course was not targeted, but was one of the three possibilities or preferences which I had the chance to take (female, Automotive Engineering)

> The choice wasn't easy. I had to choose from among very different faculties and appealing courses, including the non-scientific ones. I chose electronic engineering because it fascinated me more than any other faculty, it had the courses I liked the best, and would lead to a very interesting profession, from my point of view (female, Electronic Engineering)

These data demonstrate a more specific choice-making method than the one used by males, and also raises the issue of the influence by various people in the selection process. A large number of responses emphasized the importance of certain key figures (as mentioned in Chaps. 9 and 11), in the following order: teachers who had encouraged and supported the student, parents with degrees, tutors (where a mentoring service operated), and friends. The support of the teachers was mentioned especially by female students, who appeared to be more influenced by gender stereotypes attached to scientific or engineering professions as mentioned by a student:

> In my course it is predominantly men. I guess this is so because the figure of the engineer in society is still male-dominated (female, Electronic Engineering)

Expectation of Success

Self-image and the evaluation of one's abilities are connected with the chances of success (see Chap. 2). Hence, past experiences and future prospects were expressly cited by the respondents. The more students felt that they had acquired knowledge of a subject during their schooling years, the more probable it was that they would choose it as their subject:

> Because I acquired a basis at high school, and I like electronics (female, Mechanical Engineering)

These reasons were cited by a small number of students, but they exhibited a particular combination of elements closely connected with the other dimensions that we have examined above.

Expectations of success were, in fact, linked with a future public image, particularly for male students in Biotechnology and Engineering faculties:

> It seemed to me the degree course which would give me the best chance of entering the labour market in the future, and which was the closest to my passions (male, Mechanical Engineering)

Finally, we consider interest issues and self-evaluation elements proposed by respondents who had selected their degree course on the basis of a search for success.

> Since I was a child I feel great interest towards mechanics, particularly for automotive engineer for which I think I will get good results (male, Mechanical Engineering)

> I chose my degree course purely and simply on the basis of the pleasure I feel in studying the subject and because of the good marks obtained in the secondary school (female, Biology)

Cultural Features

This dimension includes school and educational factors (the role of the teachers, the educational background, and previous experiences), and also external factors in general (the role and expectations of family, friends, and influences outside the school environment).

For many students, the linearity of their educational background, in the sense of a continuing link between university studies and the area of interest in their secondary school, was a determining factor in their choice (see Chaps. 2 and 9). Continuity between high school and university choices can be interpreted in two ways: the perception of having acquired the necessary knowledge and basis for studying the subject and, at the same time, the desire for consistency in continuing to develop a route which had already been selected:

> Having already experienced what chemistry is at high school, I confirmed this choice (male, Chemistry)

> The subjects I took at high school were very interesting and stimulating, so I decided to continue with the same subjects I started to study at school: biology, pathology, hygiene, and all the subjects connected with the medical-scientific environment (female, Biology)

These cultural features, according to the Eccles model, contribute to shape motivations and reinforce intrinsic values such as interest, attainment and utility.

Out of school experiences were also mentioned as having a special significance. Male and female students cited the importance of activities "in the field", such as the time spent in laboratories and guided visits to research centres or museums:

> I wanted a complete change. I went to a high school which specialized in languages. After a trip, I decided I wanted to do biology (female, Biology)

As stated earlier good performance levels at high school raised students' levels of self-esteem and satisfaction and directed them towards certain subjects, which made the selection process easier:

> My decision was influenced by the good results I obtained at school in similar subjects, and by the type of courses which I would like to do (male, Chemistry)

The positive influence exerted by teachers was also linked to good performance, owing both to their teaching methods and to the advice and help received in choosing the degree course:

> I had a good physics teacher in my second last year at high school, which made me understand how great the subject was (male, Chemistry)

Along with aspects belonging to the educational system, cultural factors such as family influences are also to be cited:

> I've been interested in mechanics since I was a boy, especially in automotive engineering. My family environment helped me cultivate this interest (male, Mechanical Engineering)

> My mother works for the environment, and I want to specialize in this area as well (female, Biology)

The emphasis placed by the respondents on the family environment confirms the results of a number of studies which highlight the influence of parents on school and university choices (Aschbacher et al. 2010; Sjaastad 2011). The students also attached a positive value to the family's cultural level, as mentioned in Chap. 6, recognizing the importance of having parents or siblings with degrees in scientific subjects:

> The profession of one of my parents has aroused my interest, since I was a young child (male, Mechanical Engineering)

> Thanks to my uncle, who also studied engineering at [...] University, and to my wish to carry on the family business with a higher educational qualification than my father had (male, Mechanical Engineering)

This shows signs that university courses perpetuate the same inequalities that exist among social groups (Cavalli and Argentin 2010).

There were some interesting gender differences concerning the influence of family as well as school experiences. Female students cited restrictive family ties – sometimes explicitly, sometimes more obliquely:

> I wanted to do an arts degree, but my parents 'warmly advised' me to do biology (female, Biology)

> My parents didn't let me choose the faculty I wanted. I chose the one which was closest to it so I could realize my dream one day (female, Mechanical Engineering)

A few students, mostly males, cited family expectations as crucial for their decisions. They are expressly oriented towards the prospect of a secure job after graduation:

> Family decisions, for a secure job in the future (male, Chemistry)

> Around the table with my parents, we discussed the job which would give the best employment guarantees, and we chose engineering (male, Electrical Engineering)

Once again in the case of the male respondents, work prospects after graduation were positively understood as the continuation of their parents' career, especially their father's:

> My choice is the consequence of work experiences I've had thanks to my father, and the many discussions with artisans and engineers in the sector (male, Automotive Engineering)

> With the help of my family, I've developed a liking for scientific subjects, and my father is a pharmacist (male, Biotechnology)

Interesting differences also emerge when disciplinary areas are taken into account. Among students of Biology, Chemistry, and Biotechnology, laboratory experience and practical activities on school curricula were recurring elements, especially among females:

> At high school, I took part in a series of Biotechnology laboratory classes, and I became very interested in this degree course (female, Biotechnology)

The various channels of scientific information were used by male and female students in different ways. Women have an interest in all degree courses, whilst among men there were more cases of physics students who attached significance to participation in science-related events, the reading of scientific magazines and books, and international mathematics and science games:

> I chose physics after I re-read some articles on particle physics. I'd always had an interest in science, but during the years I hesitated over my final decision for a long time (male, Physics)

> I chose it after an astrophysics conference on a subject I'd been fond of for some time (male, Physics)

Natural Features

Some of the students' accounts raise the issue of a predisposition for science and technology. This is a set of aptitudes believed to be innate and necessary for the choice of a university degree. Predisposition orients towards choosing which degree to take and, more particularly, it makes a person shape his/her own identity. A student's understanding that he or she has an innate ability in a subject will induce him or her to express a choice or, in some cases, a real preference:

> Since I was a small girl, I've always felt a propensity for and an interest in scientific subjects, which over the years turned into a true passion and a desire to undertake research in oncology (female, Biology)

The responses included expressions like "I've always had an aptitude", "since I was a small boy/girl, I've had an inclination", "I've always felt I had a talent". These expressions highlight the importance of factors relating to identity – rather than to cultural aspects – as a part of the concept of self, self-perception and self-knowledge at an academic level (Bong and Skaalvik 2003). These factors reinforce the values which lead to a choice, as can be seen from these examples:

> I chose this degree course because I've always been curious about science, because I have a predisposition for Mathematics and Physics, and because of the satisfaction they gave me at high school (male, Chemistry)

> I chose this university course after evaluating what came closest to my aptitude, given that both Physics and Chemistry fascinated me (female, Chemistry)

Students made numerous references to beliefs and convictions supporting individual choices. These factors, when grouped together, enabled students to take their decisions, reinforce them gradually, and add value to the activity, with a view to success in their university careers:

> I chose a university course in computer science because I've always been very good with electronics, machinery, and above all with programming languages (male, Computer Science)

> I have a talent for mathematics and solving problems, so the choice was Electronic Engineering because it is the most scientific and I was most curious about it (male, Electronic Engineering)

Perception of one's innate abilities shows significant gender difference. This issue is referred to twice as often by males as by females. The differences reside above all in the content and the type of language used. Only in a few cases did female students motivate their choice exclusively on the basis of their aptitude, predisposition, or ability. Their responses frequently seemed complex: they included natural features combined with other evaluations relating to an interest in and a vocation for science or the various scientific subjects (attainment value), socio-cultural aspects, such as advice and orientation from teachers and parents (cultural features), and practical elements relating to future employment (utility value). These elements suggest that for female students the construct of identity is a composite process at the interface of individual personality, cultural context and social relationship. On the whole, a series of elements can be attributed to different dimensions:

> I have a great aptitude for science and I wanted to get a good preparation for the world of work (female, Electrical Engineering)

> I know that I am more inclined in scientific and practical subjects, but also I have the desire to enter in the Italian military police force (female, Chemistry)

Among the responses coming from male students, on the other hand, assertive elements were more frequent. They were distinct from other topics and more related to abilities and aptitudes for scientific subjects in general, or specific study subjects:

> I've always had a certain propensity for scientific subjects (male, Mechanical Engineering)

> I've always had a certain affinity for the subject (male, Computer Science)

Differences were also apparent in the language used to attribute abilities and aptitudes to scientific learning, which in some cases was a measure to distinguish awareness among students, their capacities, or their different levels of self-esteem.

Subjective statements were more common among female students ("I feel I have a talent for..."), whereas in the case of males, they were often couched in assertive terms ("I have a talent for..."):

> I have a talent for mathematics and solving problems, so the choice was Electronics Engineering because it is the most scientific one and I was most curious about it (male, Electronic Engineering)

Among female responses, we have:

> I analysed all the scientific subjects for which I felt I had a talent, and I chose the most interesting one, with the best job opportunities (female, Physics)

> I felt I had a talent for this type of subject (female, Biology)

Conclusions

The foregoing qualitative analysis enables us to make some assertions, starting from the most frequent to the less cited dimensions.

The key elements cited by male and female students were those which can be included in the intrinsic value category as proposed in Chap. 9: about half of the responses – mostly provided by women – showed passion for, pleasure in, and an attraction to the subject; those who enrolled in Biology, Biotechnology, Mathematics, Statistics, Physics, and Chemistry seemed more likely to choose their university career on the basis of intrinsic values than their male counterparts. These findings are also in line with the process of course selection an individual, personal and special choice (Holmegaard et al. 2012).

Secondly, we find numerous cultural elements cited in the list of priorities: the influence of the family context, of teachers at secondary school, and of the educational and school background. Just under half of the students – mainly males – cited cultural factors. Women's responses were nonetheless more related to elements such as the influence of their teachers, the average grade obtained at school, and support from the family. This consideration was linked – especially for female students – with the importance of identifying people who could be important points of reference along the path of choosing tertiary education, such as tutors or mentors. It is notable that those countries where this type of service has been instituted tend to have higher retention levels (Larose et al. 2011).

Males, above all those who enrolled in courses with a significant technical component (Electronic, Mechanical and Computer Engineering) cited educational background, such as the secondary school attended, among the cultural factors.

Utility values were cited less frequently than utility values and cultural elements among the motivations behind the subject choice. In this case there were gender differences within individual degree courses too: males enrolled in Engineering and Computer Science prioritized the opportunity for a satisfying career, also from an economic standpoint, professional success, and therefore a significant status. Utility aspects were equally important for male and female students, but the latter were more oriented towards professional relevance, and only in very few cases (e.g. women doing very male-oriented courses) did students cite career, success, and income aspects.

Innate-type motivations were cited mostly by male students, which confirms a male propensity for attributing innate characteristics to the difference in interests between men and women. Although present in only a very small proportion of responses, vocational motivations (attainment values) were cited by students primarily from Biology, Biotechnology, Physics, Mathematics and Mechanical Engineering.

On the whole, we may say that, on account of our results, males are more attracted by the prospect of achieving status in the techno-scientific sphere, given that traditional socio-cultural models and the institutions are nearly always represented by male figures (see Chap. 20). A perception of science as a masculine environment is therefore reinforced in their representations. These elements of self-confidence are less present among female students, who mainly rely on an interest in and a passion for scientific studies, and less on aspects related to their role and institutional careers.

Given that identity and gender identity are central to choice of higher education in science and technology (see Chaps. 3 and 4), it is relevant to consider the ways in which students have described their choice process given that we found numerous references related to identity construction. It is clear that these young people are undergoing a crucial period for their individualization processes. Their search for meaning drives them to manifest their desire to be unique, but – for males, above all – to rely on pre-established roles, which in the case of science and technology are easily available and provide them with reassurance.

References

Ackerman, P. L. (1996). A theory of adult intellectual development: Process, personality, interests, and knowledge. *Intelligence, 22*, 229–259.

Ackerman, P. L., Bowen, K. R., Beier, M. E., & Kanfer, R. (2001). Determinants of individual differences and gender differences in knowledge. *Journal of Educational Psychology, 93*, 797–825.

Allegrini, A. (2004) Donne, lavoro e tecnologie IC. Prospettive femministe/di genere sul lavoro nella società in rete del capitalismo postfordista. In *Le ricerche del progetto PORTICO. Iniziativa Comunitaria EQUAL*, Bologna, Pitagora.

Aschbacher, P. R., Li, E., & Roth, E. J. (2010). Is science me? High school students' identities, participation and aspirations in science, engineering and medicine. *Journal of Research in Science Teaching, 47*(5), 564–582.

Blickenstaff, J. C. (2005). Women and science careers: Leaky pipeline or gender filter? *Gender and Education, 17*(4), 369–386.

Bong, M., & Skaalvik, E. M. (2003). Academic self-concept and self-efficacy: How different are they really? *Educational Psychology Review, 15*(2003), 1–40.

Brotman, J. S., & Moore, F. M. (2008). Girls and science: A review of four themes in the science education literature. *Journal of Research in Science Teaching, 45*(9), 971–1002.

Cavalli, A., & Argentin, G. (2010) Gli insegnanti italiani: come cambia il modo di fare scuola. Terza indagine dell'Istituto IARD sulle condizioni di vita e di lavoro nella scuola italiana. Il Mulino.

Eccles, J. S. (1994). Understanding women's educational and occupational choices, applying the Eccles et al. model of achievement-related choices. *Psychology of Women Quarterly, 18*(4), 585–609.

Eccles, J. S., & Wigfield, A. (2002). Motivational beliefs, values and goals. *Annual Review of Psychology, 53*, 109–132.

Holmegaard, H. T., Ulriksen, L. M., & Madsen, L. M. (2012). The process of choosing what to study: A longitudinal study of upper secondary students' identity work when choosing higher education. *Scandinavian Journal of Educational Research, 58*(1), 1–20. doi:10.1080/00313831.2012.696212.

James, R. (2002). *Socioeconomic background and higher education participation: An analysis of school students' aspiration and expectations*. Melbourne: Centre for the Study of Higher Education, University of Melbourne.

Jones, M. G., Howe, A., & Rua, M. J. (2000). Gender differences in students' experiences, interests, and attitudes toward science and scientists. *Science Education, 84*, 180–192.

Larose, S., et al. (2011). Academic mentoring and dropout prevention for students in math, science and technology. Mentoring and tutoring. *Partnership in Learning, 19*(4), 419–439.

Munk, M. (2011). *Educational choice: Which mechanism are at stake?* Copenhagen: Centre for Mobility research, Department of Political Science, Aalborg University.

Osborne, J., & Collins, S. (2000). *Pupils' and parents' view of the school science curriculum*. London: King's College London.

Rosenberg, M. (1979). *Conceiving the self*. New York: Basic Books.

Rosenberg, M. (1986). Self-concept from middle childhood through adolescence. In J. Suls (Ed.), *Psychological perspectives on the self* (Vol. 3, pp. 107–136). Hillsdale: Erlbaum.

Ryan, B., Adams, G., Gulotta, T., Weissberg, R., & Hampton, R. (Eds.). (1995). *The family-school connection: Theory, research, and practice*. Thousand Oaks: Sage.

Scantlebury, K., & Baker, D. (2007). Gender issues in science education research: Remembering where the difference lies. In S. K. Abell & N. G. Lederman (Eds.), *Handbook of research on science education* (pp. 257–285). Mahwah: Lawrence Erlbaum.

Schiefele, U., Krapp, A., & Winteler, A. (1992). Interest as a predictor of academic achievement: A meta-analysis of research. In K. A. Renninger, S. Hidi, & A. Krapp (Eds.), *The role of interest in learning and development* (pp. 183–212). Hillsdale: Erlbaum.

Sjaastad, J. (2011). Sources of inspiration: The role of significant persons in young people's choice of science in higher education. *International Journal of Science Education, 33*, 1–22.

Sjøberg, S. (2000). Interesting all children in 'science for all'. In R. Millar, J. Leach, & J. Osborne (Eds.), *Improving science education* (pp. 165–186). Buckingham: Open University Press.

Strauss, A., & Corbin, J. (1990). *Basics of qualitative research: Grounded theory procedures and techniques*. London: Sage.

Wildhagen, T. (2009). Why does cultural capital matter for high school academic performance? *The Sociological Quarterly, 50*, 173–200.

Chapter 19
Being a Woman in a Man's Place or Being a Man in a Woman's Place: Insights into Students' Experiences of Science and Engineering at University

Lene Møller Madsen, Henriette Tolstrup Holmegaard, and Lars Ulriksen

Introduction

In recent years the literature within science education has been inspired by feminist theories led by Judith Butler (1993) to address gender as something students *perform* through culture (Archer et al. 2010; Sinnes and Løken 2012, see also Chaps. 4, 6 and 17 in this book). As a consequence, research challenges the assumptions that men and boys, and women and girls, belong to homogeneous gender groups, who are masculine and feminine in one particular way that is shared by either men or women (Gilbert and Calvert 2003; Henwood 1998; Phillips 2007). Rather it is suggested that research should approach gender as a complex category in which students position themselves (Davies and Harré 1990) – and that the way students position themselves changes in accordance to the cultural context and social relations they participate in. From this perspective science and engineering is not gender-neutral: 'Scientific knowledge, like other forms of knowledge, is gendered. Science cannot produce culture-free, gender-neutral knowledge' (Brickhouse 2001, p. 283).

As a consequence research in students' participation in science and engineering should focus on the relationship between the culture the students engage in, the students' ways of performing gender and how various attempts at positioning are recognized or not. Davies and Harré (1990) introduce the notion of positioning to approach the way ongoing identities are constructed and renegotiated as we engage ourselves in new social relations, draw on different discourses and participate in different cultural contexts. Hasse (2002, 2008) suggests that this process has to be studied as a learning process. She carried out an anthropological study in which she enrolled as a physics student together with other first year students. She explored

L.M. Madsen (✉) • H.T. Holmegaard • L. Ulriksen
Department of Science Education, University of Copenhagen, Øster Voldgade 3, DK-1350 Copenhagen C, Denmark
e-mail: LMMadsen@ind.ku.dk

E.K. Henriksen et al. (eds.), *Understanding Student Participation and Choice in Science and Technology Education*, DOI 10.1007/978-94-007-7793-4_19

how the students learn *to become* physics students and how they perform gender in terms of gaining recognition as 'proper' physics students within the culture of the study programme.

In the context of engineering, Tonso (2006) conducted an ethnographic study highlighting how engineering students through engineering culture and practice are required to develop '*into scientific and engineering selves*' (p. 304) in certain (gendered) ways to present themselves *as engineers* in a way that is recognised by the campus community. As with Hasse, the point of departure of Tonso is students' meeting with a certain culture (in this case engineering) combined with a particular interest in the performance of gendered identities.

> As contextualized in activity, identity production at PES [Public Engineering School] was a process through which persons' sense of themselves as engineers led to performances of engineer selves that were viewed through lenses of cultural forms for campus engineer identity, and where recognition *as* an engineer conferred belonging. (Tonso 2006, p. 303)

Further the focus of Tonso was on how gender is produced in certain ways when a minority (of female students) meet a majority (of male students):

> Women were, and to a great degree still are, considered people who are welcome only to the extent they accept the way things have historically been done'. (Tonso 1999, p. 279)

In particular, STEM and engineering study programmes face a heavy imbalance in students' biological sex[1]. This imbalance appears to influence ways of getting recognized within the study programmes. Gonsalves (2010) shows how women in doctoral physics programmes position femininity as something outside of physics, different from ordinary women with stereotypical femininity. Instead they position themselves as 'tomboys' belonging to physics. Also Due (2012) points at two competing discourses in physics which sets the scene for students' available positions; one highlights physics as a masculine discipline, and another physics as a gender neutral discipline. A similar conclusion is reached in a study comparing the discourses available within physics and biology programmes for students in their production of scientist subjectivities. It is found that a narrow range of gendered student science subjectivities are available in physical science. On the contrary student-led activities as found in biology, provides opportunities for new science identities that transcend masculine/feminine dualisms (Hughes 2001). However students from female-dominated and gender-mixed disciplines perceive men and women as being intrinsically different; more so than do students from male-dominated disciplines. This, in different ways, sets the scene for gendered ways of being recognized as a proper student. This focus on gender similarities or differences has been shown to depend on the particular culture:

[1] We distinguish between "sex" and "gender". "Sex" refers to the distinction between male and female based on biological attributes while "gender" refers to the way male and female students interpret the social and culturally embedded frames and expectations of being male or female (see Chap. 4). Most of this chapter deals with gender (the students handling the expectations), but we sometimes refer to the biological distinction (e.g., the distribution of male and female students) and then use the term "sex". To emphasise the difference, we add "biological" sex.

> If the question focuses on values, male dominated disciplines tend to highlight gender similarities. Conversely, if the question is about concrete gender equality work the rationale is based more on gender differences, differences that are often self-evident and taken for granted in everyday situations. (Haake 2011, p. 124)

But striving at study programmes with an equal balance of students' biological sex is not the solution: 'It should make us suspicious of attempts to produce a more 'balanced' science simply by increasing the number of women in it' (Gilbert and Calvert 2003, p. 875). More women in science does not necessarily change the way the knowledge-structure is gendered. To change students' access to science whatever way they perform gender we must therefore study how science culture includes certain ways of doing gender while excluding others.

The above review calls for research on students in study programmes with a heavy gender imbalance, to focus on how participation in STEM is being perceived as gendered by the students, how students position themselves within those gendered positions, and which positions they experience as being recognized and which they have to renegotiate to eventually feel they belong. This process of encountering a STEM study-programme with a heavy gender imbalance is therefore related to students' negotiations of their identities and the learning-process the students undergo when meeting their first year and strive at getting socially and academically integrated (Tinto 1993). This is the focus of the research reported in this chapter.

Aim and Research Questions

We explore how students who enter a STEM study programme within higher education negotiate their identity in their meeting with a STEM higher education study programme with a heavy imbalance of students' biological sex. We explore study programmes with both a majority of female and male students. As a point of departure, in three specific study programmes we wish to explore: how do various cultural settings affect students' construction of their identities. More particularly:

- How do students in general describe their study programme, and what do they perceive as being central for being recognised within it?
- How do the majority students perceive the minority students, and what do they highlight as being important for the minority to become recognised or not recognised within the study programme?
- How do the minority students perceive their position within the study programme and how do they relate to it?

Collecting and Analyzing Data

Three higher education study-programs have been selected based on information on numbers of female and male students. These are computer science and molecular biomedicine, both at the University of Copenhagen, and physics and nanotechnology at the Danish Technical University. Within computer science, 4–9 % female students have been enrolled in the period 2009–2011 (the number has declined throughout the period). Within molecular biomedicine the proportions are 13–24 % males from 2009 to 2011, (the number has increased throughout the period) and within physics and nanotechnology there have been 4–25 % female students in the period of 2008–2010 (the number has increased throughout the period (www.studier.ku.dk and www.dst.dk).

To address the above research questions we have applied a multi-method triangulation design where different types of information are obtained about the same theme with different methods in order to obtain an in-depth understanding (Denzin and Lincoln 2000). In the triangulation we used writing exercises, workshops and qualitative interviews.

Written exercises were used to get access to the students' considerations of their choice of education and their individual experiences within the first few years of their university education' (40 written descriptions were obtained). In the workshops the students were working in groups. Firstly, they were asked individually to make a list of themes they considered important for a student to get through their first year at university. Secondly, within groups of peers they were asked to prioritize the themes. This exercise was used to provide insights into the possible gendered negotiations among students about the themes (we held 8 workshops with 41 participating students). Qualitative individual and group interviews were held; firstly to unfold the students' narratives of entering the selected educational programs and their experiences during the first year, secondly to gain insight into how the students negotiate their gendered power position. Twelve qualitative interviews were performed involving 31 students, consisting of five individual and seven group interviews. All student names in this chapter are pseudonyms as to provide the students with anonymity.

The analysis presented in the Results section is divided in two parts. The first part offers an analysis of all students' description of their study programme. From these descriptions we extract a general discourse about the study programme, although we do not include all the variations in the descriptions. As a consequence not all of the participating students would recognise their own perceptions in reading the extract. In the second part of the analysis we show how the minority is seen by the majority of students and further how minority students perceive the cultural setting and position them in relation to it.

Results

Students' Experiences of Their Programme

The descriptions presented below result from our analysis of the students' experiences within their first year of studies. The focus is on the students' negotiations of their identity in relation to the subject they encounter. The purpose is to show the discourses the students draw on when describing their study programme. The second part of the analysis shows how the students position themselves in relation to the discourses presented below.

Physics and Nanotechnology

The students describe the study programme physics and nanotechnology as requiring the students to be in love with physics. Although all students are interested in science and some in nanotechnology, it is the physics that defines the students belonging to the educational program. As one student puts it: *'it is like other subjects [*i.e. *other than physics] don't quite reach the depth of the world's content'*[2]. Another strong signifier among the students is a common experience that now they are (finally) challenged intellectually. The students explain how the study programme is characterised by high standards; something which is much appreciated and that they have missed in their previous educational experiences. The students describe how they compete in reaching the high standards of the study programme, which they strive to match. The students also identify social integration as important and something that is worth investing in, often in combination with the heavy workload of the study programme.

Computer Science

The students' descriptions of their attachment to computer science are rather vague and they describe a diverse array of ways to become a computer science student: two of the common attachments can be broadly defined as liking computers and wanting to do programming. The students describe a strong student community as central for their access to learning to think like a computer scientist. They define themselves in relation to this community either as being a member or by recognizing its existence. Older students play an important role in the integration and inclusion of new students into the community during the first years of study. Inclusion is social but also to a high degree academic, as one student expresses it: *'One of the important things, both in order to complete and to 'keep it' is to have a good social network'*. The students describe the computer science community as distancing itself from the official university, due to an experience of lack of structure, technical problems and messiness.

---

[2] Please note that the interviews were made in Danish. We have chosen verbatim translations rather than linguistically correct ones.

Molecular Biomedicine

The students describe becoming a molecular biomedicine academic not on its own but in relation to other professions. First of all, being a molecular biomedicine academic is described as not being interested in having patients like a medical doctor although fascinated by medicine. Secondly, it is described as not being interested in animals and plants although being interested in biology. The students define being a molecular biomedicine academic as something it is not – as a residual. Dealing with becoming a molecular biomedicine academic is hard. One student says: *'many of them [fellow students] are really ambitious; it affects me and pressures me'*. Intellectually the students describe a culture where high grades are demanded and the label 'elite' is put forward as a requirement the students need to meet to belong. Also, socially the students describe a study programme with many activities which they perceive being important to participate in to gain belonging. The majority of girls are put forward as the explaining factor of the ambiguous, uniform, high pressure, and non-relaxed possibilities. In this way social relations act as a way to cope with the pressure.

These different descriptions of the three study programmes clearly give different frameworks for all the students to negotiate within in their process of constructing an identity. Within physics and nanotechnology you need to be fascinated by physics, and a very high level of performance is expected. However, the students indicate this as a relief; finally they are been challenged. The study has a strong common identity and is competitive. Within computer science, the students only have a vague attachment to the content, there exists a very strong academic and social study environment including both new and older students, and the study programme is characterised by many sub-cultures making the horizontal cohesiveness among first-year students vague. Within molecular biomedicine, the deselection of medicine and biology unites the students. There exists a high level of performance that stresses the students in various ways, and both male and female students ascribe gender significance to their negotiation of identity.

The result of the workshops shows that across all the three educational programmes it is the social inclusion that is central for the students. Almost all of the groups prioritise social integration in various forms as the most important topic for surviving the first year in their study programme. This appears to be in opposition to the students' descriptions of the three studies as very different, as described in the following. One interpretation is that the social dimension overrides the different gendered cultures the students are negotiating within. Another interpretation is that the students ascribe different meanings to what being social means in different cultural settings.

In the following we wish to combine the students' descriptions of their study programme, with quotes from the students of the majority biological sex about their perceptions of the study programme and how to navigate to become a proper student at the study programme, and also quotes from the students with the minority biological sex illustrating how they perceive themselves belonging to the study programme. By bringing together these perceptions of being a student, and in particular a student possessing a minority biological sex, we aim to analyze what is recognized within each programme and how students negotiate their identity and gender to fit in.

How to Become a Physics and Nanotechnology Student When Being the Female Minority

Within physics and nanotechnology, a group of male students describe what it must be like to be a girl on the study programme:

> Allan: 'I think they easily become one of the boys – they need to adjust when there is such a huge [gender imbalance] it's going to be a male culture' [no matter what].
> Christian: 'If you don't fit in, then you stand too much out, then you are not part of the club and then it is not possible to be here.'

As the male students in this quote indicate, the girls need to be *one of the boys* to fit in. They need to fully assimilate into the *male culture*. This is echoed across the data from physics and nanotechnology:

> Allan: 'A physicist as a woman – those two things do just not fit well!'
> Christian: 'No, that does not fit with the picture'

To belong within physics and nanotechnology, the minority (girls) need to ascribe to a particular non-feminine culture in their negotiation of identity in order to get recognized as a full-blooded physicist. To some of the girls this requires a negotiation of who they perceive themselves as being. The students ascribe being a student of physics and nanotechnology as a place for high level and pace, and this culture requires the girls to perform their gender in particular ways to be recognised as physicists. As one group of male students explained, if one of the male students faces difficulties in keeping the pace, he can still get recognised by fellow students if he involves himself in the social part of the study programme. In contrast, if a female student does not keep the pace, it seems incompatible with being *one of the boys*, her only way to stay within the study programme is, as explained by the group of male students, to be good looking. Good looking, though, is not being related to 'being one of the boys' – therefore being in love with physics requires of girls not to be too girlish and to keep the pace. An example of keeping this balance of being one of the boys is described by Louise:

> Sometimes the male students say things they do not mean seriously, For example me and my fellow student walked together at campus, and saw one of the older female students, and he says *'Karen is just the only pretty girl here at physics'* after a while of silence I reply: *'Thanks William'*, and he was like: *'God no, no, no, I am not...'*

It is clear from the quote that the female student is perceived as 'one of the boys' by the male student. But as there exists an opposition between being girlish and being a physicist – the female students cannot be recognized within both categories at the same time. In this case, Louise has succeeded in becoming like the majority, in this case a masculine one, so much that she is no longer perceived as female. But this balancing how to position one's gender in a way that is recognised as belonging to physics, without being in danger of getting feminized and thereby excluded from being recognized as a proper physicist, can be a difficult balance to keep for some of the female students:

> Brian: 'Consider how much she [one of the female students] is getting bullied with she is going to take a shower [at the retreat[3]]. Rasmus could take as long a shower as he would, but Laila was instantly bullied with expressions as: you have 10 minutes' [indicating that women take long showers]

It seems that the girls within physics and nanotechnology are running the risk of being feminized within the study programme, and this is not only affecting their social integration within the study programme but also their academic integration.

> Olga: 'We watched Myths Busters [a science programme on the Discovery TV channel] in one of the lectures and then our teacher had found some mistakes in the programme which we should identity. And then, I don't know, it became very boyish like 'girls cannot do this'. . .and then we thought 'yes we can, we are actually some right here'
> Interviewer: 'How do you experience it when such things happen?'
> Olga: 'I think it is the other girls, not me I do not take it on me, as it is me. I know that if you take 100 girls they cannot – I do not find it to be something personal'.

The example shows how the female students do not internalize and recognize themselves within the offered position, but rather exclude it as something that concerns other girls.

The desire to 'be one of the boys' has some consequences for the female students. Louise explains how she does not have any relationship with the other girls on the study programme, but belongs to a group of male students – a group that was formed by the institution in the beginning of first year with the purpose of introducing the students to the university. At the beginning of the year an older student was attached to the group as a tutor, but after the formal meetings ended the group have kept on meeting. She further explains some situations where she feels herself being positioned as feminine; when she is participating in social activities she is getting a lot of attention, which she finds to be an advantage for 'boosting her self confidence', but sometimes in her study group she find it hard to 'be one of the boys' when the talk centers around 'toilet-habits' or 'computer games', two topics often debated in the group. Another student, Sarah explains how you need to accept the jargon to be part of the study programme:

> Sarah: 'At some point you are just so much one of the guys. I joke with it myself' [that she is one of the guys]
> Interviewer: 'But do you experience that it is necessary for you to have this jargon or to be one of the boys in order to be here – do you understand what I mean?'

---

[3] At many Danish Universities older students arrange a retreat, typically a weekend or a whole week, where the new students are taken to a summerhouse. The intention is social and to introduce the students to the study life through the experiences of older students.

Sarah: 'Yes, yes, yes I have really not thought about it, so not 100 % but in some point yes, I think you really need maybe not to use the jargon yourself but you need to be able to accept it.'
Interviewer: 'Do you also use the jargon yourself?'
Sarah: 'More than I did before'
Interviewer: 'Is it a jargon that you only use at [the particular place for the study programme]?'
Sarah: 'I use it when I am together with the guys out here...it is not something I go and think so much of.'

Sarah ascribes herself to the dominant culture at the study programme by both accepting its existence and premises and herself using a particular language used by the male students. She has adjusted her behaviour in a way that is legitimate as a physics and nanotechnology student and she has internalized it in a way that she finds makes it legitimate to be female too. In that respect she is fully integrated as a physicist, however through an adaptation of practices within the community.

These examples show how gender is negotiated in order to become like the majority gender, in this case a masculine one. These findings are echoed in the literature. Danielsson (2009) finds that female physicists balance the norms for being a women and being a physicist by positioning themselves as different from other women. Another study among engineering students shows how female engineering students needed to perform their gender in particular ways to gain recognition and hence to apply certain coping strategies such as acting like one of the boys, accepting gender discrimination and adopting an 'anti-woman' approach (Powell et al. 2009):

> In 'doing' engineering, women often 'undo' their gender. Such gender performance does nothing to challenge the gendered culture of engineering, and in many ways contributes to maintaining an environment that is hostile to women. (Powell et al. 2009, p. 411)

In a study of students in the social sciences Søndergaard (1996) shows how academic prestige is linked to masculinity. Her study explored the construction of gender in academia in relation to different aspects, including the academic practice, but also how the students expressed themselves through their clothing and in their sexual encounters. These different practices are all gendered and linked to interpretations of the individual students' practices. Hence, not only is academic competence and prestige linked to masculinity, but some subfields of the discipline are considered masculine (and hard) while others are linked with femininity. In other practices, there are expectations concerning how males and females should act, for instance, who should be the active partner in a sexual encounter.

Søndergaard uses the concept of a matrix to explain how this web of gendered expectations and interpretations works. Importantly the gendered practices do not prevent female students from entering a subfield with masculine connotations, or to make the first approach on the dance floor. What they do, however, is that male and female students are met with particular expectations and interpretations based on what Søndergaaard coins as 'the sign on the body', that is, the biological sex that is visible to the outside world. A female student entering the discipline will therefore be tacitly expected to be less competent than the male students, just as she will be expected to have a preference for the subfields that are connoted as feminine. Female students entering fields that are considered masculine are, so to speak, behind on points, because their biological sex is interpreted as a sign of less competence.

The idea of the matrix is that the individual students' practices in different contexts affect each other. If, for instance, a female student wishes to enhance her status as competent she could downplay or neutralise her clothing, hairstyle, use of make-up, etc., in order to appear less female and through this evade the interpretation as less competent. This practice, neutralising her appearance and entering a masculine subfield, will make her appear less feminine in other contexts as well, for instance, in the emotional and sexual encounter. Conversely, if the female student maintains a distinct feminine appearance in order to be recognised as feminine in social contexts she runs the risk of being interpreted as less competent. Students therefore need to balance how they 'score' at the different matrices in order to be recognised in different social contexts, but the opportunities for male and female students will be different from the outset.

As stated in the first part of the analysis the culture within physics and nanotechnology also affects the male students, who need to position themselves as clever and in love with physics. Having a male sign on the body, in other words, still requires a practice that is considered legitimate and recognisable within the disciplinary culture.

How to Become a Computer Science Student When Being the Female Minority

Within computer science the findings are more complex than the homogeneous picture within physics and nanotechnology. As stated earlier, the students in general describe a diverse array of ways to attach to the study programme, and hence there seem to be more diverse perceptions of being the female minority:

Kenneth: 'I do not think about it'
Lars: 'It's no problem'
Søren: 'Whether it is a boy or a girl doesn't matter'
Ryan: 'The girls have the same terms as the rest of us'......
Kenneth: 'The girls that have been best adapted are the ones that are most masculine'......
Søren: 'The girl [in our group] did put more into the layout of the assignment'
Ryan: 'I have a longhaired boy in my group: he does things like that'

These quotes from male computer science students show the various ways that they ascribe meaning to being a girl on the programme. The girls need to be one of the boys, the girls are not different from us and some boys are like the girls. Based on our interviews and the other material we have collected within the computer science programme our interpretation is that this is the result of sub-cultures within the study programme being gendered in different ways. It is however, beyond the scope of this chapter to unfold this complexity in greater detail.

A central question seems to be how this affects the female students' social and academic integration strategies. In the narrative of Emily, she tells how the social integration requires a certain vocabulary:

> The way you speak to each other; the content, the terminology, and the way of competing is very excluding. Girls get frightened about it – and it's hard to get into it (...) But very identity-building (...) It is cool to be the one who knows slang if you are a part of the club, then it's nice, you get recognized

Emily explains how the very technical jargon has been a part of her vocabulary due to her older brothers. Furthermore she explains that she is different from other girls since she has a high IQ which she feels helps her get recognized, because she does not struggle much with the academic content:

> I think [the jargon] is unpleasant for other girls, and they feel more stupid than they necessarily are. That's part of the game, to make people who do not know it [the jargon] feel stupid

Emily explains how people without high self-confidence will find it *'a torment to be here'*, and she explains how in particularly the female students are vulnerable to this because they have a tendency to underestimate their own abilities. Concerning the academic integration she explains how the male and female students have different learning strategies, and that the male way of learning is enhanced by the teaching on the study programme:

> Boys try again and again and again to find a solution. They search for information, read books – try to fix it in some way or the other. Most girls meeting a problem think, I do not know what to do, I will ask a teacher or a fellow-student (...) The learning process is more social for most girls.

But by asking questions without having tried everything out on her own, there is a danger of a girl being positioned as un-intelligent, and the girl may feel that it enhances the picture of 'girls cannot do computer science'. To counteract that she might try harder on her own before asking. And a large part of the culture at computer science, as Emily describes it, has to do with being intelligent: *if you are smart you are recognized no matter how you behave*. According to her you need to learn to learn in a new way, and this way is implicitly enhanced by the teaching:

> Emily: 'Large parts of the content that was included in the exam paper...was content that you haven't heard more about before. [During the course the teacher said]: there exist this programme, play with it, take it home with you...you will figure it out. At the exam almost none of the girls knew how to solve the problem related to that programme.'

This 'take it home and play with it' presupposes that the students by themselves find not only the solution but also learn different ways of getting to it, and according to Emily this is experienced as *anxiety-provoking* by most girls. From the example it seems that the study of computer science presupposes gendered experiences and practices that do not necessary reflect competences but different ways of approaching a problem.

In Sofie's narrative her perception of the social culture is different from that of Emily. Sofie explains how you always can find help if you ask other students, and they will be happy to help you. *'Forget your pride, and ask for help'*, she explains – failing is quite normal, and you need to ask for help to get through. Only very few students pass the exams within the estimated study period, she explains. One interpretation of the two girls' differing descriptions is that there are distinct subcultures, gendered in different ways. Emily with her self-reported high IQ might be a part of the competition of being a smart student, while Sofie apparently is more focused in how to get through. This seems to have an influence on how they position themselves and perform their gender, and whether or not they find themselves exposed when asking for help.

Emily's descriptions of the learning culture within the computer science programme is supported within the literature. Hasse (2002) describes how girls in physics are confused because they did NOT (like they were used to from school) get credit for "following the instructions". Rather what was recognized was being playful and trying out own ideas. Our data point towards similar findings, although the gendered learning cultures within computer science programmes require further study.

How to Become a Molecular Biomedicine Student When Being the Male Minority

Very few boys enter molecular biomedicine. Contrary to the two other study programmes, this student minority is expected to construct their own community together, as stated by these female molecular biomedicine students:

Karen: 'You need to be a loud boy and good at the social – not isolate oneself – that does not work'
Susan: 'It must be difficult because there are not so many they can hang with' [implicit that you need to be able to hang with someone from our own sex]
Pernille: 'They [the boys] are good at doing something with the boys at other levels – they stick very well together'
Fanny: 'You have a need for talking to someone with the same biological sex'

The boys are recognized as belonging to molecular biomedicine but perceived as different from the girls and gendered in a way that makes it difficult for the girls to hang out with them. This affects the male students' social integration.

The male students also express their need to create a space of their own masculine setting together with other male students on the study programme. As a consequence they form a collective group, or gang, of all male students regardless of the number of years studied at molecular biomedicine. This gang is organized by the students themselves. Two male students describe the gang in this way:

Peter: 'We have a gang away from [the study programme], such a male thing, where the politically correct stuff like saying that you don't need to drink, just vanishes'
Will: 'But there we also know that everybody thinks it is fun'
Interviewer: 'Could you tell a little more about this gang?'

Will: 'We went to this gang inauguration with extreme drinking, different games – really masculine, like being in a sauna and drinking booze and then going out and running around naked'

Peter: 'It is also a kind of natural isn't it – when we are boys we need to find some way to stick together'

For these boys the negotiating of gender concerns their ability to deal with being with a feminine majority and a way to deal with this is to create a space of their own masculine setting together with other male students on the study programme. From this perspective it seems that molecular biomedicine is gender-segregated, that makes it is hard for the boys to become one of the girls and furthermore, that it is perceived as being unattractive to boys. Being together with fellow students sharing the same gender is described as a relief:

Peter: 'You kind of sometimes miss boys, it can be very girly I think...cause you are together with girls all the time – you can miss being together with boys occasionally'

Will: 'We have experienced that when you finally get out and it is only boys from the study programme, then it almost like a, you get really relaxed and talk about things which you have been left alone with'

Peter: 'Totally relief, yes that is real enough'

Will: 'Then we almost talk ourselves as girls because there is so much to talk about'

[Both students laugh]

'To talk like girls' is an expression that is mentioned several times, and which also seems to set the scene for the male students positioning themselves within molecular biomedicine. They are required to perform a certain kind of masculinity to feel recognized as biomedicine students. In the above quotes the female students articulate this as the male students are required to be loud and good at socializing, and this way of getting recognized seems to be a challenge to some of the male students:

Peter: 'Actually, I tried it yesterday: wow, how girls are good a small-talking. When I come into this room – I just freeze, but for the girls it only takes 30 seconds, then they have a conversation going on'.

Will: 'Yes, in the beginning of the study you could sit and really feel outside'.

At the same time as being loud and social expectations seem to be a perception of the male students, the male students negotiate how to become recognized as a molecular biomedicine student without it being on the premises of the majority gender. An example is one of the boys telling how the boys position themselves as something else than what they describe as '*the calendar girls*': this is used as an expression by the male students to denote female students reaching for their calendars as soon as some information is given. It covers how the female students are well organized, have a high self discipline and work ethic. This behavior is perceived by the male students as a symbol of a very organized and controlled life which they do not want to adapt themselves to, and they describe it as a competence possessed by the girls. Instead the male students negotiate what is described as a very ambitious study culture with a high performance pressure. '*We do not show in class that we are wise although we are*' [like the women tend to do according to the interviewed men] and '*we do not need to have everything under control*'. The male students thereby position themselves in opposition to being a *calendar girl* which affects their interaction with the study programme.

The male students within molecular biomedicine are perceived as something different from the female majority by both the female and male students themselves. They cannot hang out with the girls all the time and are expected to prefer each other's company. They are required to perform a certain kind of masculinity, being loud and having a good social life, which sets the frame for their positions and way of belonging to the study programme. Also in an academic context the male students positions themselves as different from the well organized *calendar girls* who they perceive to be controlled. They do not find it necessary to present and position themselves as in control and clever to get recognized, as they describe the girls do. Compared to the girls within physics and nanotechnology, in one perspective the males within molecular biomedicine have room to be a molecular biomedicine student in different ways than the majority within the study programme. In another way they are expected to be different from the female majority– where the minority of female students in physics and nanotechnology are expected to become one of the boys. These differences are discussed below.

Discussion

The task of negotiating one's identity is a project for all students entering a higher education programme (Holmegaard et al. 2014; Ulriksen et al. 2013). In the present analysis we show how both male and female students, being a minority in their study programme, also need to engage in a gendered identity negotiation-process in struggling to belong and become socially and academically integrated into their new study programme.

The results show how students apply different gendered strategies for being recognized within the three study programmes; computer science, molecular biomedicine and physics and nanotechnology. They range from striving to become like the majority to explicitly maintaining one's differences. Whereas the female students have different strategies for 'being as' within computer science and physics and nanotechnology, the male students in various ways struggle to 'fit in' within molecular biomedicine.

The female students in their narratives of being the minority gender within both physics and nanotechnology and computer science relate themselves to being more *masculine* than other girls, in telling an individual history of how they previously have belonged to a man's world. For example they have only male friends, they have only played with boys in their childhood, or they are used to being in a male dominated environment during earlier educational settings. The possible strategy they see in order to be recognized as respectively a computer scientist or a physicist is to ascribe masculinity into their identity in various degrees. They are required to modify their gender within the negotiation process of entering a new study programme and getting recognized as a proper student within it by assimilating.

However, it also seems from the analysis that there is slightly more room for doing gender within computer science for the female students than within physics and nanotechnology. We ascribe this to our notion of different subcultures within computer science that appears to give the female students a range of possibilities for gaining recognition.

Considering molecular biomedicine, where the male students are the minority, we found a different pattern. The male students we interviewed expressed themselves not with a 'being as' but with a 'fitting in'. They denote themselves as different from the majority gender and not by trying to behave as one of them. None of the male students had a history of being in a girls' world or defined themselves as a girl-boy. From the analysis it seems that the male students were able to negotiate an identity without discarding their masculinity and still be legitimate members of the culture. Yet they are still required to perform a certain kind of masculinity and position themselves as something different from the girls. Thus, their negotiation strategy to become integrated into their study programme could be labelled 'segregation'.

Another aspect of the strongly dominant gendered culture is that for the majority of students certain ways of doing gender is perceived to be legitimate, leaving out other ways. It is relevant to assume that certain male students' positions within computer science and physics and nanotechnology are required to gain recognition and hence that other attempts are being marginalised. However, it is beyond the scope of this chapter to unfold this perspective.

Overall, our analysis shows that the different higher education programmes provide different but well defined and narrow frameworks for what ways of doing gender are legitimate and recognizable, and that the students' negotiations deal with which forms of doing gender the students experience as acceptable. From the analysis it seems that female students are urged to position themselves as non-feminine whereas male students are restricted to positioning a certain kind of masculinity to become recognized. The female students aim at positioning themselves as aligned with the male majority, and in doing so they cannot be too girlish. Rather they struggle to become one of the boys. For instance, none of the interviewed female students talked about a feminine sisterhood in any way. If this positioning fails our results suggest that only the female students are in danger of being excluded as not clever enough. Following Søndergaard (1996) this is related to the way gender and competence are related in the matrix. When academic competence is related to masculinity, female students need to understate their gender in order to be recognised as competent but at the same time they need to balance this performance with other practices if they still wish to be recognised as female. On the other hand, male students are expected to perform masculinity together, which is approved and encouraged by the majority of girls to be legitimate members of the culture. However, in an academic context where masculinity initially is convergent with competence male students seem to have a broader range of positioning possibilities in order to become recognised.

References

Archer, L., DeWitt, J., Osborne, J., Dillon, J., Willis, B., & Wong, B. (2010). Doing' science versus "being" a scientist: Examining 10/11-year-old schoolchildren's constructions of science through the lens of identity. *Science Education, 94*(4), 617–639.

Brickhouse, N. W. (2001). Embodying science: A feminist perspective on learning. *Journal of Research in Science Teaching, 38*(3), 282–295.

Butler, J. (1993). *Bodies that matter. On the discursive limits of "sex"*. New York: Routledge.

Danielsson, A. T. (2009). *Doing physics – Doing gender*. Uppsala: Uppsala Universitet.

Davies, B., & Harré, R. (1990). Positioning: The discoursive production of selves. *Journal for the Theory of Social Behaviour, 20*(1).

Denzin, N., & Lincoln, Y. (2000). *The handbook of qualitative research*. Thousand Oaks: Sage.

Due, K. (2012). Who is the competent physics student? A study of students' positions and social interaction in small-group discussions. *Cultural Studies of Science Education*, 1–19.

Gilbert, J., & Calvert, S. (2003). Challenging accepted wisdom: Looking at the gender and science education question through a different lens. *International Journal of Science Education, 25*(7), 861–878.

Gonsalves, A. J. (2010). *Discourses and gender in doctoral physics: A hard look inside a hard science*. Montréal: McGill University. Doctor of Philosophy.

Haake, U. (2011). Contradictory values in doctoral education: A study of gender composition in disciplines in Swedish academia. *Higher Education, 62*, 113–127.

Hasse, C. (2002). *Kultur i bevægelse – fra deltagerobservation til kulturanalyse – i det fysiske rum*. Frederiksberg: Samfundslitteratur.

Hasse, C. (2008). Learning and transition in a culture of playful physicists. *European Journal of Psychology of Education, 23*(2), 149–165.

Henwood, F. (1998). Engineering difference: Discourses on gender, sexuality and work in a college of technology. *Gender and Education, 10*, 35–49.

Holmegaard, H., Madsen, L., & Ulriksen, L. (2014). A journey of negotiation and belonging: understanding students' transitions to science and engineering in higher education. *Cultural Studies of Science Education*, *9*(3), 755–786. doi: 10.1007/s11422-013-9542-3.

Hughes, G. (2001). Exploring the availability of student scientist identities within curriculum discourse: An anti-essentialist approach to gender-inclusive science. *Gender and Education, 13*(3), 275–290

Phillips, A. (2007). Re-inscribing gender binaries: Deconstructing the dominant discourse around women's equality in science, engineering, and technology. *Sociological Review, 55*, 768–787.

Powell, A., Bagilhole, B., & Dainty, A. (2009). How women engineers do and undo gender: Consequences for gender equality. *Gender, Work and Organization, 16*(4), 411–428.

Sinnes, A., & Løken, M. (2012). Gendered education in a gendered world: Looking beyond cosmetic solutions to the gender gap in science. *Cultural Studies of Science Education*, 1–22.

Søndergaard, D. M. (1996). *Tegnet på kroppen. Køn: Koder og konstruktioner blandt unge voksne i akademia*. København: Museum Tusculanums Forlag, Københavns Universitet.

Tinto, V. (1993). *Leaving college. Rethinking the causes and cures of student attrition* (2nd ed.). Chicago/London: The University of Chicago Press.

Tonso, K. L. (1999). Engineering gender – Gendering engineering: A cultural model for belonging. *Journal of Women and Minorities in Science and Engineering, 5*, 365–404.

Tonso, K. L. (2006). Student engineers and engineer identity: Campus engineer identities as figured world. *Cultural Studies of Science Education, 1*(2), 273–307.

Ulriksen, L., Holmegaard, H. T., & Madsen, L. M. (2013). Weaving a bridge of sense: Students' narrative constructions as a lens for understanding students' coping with the gap between expectancies and experiences when entering higher education. *European Educational Research Journal, 12*(3), 310–319.

Chapter 20
Italian Students' Ideas About *Gender* and *Science* in Late-Modern Societies: Interpretations from a Feminist Perspective

Alessandra Allegrini

Introduction

This chapter draws on a selection of qualitative findings from the Italian IRIS survey, which allow us to explore different aspects shaping the representation of gender and science among young women and men enrolled in different scientific courses: biology, biotechnology, physics, chemistry, mathematics and statistics, computer sciences, mechanical engineering, electronic engineering, chemical engineering. The relevance of this issue within feminist approaches to STEM educational choices is specified in Chap. 4, where it is described how science might be viewed as part of a gender discourse, invested with gendered attributes that can impact on the choice process in science.

Several data from the Italian IRIS survey, both qualitative and quantitative, are relevant to this issue, in particular the answers to three open questions: (1) Do you attend a course where one gender is over-represented? If so, why do you think this is the case?; (2) Do you see any reason why the situation described above should change – and if so, what do you think could be done to change it?; (3) Describe how you came to choose this course. These questions are part of a larger questionnaire which was distributed to students attending their first year in 45 Italian universities, with 2,667 valid cases collected in spring 2010. 2,203 students answered the first question, 1,506 answered the second question and 2,135 answered the third one (see Appendix).

Although the overall qualitative data obtained through these three questions offered evidence to formulate hypotheses about students' representation of gender and science, in this chapter I mainly focus on the answers to the first question, which was specifically formulated to study students' perceptions and ideas about

A. Allegrini (✉)
Observa Science in Society, Viale A. Fusinieri, 65, 36100 Vicenza, Italy
e-mail: allegrini_a@yahoo.it

E.K. Henriksen et al. (eds.), *Understanding Student Participation and Choice in Science and Technology Education*, DOI 10.1007/978-94-007-7793-4_20

the gender dimension of the scientific course attended at university. To sum up, the analysis of the responses to this question provided:

- An understanding of Italian male and female students' perceptions about the gender dimension of the scientific course attended;
- An understanding of the way Italian male and female students symbolically represent gender as a concept, and the way they conceive it both in relation to science and to the different scientific subjects;
- A description of some emerging features in the gender and science imagery of Italian males and females in a late-modern society.

Research Methods

Data were first analysed by means of *content analysis*, an inductive process which starts from the texts, and then assigns to every answer one or more proper codes. Through a selection and assessment activity carried out by a research team, student responses were coded into categories, in order to build up a more restricted body of data available for syntheses and comparisons (Krippendorf 2007). Atlas ti v.5 software was used in order to select and study the written texts which were initially structured as single parts, then assigned to several codes and further interpretative dimensions.[1]

Three interpretative dimensions have been formulated on the basis of key concepts employed in feminist theories: *culture/nature; male hard sciences/female soft sciences; equality/difference*. These binary concepts are complex and open to different interpretations, and their meanings are intertwined. In the Italian IRIS survey they structured a feminist interpretative framework, philosophically and historically oriented, which helped to understand the way gender and science are represented within the imagery of the respondents. To be more precise, these polarisations have been employed as test categories, thus helping to enquire whether this imagery presents continuity or discontinuity aspects with reference to a traditional gendered discourse.

A brief clarification of the way these concepts have been specifically understood and employed in the Italian IRIS survey is further necessary here. A more detailed explanation of them can be found in Chap. 4.

The polarisation *culture/nature* lies at the root of the way the gender relationship has been figured out throughout the history of Western thought. In classical times, it started to be essentially conceived as a female subordination to the male being. The

[1] Giuseppe Pellegrini and Chiara Segafredo outlined this preliminary content analysis, discussing it with Alessandra Allegrini who analysed and interpreted the collected data from a feminist perspective. Besides the feminist interpretative dimensions described in this chapter, Eccles' model categories have been employed for analysing qualitative outcomes about students' motivations towards scientific subjects (Chap. 18).

naturally given female difference has been negatively seen as lacking compared to male sameness and uniqueness – neutral, abstract and universal (Irigaray 1974, 1977; Cavarero 1987, 1990; Fraisse 1991, 1996). In the modern age, this symbolic order also had a social meaning and value, having an impact on what is commonly known as "the gender (or sexual) division of social roles". In the shift to modern society, it started to be the organizational principle at the bottom of the long-lasting division between the "public/productive male" sphere and the "private/reproductive female" sphere, also known as the "male breadwinner/female caregiver" model in the Anglo-Saxon literature (Ferber and Nelson 2003; Picchio 2003).

The binary concept *male hard sciences/female soft sciences* might be considered a specific derivation of this gendered discourse, with a double-edged significance: a symbolic one (gender and science representations) and a material one (gender heterogeneity in science course choice). The material dimension of this polarity is largely documented in national and international research on the gender gap in STEM studies and educational choices: the number of women enrolling in life and health sciences has been increasing, while women are still a minority in technologically-orientated sciences, such as engineering and computer science, but also in physics and mathematics. In the Italian IRIS survey, the sample composition confirms this trend: male students form the large majority in engineering, computer science and, to a lesser extent, in physics and mathematics, while females are in the majority in biology and biotechnology.

Women have repeatedly attempted to overcome this dichotomous and oppositional order of discourse, a process manifested throughout the three waves of feminism. From the First-wave to the Second-wave, and until nowadays, *equality/difference* have been the two conceptual terms in which they located their subjectivity. On the one hand, women need to be equal to men – as far as free opportunities in thought, word and action are concerned. On the other hand, they want to be free to be different, assuming that this difference is not a negative and subaltern, rather a positive and exceeding concept (Offen 1988; Scott 1988; Groppi 1993; Saraceno 2008).

Before presenting and interpreting results from IRIS, it is worth addressing the structural limits of the research method adopted here, that is a qualitative analysis based on written texts. Differently from oral narratives, written texts are not collected within a face-to-face relationship between the interviewer and the interviewed, so that they do not really allow to deeply figure out the inner subjective notions and views behind the terminologies employed by each respondent. This is why the interpretations of the overall results that I will offer in this chapter must be regarded with particular care, especially since they emphasize a strong stereotypical imagery of gender and science. The question wording in itself might even have called forth such an imagery, which the respondents may not personally hold. At the same time, these interpretations are based on a very high number of responses, which enables observing and registering the most recurrent and specific types of representations held by the respondents. This is an undeniable advantage that any oral research method would hardly provide. Moreover, even if the emerging imagery might have been influenced by the questions, so that it tends to reflect the responses students believe they were expected to give, this is in itself an indication of the kind of perceptions about gender and science that are prevalent in the present society and continue to shape students' choices in STEM.

Male Hard Sciences/Female Soft Sciences

This gendered polarisation is highly relevant for interpreting the Italian IRIS findings. Besides sorting students' STEM choices materially, thus outlining the gender gap underpinning them, it deeply shapes students' gendered representation of the scientific subjects they study, and more widely the representation of gender and science. Indeed, it has been observed to be a rather recurrent tendency to sharply distinguish between "male sciences" and "female sciences", and more specifically among "what is male" and "what is female" in the interests, attitudes and abilities required to learn scientific subjects, as well as in students' conceived future work scenarios after university. The gendered distinction male hard sciences/ female soft sciences therefore keeps a significant symbolic value in the imagery of the respondents, confirming that the gender gap in STEM course choices has a significant counterpart also in the gender and science representations.

Not surprisingly, on the one hand, life and health sciences, along with humanistic subjects, are frequently connected to care-giving attitudes or ideals, which are considered typically female.

> There are many male students because it is hard that a woman likes these things, compared to law or medicine, where she can express in the best way her will to help others (Male, Mechanical Engineering)

> Unfortunately there is a high prevalence of females, because women are more sensible towards humanitarian problems (Female, Biology).

As it is underlined in Chap. 18, also the analysis of the responses to the above mentioned question "describe how you came to choose this course" pointed to the same association between life and health sciences, and female care-giving aspirations, which has been recurrently observed behind students' motivations for choosing their university course. Female students often explain their choice by a sensitivity to helping other people and dealing with health problems.

> I have chosen this course because of my desire is to help the others, and research seems to me the most proper field to realise my passions (Female, Biology)

> I have chosen this course because I feel the necessity to give my own contribute to society, my personal contribution to "improve life conditions in the world" (Female, Chemistry).

On the other hand, technical subjects are most of time connected to male prerogatives, interests and abilities, both by male and female students, and especially among those enrolled in physics, mechanical engineering and electronic engineering courses.

> The course is very homogeneous, although I thought informatics was more oriented to a male public (Male, Computer Science)

> The participation in the course is predominantly male, because there is a common belief that studying informatics means studying "the way to adjust a PC" (Male, Computer Science)

> In my course there is a male prevalence, probably because of the shared conviction that mechanical engineering is a study course for males. Actually in this course we don't only make machines (Female, Mechanical Engineering);

> Male prevalence. Male sex is more willed to follow this study course because he has a major contact with material (e.g. video-games), but also because of some stereotypes. (Male, Computer Science)

The practical dimensions of a subject are often perceived as interconnected with technical aspects, once again frequently associated with men. On the contrary, theoretical interests and attitudes are more often seen as female prerogatives, especially the theoretical dimension of mathematics.

> Mathematics is more theoretical and less practical than for instance physics, this is why it attracts more girls. (Female, Mathematics)

> There is a female prevalence, because mathematics is the scientific subject most suitable for girls' attitudes, since it is the less technical subject. (Female, Mathematics)

At first sight, ascribing the theoretical features of mathematics to women's interests and inclinations could be evaluated as a non-traditional aspect, that challenges the historical association of science with masculinity attached to the traditional distinction male hard sciences/female soft sciences. Actually, this is not the case if we consider other traditional features that draw this association. First, to a larger extent theoretical aspects are not uniquely attributed to mathematics but also to research and study, as evidenced by students enrolled in biology and biotechnology as reasons for the female prevalence in these courses.

> There is a female prevalence; the reason is that nowadays girls are the most motivated to continue to study, much more than boys (Male, Biotechnology)

As I will clarify in the next paragraph, students often remark a natural female tendency to methodical study, meant as a compensatory effort to the lack of male innate abilities in learning scientific subjects. A second aspect to consider is that a high number of students report that women decide to study mathematics because they wish to enter school teaching, which can undoubtedly be seen as a typically female job, traditionally connected with care-giving ideals.

> There's a female prevalence in my course. I do not know why, but doing "the mathematics teacher" is a quite common female ideal (Male, Mathematics)

> Female prevalence, because of the main opportunity in future teaching jobs, which attract more women (Female, Mathematics)

Overall, the findings presented here clearly pinpoint a rather traditional trend in the students' gendered imagery of science, since they point out a reiterated association of the "hard sciences" with men, and the "soft sciences" with women.

Among these findings, the most important and emerging one is the highly frequent association between technical subjects, technology, technical aspects and men, while these aspects are never connected with women. This finding suggests the primacy of the male-gendered connotation of techno-science over science in the students' imagery. On the one hand, physics or mathematics are not among the subjects mostly perceived as "male hard sciences", compared to engineering and computer science, that – compared to physics and mathematics – also have the highest number of male students enrolled. On the other hand, besides biology, health sciences and humanistic disciplines, biotechnology tends to fit in the same trend: next to the other traditional "female soft sciences", also biotechnology is symbolically associated with typical female aspects, so that it tends to be represented as a "female soft science".

As already underlined in Chap. 18, the case of biotechnology is indeed rather emblematic. Students perceive a strong association between biotechnology and engineering, in so far as they frequently compare one with the other, more precisely describing them as gender connoted opposite fields.

> There is a female prevalence, probably because males prefer engineering or informatics (Female, Biotechnology)

> The female component is more consistent, because there is a high percentage of males enrolled in engineering or polytechnics (Male, Biotechnology)

Although engineering and biotechnology are often perceived as related disciplines, it is remarkable that technology – which is supposed to be a central feature of biotechnology – is not considered a distinguishing aspect of this field of study, while on the contrary, biology, medicine, laboratory activities and the typically female inclinations ascribed such as care-giving and helping others are perceived to characterise this field. As many students attending biotechnology assert:

> I attend a course where there is a female prevalence: women are more patient in laboratory (Female, Biotechnology)

> There is a slight female prevalence, because they are more willing to engage in laboratory activity (Male, Biotechnology)

> Female prevalence. Maybe because this scientific subject, useful to help others or contribute to society and people, attracts more a female sensibility (Female, Biotechnology)

> There is a female prevalence, maybe because women are more fascinated by the biological or animal field than males (Male, Biotechnology)

The association of technology with men is also found in several responses stressing a major female proximity to scientific disciplines, which implies a sharp distinction between scientific disciplines – associated with women – and technological disciplines – associated with men.

> There is a female prevalence. In my opinion the reason is that humanistic-scientific disciplines are considered nearer to a female personality than a male one (Female, Biology)

> This course is more attended by females, maybe because biology attracts more girls, since we have a major predisposition towards scientific subjects (Female, Biology).

By means of seeing the terms 'scientific-humanistic' and 'theoretical' as alternative and opposite concepts to 'technical' and 'practical', these responses can moreover be interpreted in the light of the considerations offered above about the connections of theoretical aspects of scientific subjects, mathematics in particular, to female prerogatives.

The primacy of the male-gendered connotation of techno-science over science finds further evidence in other Italian inquiries into the gender and science representations in secondary school educational contexts, indicating that the gendered polarisation male hard sciences/female soft sciences still plays a central role in the students' imagery, although partly reworked into major techno-scientific meaning (Allegrini 2009).

Culture/Nature

Students cite a wide range of factors as the main reasons for the gender composition of the course they attend. These factors can be interpreted in the light of the culture/nature polarisation. Different social, cultural and historical elements, both internal and external to the educational system, are explicitly offered as cultural reasons for the gender imbalance observed, or for the prevalence of one gender over the other. Moreover, they unveil a historical-cultural idea of gender as a concept. Other factors, internal and external to the educational system, are suggested as natural reasons for the gender composition of the courses attended, especially for the gender imbalance perceived in several disciplines. As I will further clarify in the next paragraph, they explicitly or implicitly assume a deterministic and essentialist view of sexual difference, both from a biological and psychological point of view.

Although natural aspects are detected in the responses of students of both genders, the analysis of the overall outcomes show that male respondents, more than female, remark this type of features as reasons for being dominant in the course they attend. Students enrolled in male-dominated courses, or, more precisely, the courses where they perceive to be the majority – namely engineering, computer science and, to some extent, physics – especially share this naturalised conception of sexual difference, as a naturally given difference among males and females, as far as skills, abilities, capacities and cleverness are concerned.

> I suppose that in my course there is a prevalence of male students because these subjects fit more with male abilities and capacities (Male, Mechanical Engineering)
>
> There is a male prevalence because women are afraid of the word 'mathematics' (Male, Mechanical Engineering)
>
> Male students, because I still have to find a really clever girl. (Male, Computer Science)

Attitudes, interests, abilities in technical subjects, most of all computer science, are especially evaluated as male attributes, often by virtue of a natural reason.

> There is definitely a male prevalence, due to a way of thinking and reasoning, which is closer to the one of a machine (Male, Computer Science)

> My course is male dominated because it is a purely technical-working study course, which is unsuitable for females. (Male, Mechanical Engineering)

Besides the technical dimension of a study subject, its practical relevance, and the overall concrete dimension of a study course, are often male-connoted on the basis of an essentialist difference among sexes, most of all among students in engineering and computer science courses.

> Engineering is a very practical subject (. . .), that is a male characteristic, more than female (Male, Computer science)

> The prevalence is male because this is a very practical study course (Male, Computer Science)

In male-dominated courses, engineering and computer science in particular, women give a larger spectrum of cultural reasons to explain the gender imbalance. Different elements describing the sociocultural background and structuring cultural beliefs, such as gender stereotypes and prejudices, or historical factors such as the history of the gender relationship in the past, are mostly named by these female students, whose remarks often turn into forms of denunciation and social critique.

> Male students are prevalent, because we live in a male chauvinist society, where there is the conviction that women are unsuitable for these things (Female, Computer Science)

> There is a prevalence of male students. It is a cultural inheritance, because men more than women have always undertaken scientific studies (Female, Physics)

> I believe to attend a course with a male prevalence because, according to the current stereotypes, my interests are those of boys (Female, Mechanical Engineering)

> There are many male students because there are still prejudices against women, who are not considered to do much better than men. (Female, Mechanical Engineering)

Women enrolled in engineering also underline gender stereotypes attached to the engineer as a male professional figure.

> In my course there is a prevalence of male students. I suppose this is the case because in our society the engineer is a purely male figure (Female, Electronic Engineering)

It is worth noticing that the tendency to represent gender and science in a stereotypical and essentialist way also appears in the responses to other open questions, in particular the question "do you see any reason why the situation described above should change – and if so, what do you think could be done to change it?". The analysis of the students' responses to this question highlighted that this trend is once again largely shared among male students enrolled in engineering and computer science. They represent the majority of all the respondents who do not consider a change in the gender composition of their study course to be feasible. They justify this belief on the basis of naturally deterministic reasons, often employed to defend a status quo, thus to reaffirm a strong association between masculinity and technological hard sciences.

> No, this situation will not change. Females are like they are, because of their nature (Male, Mechanical Engineering)

> If this situation has not changed yet, it will never change. While mathematics is getting better, informatics will stay like it is now. Although there have been some, I cannot imagine women contributing to innovation in this field. (Male, Computer Science)

The tendency to offer naturalised reasons, which is prevalent among male students attending male-dominated courses, is also frequent among female students attending female-dominated courses. Nevertheless, in both cases, men more than women are those who report a wider range of naturalised factors. While in male-dominated disciplines the majority of male students justify this situation by remarking their own ability and innate attitude towards science, specific scientific subjects – especially technical subjects – in this case, the innate aspects attributed to women are mainly behavioural inclinations towards care-giving and human relationships. As already mentioned in the previous paragraph, these female qualities are mainly associated with humanistic subjects, life and health sciences, including biotechnology. Women enrolled in biology and biotechnology particularly underline these aspects.

> It is a course prevalently attended by female students, because women have an innate attitude to be interested in all the problems connected to our planet. (Female, Biology)

In disciplines such as mathematics, statistics, physics and chemistry, where more women are enrolled and the perception of the scientific field is more gender balanced, and in several cases even female-dominated, the idea of a typically female care-giving prerogative disappears, leaving instead space to the belief of an innate female capacity for studying and learning. More specifically, methodical study, often described as a natural inclination to make efforts, is the most cited female capacity in these disciplines.

> There is a female prevalence, because girls are more inclined towards studying and effort (Female, Physics)

> There is a female prevalence because, in my opinion, female sex is more inclined towards and constant in studying (Female, Mathematics)

> There is a slightly female prevalence, probably because women are more willing to make sacrifices. (Female, Biology)

Sometimes, this typically female predisposition for effort is also related to laboratory activities, most of all in biotechnology courses.

> I attend a course with a female prevalence, because I believe women are more patient within laboratories (Female, Biotechnology)

It should be noticed that these are not really innate abilities or capacities towards scientific subjects, rather innate behaviours in the learning style adopted in the study process. In Italy, several qualitative enquires on didactic-pedagogical issues from a gender perspective have shown that this characterisation of a typically female learning style begin already in secondary school (Mapelli 2004; Tamanini 2007a, b; Padoan and Sangiuliano 2008). Teachers play a major role in reinforcing this characterisation, negatively evaluating the methodical female learning style as a compensatory effort to the lack of male innate abilities and cleverness in learning scientific subjects (Allegrini 2009, 2012).

Equality/Difference

The interpretative dimension equality/difference highlights other significant elements which structure students' imagery of gender and science, more specifically how the gender relationship is configured in relation to science and the different scientific disciplines.

The great majority of respondents conceptualise the gender relationship as women's equality with men, in regard to interests, capacities, abilities and inclinations towards science and specific scientific subjects. This idea of equality is mainly meant as absence of difference, whereas difference is synonymous of gender stereotypes, prejudices, discriminations affecting women and not men in science, in culture and society. Difference is thus conceived as a negative concept in opposition to equality. This idea of equality is explicitly underlined by students who observe a gender balance in their study course.

> There is no difference in my opinion: this course is attended only by who has interest in the subjects to be studied (Male, Electronic Engineering)

Among the respondents, women are those who especially support a gender balance, compared to the large majority of men who perceive to be dominant in their study course. They particularly assert a gender balance by underlining the absence of gender stereotypes and prejudices in their study course. In their words, difference means discriminations and prejudices against women, no longer affecting present society and science studies.

> We are balanced; nowadays there is not any more so much difference or prejudices (Female, Chemistry)

Also women perceiving a male prevalence in male-dominated courses, most of all engineering, assume the same idea of difference as a negative concept: an idea of equality of women with men in the sense of absence of differences meant as discriminations. As already underlined, these young women are the most critical towards gender stereotypes and willing to denounce them.

Most of the students perceiving gender balance address equality with a slightly different meaning: as in-difference or gender neutrality, that is considered a distinguishing feature of the interest in science.

> There is no difference in my opinion: this course is joined only by who is interested in the subjects to be studied (Male, Electronic Engineering)

> I do not think there is a female or male prevalence, because the choice to study a subject does not depend on sex, rather on interests (Female, Biology)

> No gender prevails on the other, and this is because scientific subjects are a shared interest common to everyone (Female, Biotechnology)

> There is not a prevalence of male or female students; the course is heterogeneous. I think this is because the study subject is interesting regardless of a student's sex (Male, Chemistry)

Nevertheless, I would suggest that the same idea of difference as a negative concept, opposite to equality, is assumed also in these cases, since gender neutrality is conceptually and historically linked to this way of conceiving difference. Further details concerning this issue can be found in Chap. 4.

In some other cases, it is science itself, or a particular scientific subject, that is considered neutral, not influenced by external elements such as gender.

> The percentage of males and females is the same. Probably because mathematics is a hybrid subject (Female, Mathematics)

Science is not only considered neutral from the point of view of the subjects who study and practice science; it is rather neutral as far as its own methods, categories and approaches are concerned. As it is pointed out in Chap. 4, this can be considered a sign of a historical heritage coming from a positivist idea of science: a pure science, that is free from external elements influencing it. From this perspective, equality among men and women – meant as gender neutrality – is linked to an idea of scientific objectivity as neutrality. Over the last 40 years, feminist theories on science have contested this idea of neutrality, trying to undermine the internalist conception of science and its objectivity that, in short, represents the original male association with science behind the ideal of neutral objectivity (Bordo 1987; Keller 1983, 1985; Harding 1991, 1993; Haraway 1988; Longino 1990, 1996).

Apart from being understood as gender equality, or gender in-difference, the gender relationship is often elaborated in an essentialist way, as a naturally given difference. As it as been already described in the previous paragraph, a large proportion of respondents indeed share a widespread tendency to naturalise difference among women and men, especially men in male-dominated disciplines, who recurrently consider this natural difference a reason for justifying to be the majority in their course. In line with what is described in Chap. 4 about gender identity as a performative practice (Butler 1990, 1993), it is possible to suggest that both young men and women need to ascribe their belonging to a certain gender, reinforcing their gender identity as a personal and social role through a reiterated practice. However, while the former perform their gender by over-stressing gender difference, the latter often perform their gender by denying their difference. If that is the case, difference is meant as discrimination, and urges girls' need to be equal to boys, most of all in male-dominated sciences. An analogous interpretation is offered in Chap. 19, where it is described the way female students enrolled in male-dominated university courses, such as computer science, physics and nanotechnology, perform their gender by trying to be "more masculine" than other women, and assimilating with the majority gender group. On the other hand, male students in the female-dominated course of molecular biomedicine tend to denote themselves as different from the majority gender, therefore not trying to become as them. By employing another key concept introduced in Chap. 4, "hegemonic masculinity" (Connell 1987; Connell and Messerschmidt 2005),

I would also suggest that both women and men who differ from "hegemonic masculinity" might perceive to be devalued.

Recently, other Italian surveys have attested this double tendency in the imagery of the gender relationship: the alternation between an essentialist conception of sexual difference and a conception that denotes in-difference or gender neutrality (Contarello et al. 2008, 2009; Allegrini 2009, 2012). Although further analysis should be required, from a feminist perspective it is possible to argue that this is not really a conceptual alternative, but rather the same way of seeing the gender relationship: as equality among maleness and femaleness, among men and women. Indeed, as it is clarified in Chap. 4, both these ideas might be traced back to a historical-conceptual perspective that traditionally conceives sexual difference as a female lack by virtue of a naturally given difference, as something negative to remove in favour of an ideal of equality, often meant as homogenization of embodied differences to a neutral-male universal representation.

Italian Students' Ideas About Gender and Science in Late-modern Societies: Continuity or Discontinuity with the Past?

This chapter has explored different symbolic aspects outlining the representation of gender and science within the imagery of Italian students who are enrolled in different scientific courses in their first year of university. Three historical-conceptual dimensions, based on key notions particularly relevant in feminist theories, have been employed for this purpose: culture/nature, male hard sciences/female soft sciences, equality/difference.

More precisely, these interpretative categories have specifically allowed us to explore the relationship between traditional and non-traditional features characterising students' representation of gender and science and different sciences. Among the traditional aspects, the most significant one is the recurrent inclination to stereotype in a naturalised and essentialist way "what is male" and "what is female" in the different scientific disciplines, in the interests, attitudes and abilities required to study these subjects, and in the job opportunities after university. Although a number of students, especially women in male-dominated study courses, appear to be oriented to a cultural approach to gender, this being behind their motivation to be a minority in the male-dominated course they attend, the largest proportion of respondents mainly share an idea of sexual difference meant as a naturally and essentially given difference among sexes. This reiterated assessment of male and female traits reproduces gender polarities to be properly framed in the binary concept male hard sciences/female soft sciences that in students' imagery still has a powerful symbolic meaning, in that it appears to orient their choices in science studies.

It is especially in the light of this last interpretative dimension that non-traditional aspects have been remarked, mainly the primacy of the male-gendered connotation of techno-science over science. As it has been noticed, technically-oriented disciplines, such as engineering and computer science – that have the highest number of male students enrolled – are the fields mostly represented as "male hard sciences". On the contrary, biotechnology, although perceived as related to engineering, is largely represented as a "female soft science", since technology is not considered a distinguishing aspect of this field of study. I have underlined that biology, medicine, and specific inclinations ascribed to women, such as care-giving and helping others, are instead considered the distinctive features of biotechnology.

How can we further interpret these main findings, here briefly summarised, within the socio-historical context of late-modern societies?

Modern age, or first modernity, is defined through several peculiar features which shaped a patrimony of ideas that, despite originating far back in time, visibly materialised within industrial Europe, after the Second World War. Among these features, we should mention the national-state organisation of economies in each single country; the class hierarchies between the bourgeoisie and proletariat, experts and profanes, on the basis of knowledge monopolies that were professionally produced and controlled; the "natural" territorial bond between production, cooperation and enterprise (Harvey 1993; Bologna and Fumagalli 1997; Marazzi 2001). Also the long-lasting "natural" principle that has ruled and controlled the exclusion of women from the public sphere has been a central feature. The latter expressed itself through the division among "productive male labour" and "reproductive female labour", which defined nuclear families as reproduction contexts for male salary workforce, by using biomedical knowledge in order to maintain male and female "natural" foundational principles, and translating them into a social, political, economical order (Allegrini 2004). This is in short the reason why gender roles, defining a gender or sexual division of labour, fundamentally characterised the socio-cultural order of Western societies during the first modernity.

In the transition to the so-called second-modern or late-modern societies, different social, cultural and economical changes have occurred through a number of events and processes in the last 15–20 years (Giddens 1991; Beck et al. 1994; Beck 1999; Castells 2000; Bauman 2001; Beck and Beck-Gernsheim 2002). In the IRIS project specific attention is paid to understanding youth identity-building dynamics in late-modern societies (Boe et al. 2011; Chaps. 2 and 3).

As mentioned in Chap. 4, Second and Third-wave feminism, along with the increasing feminisation of the public/productive sphere in the last decades, have played a major role in the de-traditionalisation process which has characterised late-modern societies. Late-modernity, or rather post-modernity, might indeed be considered an important step for feminism, in that it offers a way out from modernity and its values, which are inextricably linked to a traditional gender order. Feminist scholars belonging to the Third-wave feminism have conceived feminism itself as a theoretical-political instance aimed at transforming social reality and symbolic orders, as a powerful weapon disaggregating the foundational categories shaped in the modernity age (Braidotti 1992, 1994). In this transition phase, principles considered for a long time as "natural laws", and the social conditions defined as universally and naturally given, have finally lost their relevance. The rigid crystallisation of gender roles has turned into opening up opportunities not only for women but also for men.

Nevertheless, traditional features, traces of modernity, can be still noticed today. Gender roles have not completely disappeared: although they are no longer social and work roles, they still act as stereotypes, conventions, constraints that have a normative but invisible power, so that specific behaviours, expectations, competences, emotions, abilities are still conventionally ascribed to women and men. Some thinkers believe that these traditional aspects, or continuity factors with the first modernity, are now back again, with an even stronger impact on younger generations. "Re-genderisation trends", that is a tendency to come back to traditional gender roles, together with a tendency to re-actualize a naturalization of sexual difference, can be detected (Lipperini 2007). Living in a crisis situation characterised by a lack of reference points as well as widespread economical, cultural and existential precariousness, these generations tend to restructure traditional social and cultural models.

This socio-cultural trend has also been noticed in the context of scientific and technical education, within a larger pedagogical-educational frame that, visibly in Italy, tends to maintain a very traditional asset, rather distant from the complex reception and governance of changes that are affecting our present, also within the field of science (Allegrini 2009). According to some scholars, there is a "substantial stability of educational and cultural models" that "has an inevitable impact on intergenerational transmission, perpetuating – from parents to sons, and from teachers to pupils – characters, specificities, but also social expectations in regard to male and female roles" (Zajczyk 2007, p. 159).

Returning to the Italian IRIS findings, I would suggest that the traditional aspects describing the students' representation of gender and science can be further understood with reference to these sociological issues. The recurrent tendency to reproduce gender polarities among "male sciences" and "female sciences", frequently grounded on an essentialist view of sexual difference, can be effectively seen as a result of re-genderisation trends re-emerging in the young generations of late-modern societies and particularly enforced by the traditional asset of the educational system.

What about the non-traditional aspects, more specifically the primacy of the male-gendered connotation of techno-science over science?

An understanding of this factor might be provided by drawing attention to other features characterising late-modern societies, such as the widespread diffusion of technology and technological objects in daily life, which, as a matter of fact, undoubtedly has a significant influence on the imagery of the late-modern societies' young generations.

Actually, the relevance of techno-science is not only a distinguishing feature of late-modern society. It is also at the core of the transformation process that has occurred in science itself over the last 15–20 years. In this process of change, science becomes more tightly connected with technological research. More precisely, science's transformations are mainly oriented towards and informed by new technologies, which overall reconfigure science from several perspectives. The international literature on this issue is so diverse that it is rather difficult to offer a brief summery here. However, the well-known book *Real Science: What It Is and What It Means* (Ziman 2000) can surely be mentioned as a shared reference. In this book, physicist and sociologist John Ziman describes the way contemporary scientific research is increasingly mediated by communicative processes carried out by information and communication technologies, which strongly modify scientific features and epistemic categories which were earlier considered unchangeable through the passage of time.

Some specific aspects underpinning science transformations have recently been discussed by Italian philosopher Elena Gagliasso and biologist Flavia Zucco, such as for instance the difficulty to sharply distinguish between science and technology, the former increasingly depending on virtual simulation practice in several scientific research fields. Considering the large amount of data computers are able to provide for formulating hypotheses, in many cases this process has been able to change the parameters of the hypothetical-deductive and experimental approach implied in the scientific method (Gagliasso and Zucco 2007, p. 7).

In the Italian IRIS survey, not only the primacy of techno-science over science has been largely observed to be deeply represented by the students, rather it is the gender connotation of this process that has been noticed to be clearly represented, especially by looking at the type of disciplines the students repeatedly polarise into male hard sciences/female soft sciences, along with the typical gendered attributes ascribed to them. In so far as this gendered polarisation still plays a pivotal role in the Italian students' imagery on gender and science, I would finally suggest that the historical association of science with masculinity is not overcome in the context of late-modern society, rather it is reconfigured in a new techno-scientific dimension.

References

Allegrini, A. (2004). Donne, lavoro e tecnologie IC. Prospettive femministe/di genere sul lavoro nella società in rete del capitalismo post-fordista. In *Associazione "Orlando" (a cura di), Le ricerche del progetto Portico* (pp. 129–202). Bologna: Pitagora.

Allegrini, A. (2009). Le Altre stelle. La dimensione di genere dei contesti educativi tecno-scientifici. Un'indagine conoscitiva. Ricerca promossa dalla Consigliera di Parità della Provincia di Verona, in collaborazione con l'Ufficio Scolastico XII di Verona e la Consulta provinciale degli studenti. In *Associazione Donne e Scienza*. www.donnescienza.it; 2012, febbraio (pp. 9–13 e 25–103). Verona: Grafiche Marchesini.

Allegrini, A. (2012). Genere e scienza nella contemporaneità: uno sguardo storico-concettuale. In S. Badaloni e & A. Contarello (Eds.), *Genere e Cambiamenti. Dalla sottorappresentazione delle donne a nuovi scenari emergenti* (pp. 85–90). Padova: Padova University Press.

Bauman, Z. (2001). *The individualized society*. Cambridge: Polity Press.

Beck, U. (1999). *What is globalisation?* Cambridge: Polity Press.

Beck, U., & Beck-Gernsheim, E. (2002). *Individualization: Institutionalized individualism and its social and political consequences*. London: Sage.
Beck, U., Giddens, A., & Lash, S. (1994). *Reflexive modernization. Politics, tradition and aesthetics in the modern social order*. Stanford: Stanford University Press.
Bøe, M. V., Henriksen, E. K., Lyons, T., & Schreiner, C. (2011). Participation in science and technology: Young people's achievement-related choices in late-modern societies. *Studies in Science Education, 47*(1), 37–72.
Bologna, S., & Fumagalli, A. (Eds.). (1997). *Il lavoro autonomo di seconda generazione. Scenari del postfordismo in Italia*. Milano: Feltrinelli.
Bordo, S. (1987). *The flight to objectivity. Essays on Cartesianism and culture*. Albany: State University of New York Press.
Braidotti, R. (1992). *Patterns of dissonance. A study of women in contemporary philosophy*. New York: Routledge.
Braidotti, R. (1994). *Nomadic subjects. Embodiment and sexual difference in contemporary feminist theory*. Cambridge: Columbia University Press.
Butler, J. (1990). *Gender trouble: Feminism and the subversion of identity*. New York/London: Routledge.
Butler, J. (1993). *Bodies that matters*. New York: Routledge
Cavarero, A. (1987). L'elaborazione filosofica della differenza sessuale. In M. C. Marcuzzo, A. Rossi-Doria (a cura di), La ricerca delle donne. Studi femministi in Italia (pp. 173–187). Torino: Rosenberg & Sellier (English translation (1991). The need for a sexed thought. In S. Kemp & P. Bono (Eds.). *Italian feminist thought: A reader*. Oxford: Blackwell.
Cavarero, A. (1990). Nonostante Platone. Figure femminili nella filosofia antica. Roma: Editori Riuniti. English Translation (1995). *In spite of Plato: A feminist reading of ancient philosophy*. Foreword by Rosi Braidotti. Cambridge: Polity Press.
Castells, M. (2000). *The rise of the network society. The information age: Economy, society and culture* (Vol. 1). Oxford: Blackwell.
Connell, R. (1987). *Gender and power*. Cambridge: Polity Press.
Connell, R., & Messerschmidt, J. (2005). Hegemonic masculinity: Rethinking the concept. *Gender and Society, 19*(6), 829–859.
Contarello, A., Badaloni, S., Brondi, S., & Manganelli, A. M. (2008). *Studentesse e studenti di fronte al "mito" della scienza. Un'indagine presso l'Università di Padova*. Padova: Università degli Studi di Padova.
Contarello, A., Badaloni, S., Brondi, S., Manganelli, A. M. (2009, 19–20 febbraio). Il genere della scienza. Una lettura psicosociale della sottorappresentazione delle donne nell'universo scientifico e tecnologico. In *Le questioni sui generi in psicologia sociale*. Abstract book (pp. 125–137). Parma: Uni.Nova.
Ferber, M. A., & Nelson, J. A. (Eds.). (2003). *Feminist economics today*. Chicago: The University of Chicago Press.
Fraisse, G. (1991). Dalla destinazione al destino. Storia filosofica della differenza dei sessi. In G. Fraisse & M. Perrot (Eds.), *Storia delle donne in occidente. L'Ottocento* (pp. 89–123). Roma: Laterza.
Fraisse, G. (1996). *La difference des sexes*. Paris: Presses Universitaires de France.
Gagliasso, E., & Zucco, F. (2007). *Il genere nel paesaggio scientifico*. Roma: Aracne.
Giddens, A. (1991). *Modernity and self-identity. Self and society in the late modern age*. Cambridge: Polity Press.
Groppi, A. (1993). Le radici di un problema. In G. Bonacchi e & A. Groppi (a cura di), *Il dilemma della cittadinanza. Diritti e doveri delle donne* (pp. 3–15). Roma/Bari: Laterza.
Haraway, D. (1988). Situated knowledges: The science question in feminism and the privilege of partial perspectives. *Feminist Studies, 14*(3), 575–599.
Harding, S. (1991). *Whose science? Whose knowledge? Thinking from women's lives*. Buckingham: Open University Press.

Harding, S. (1993). Rethinking standpoint epistemology. "What is strong objectivity?". In L. M. Alcoff & E. Potter (Eds.), *Feminist epistemologies* (pp. 49–82). New York: Routledge.
Harvey, D. (1993). *La crisi della modernità*. Milano: Il Saggiatore.
Irigaray, L. (1974). *Speculum. De l'autre femme*. Paris: Les Editions de Minuit.
Irigaray, L. (1977). *C'è sexe qui n'est pas un*. Paris: Les Editions de Minuit.
Keller, E. F. (1983). *A feeling for the organism. The life and work of Barbara McClintock*. New York: Freeman & Co.
Keller, E. F. (1985). *Reflections on gender and science*. New Haven: Yale University Press.
Krippendorf, K. (2007). *Content analysis: An introduction to its methodology*. Thousand Oaks: Sage.
Lipperini, L. (2007). *Ancora dalla parte delle bambine*. Milano: Feltrinelli.
Longino, H. (1990). *Science as social knowledge. Values and objectivity in scientific inquiry*. Princeton: Princeton University Press.
Longino, H. (1996). Cognitive and non-cognitive values in science: Re-thinking the dichotomy. In L. H. Nelson & J. Nelson (Eds.), *Feminism, science and the philosophy of science* (pp. 39–58). Boston: Kluwer Academic.
Marazzi, C. (2001). *Capitale & linguaggio. Ciclo e crisi della new economy*. Soveria Mannelli: Rubbettino.
Mapelli, B. (2004). *Nuove virtù. Percorsi di filosofia dell'educazione*. Milano: Guerini Studio.
Offen, K. (1988). Defining feminism: A comparative historical approach. *Signs: Journal of Women in Culture and Society, 14*(1), 119–157.
Padoan, I., & Sangiuliano, M. (Eds.). (2008). *Educare con differenza. Modelli educativi e pratiche formative*. Torino: Rosenberg & Sellier.
Picchio, A. (Ed.). (2003). *Unpaid work and the economy*. London: Routledge.
Saraceno, C. (2008). Tra uguaglianza e differenza: Il dilemma irrisolto della cittadinanza. In Fondazione Ermanno Gorrieri. www.fondazionegorrieri.it
Scott, J. (1988). Deconstructing equality versus difference, or, the uses of poststructuralist theory for feminism. *Feminist Studies, 14*(1), 33–51.
Tamanini, C. (a cura di) (2007a). *Maschi e femmine a scuola: stili relazionali e di apprendimento. Una ricerca sul genere e percorsi formativi*. Trento: Iprase Trentino - Provincia Autonoma di Trento Ed.
Tamanini, C. (2007b). Segregazione formativa e cultura di genere degli insegnanti. In B. Poggio (Ed.), *L'isola che non c'è. Pratiche di genere nella pubblica amministrazione tra carriere, conciliazione e nuove precarietà* (pp. 171–189). Trento: Edizioni 31.
Zajczyk, F. (2007). *La resistibile ascesa delle donne in Italia. Stereotipi di genere e costruzione di nuove identità*. Milano: Il Saggiatore.
Ziman, J. (2000). *Real science: What it is and what it means*. Cambridge: Cambridge University Press.

Part V
Conclusions and Recommendations

Chapter 21
Understanding Student Participation and Choice in Science and Technology Education: The Contribution of IRIS

Jim Ryder, Lars Ulriksen, and Maria Vetleseter Bøe

Introduction

This chapter considers the contribution of the studies in this book to our understanding of students' educational choices. This is done across five themes: theoretical perspectives; choice as a continuous process; the role of identity and social structure; gender; and methodological insights. The chapter ends with suggestions for a future research programme exploring student choice and participation.

Theoretical Perspectives

Part I of this book presents the theoretical perspectives drawn upon in the Interests and Recruitment in Science (IRIS) project. Chapter 2 focuses on the Eccles et al. expectancy-value model of achievement-related choices, which posits that educational choices can be explained by young people's beliefs about how well they will do in, for example, a study programme, and by the value they attach to the programme in question. This subjective value represents how interesting the student expects the programme to be, how easily it is negotiated into the student's identity

J. Ryder
School of Education, University of Leeds, Leeds LS2 9JT, UK
e-mail: J.Ryder@education.leeds.ac.uk

L. Ulriksen
Department of Science Education, University of Copenhagen, Øster Voldgade 3, DK-1350 Copenhagen C, Denmark
e-mail: Ulriksen@ind.ku.dk

M.V. Bøe (✉)
Norwegian Centre for Science Education, P.O. Box 1106, Blindern, N-0317 Oslo, Norway
e-mail: g.m.v.boe@naturfagsenteret.no

E.K. Henriksen et al. (eds.), *Understanding Student Participation and Choice in Science and Technology Education*, DOI 10.1007/978-94-007-7793-4_21

construction, how useful it will be for reaching other goals, and how much it will cost in terms of time and effort. According to the expectancy-value model, individuals' values and beliefs are shaped by a range of social and psychological factors, such as cultural surroundings, personal goals and self-perceptions. Chapter 2 also describes sociological theories on late-modernity, adding an understanding of how young people's expectations and values may be influenced by a late-modern zeitgeist. Such theories argue that rich, developed societies emphasise late-modern values such as self-realisation and personal well-being for the individual. This should be understood as a way of coping with changes in society leaving the individual as apparently less bound by tradition and to a large extent free to construct their own identity through life choices. This, firstly, leaves the individual with an obligation to handle these options and to be able to release his/her potential in a way that appears authentic and fitting to the individual's sense of self. Secondly, this constitutes a sense of ambivalence between being free and being at risk. Furthermore, since social class, gender, and other social categories are still affecting what is possible, the student also needs to handle the contradiction between what appears to be a condition of liberty but also involves significant limitations.

In IRIS, these perspectives have influenced instrument development and methodological choices and have aided interpretation of results. For instance, the strong focus on personal interest in the accounts IRIS respondents give of their choice (Chaps. 9, 18 and others) may be interpreted in the light of the importance of individual self expression and the influence of late-modern perspectives on youth.

Chapter 3 presents narrative theories, in particular narrative psychology, as a framework for understanding how students negotiate their educational options as part of their identity construction. This approach studies how young people construct narratives in ways that are recognised as compatible with how they see themselves and how they are perceived by others. Importantly, identity construction is viewed as an on-going and constantly changing process that, at the same time, tries to maintain a stable sense of self embedded in the surrounding culture. Elsewhere, Tinto's perspectives on social and academic integration are used in Chaps. 13 and 15 to add to our understanding of how students negotiate their choice narratives and their identity when they have started a new study programme. Perspectives on gender, presented in Chap. 4, are discussed in a later section of this chapter.

The theoretical perspectives presented in Chaps. 2 and 3 have been employed in various ways throughout the book. For example, the expectancy-value model is used as a theoretical and/or analytical framework in Chaps. 9, 11, 16 and 18. The use of narrative theories in this book varies in form, as elaborated in Chap. 3. Chapters 7 and 15 (and empirical parts of Chap. 3) use narrative psychology both as the underlying conceptual framework for understanding the notion of identity, and as a guide for the construction and analysis of interview data. Chapter 13 employs a narrative psychology conception of identity in its review of research on drop out from higher education, and Chap. 9 uses these perspectives in an analysis of short statements from students about how they came to choose their study programme.

The contribution of the present book is not so much to add to, or challenge, the above theories but to bring them together and in some cases to apply them to new contexts and settings. Theories of narrative psychology have rarely been used to inform research on STEM participation. The expectancy-value model has been used to examine choices of science and mathematics courses, but was developed as a more general model for achievement-related choices and performance. Late-modern perspectives and narrative theories are relatively undeveloped in a science education research context. Schreiner (2006) drew on broader sociological theories of late-modernity in her study of young people's orientations to science, thereby suggesting that the perspectives could be relevant also for understanding participation in STEM. The IRIS project has attempted to take these multiple theoretical perspectives and consider the extent to which, taken together, they provide fruitful insights into our understanding of STEM-related educational choices in particular.

The work of late-modernity theorists such as Giddens, Beck and Bauman has been criticized for implying that social structures such as class have lost their relevance to studies of young people's choices and behaviour (Atkinson 2008; Furlong 2009). The IRIS project has found late-modernity perspectives useful for understanding how young people regard an educational choice as their own individual project, with both a personal freedom and a personal responsibility to choose. However, as is stated in Chap. 2, the IRIS project distinguishes between young people's idea of having a free choice and the actual limitations to their freedom that are imposed by social structures such as gender and class (Archer et al. 2012).

We have found it fruitful to bring together aspects of the Eccles et al. model and narrative perspectives in understanding student choice. The arrows in the figure of the Eccles et al. model given in Chap. 2 might be interpreted to mean that choices happen at an instant in time, and are the outcome of a series of influences (working from left to right in the model). However, the empirical evidence in Chap. 3, for example, using the narrative approach, demonstrates how such an interpretation of the choice process is rather limited. Narrative theory emphasises how the influence of culture, family and peers interact constantly with the choice process. In the Eccles et al. model, the iterative nature of the influence of the parts of the model is indicated by a dotted arrow going from the choice and back to the rest of the model. However, underestimating this dotted arrow and interpreting the model in a strict left-to-right sense will restrict the possibilities to look at the on-going dynamics of a choice process. Though valuable information is provided by studies measuring either the predictive power of expectations of success and subjective values for STEM choices (Eccles 2007; Eccles et al. 2004) or students' own retrospective reports of the influence of expectations and values (Bøe 2012; and Chaps. 9 and 18 of the present volume) more and different information is needed to fully understand young people's STEM choices. Eccles and colleagues have themselves included interview components alongside questionnaire data in longitudinal studies of how expectations and values develop (see for example Fredricks and Eccles 2002; Jacobs et al. 2005). By specifically using narrative theories in qualitative investigations, IRIS demonstrates an effective way of studying how the process of making meaning of a STEM choice takes place over time – before, at, and after specific decision points.

Choice as a Continuous Process

As discussed above, one of the starting points for the work of the IRIS project was the view of student choice as a continuous process of activity and reflection by the student rather than a decision made at a specific point in time. Thus, we have made a distinction between 'decision points' and the 'choice process'. One example of a decision point would be enrolling on a specific university course (e.g., chemical engineering) following the end of upper secondary schooling. Such a decision point might be preceded by periods of reflection on future courses by the student over several years, and perhaps also (intermittently) activities around this choice such as talking to careers specialists, parents, friends and conducting internet searches about courses and careers. Furthermore, work within IRIS has also highlighted the ongoing nature of the choice process beyond key decision points. Thus, a student's formal educational experiences are characterised by a continuous choice process, punctuated by key decision points. Here we elaborate on the extent to which this perspective is supported by the research reported in this book.

The examination of school science students' retrospective reflections on the decision point of choosing post-compulsory courses in two schools in England (Chap. 7) demonstrated both the extent of the pre-decision choice process and its varying nature. Students' reflections included reference to experiences from primary schooling onwards that were seen as influencing their final decision. Furthermore, these reflections could be characterised in varying ways, for example: early commitment to specific courses followed through to the decision point; ongoing uncertainty up to the decision point; periods of commitment to specific courses interspersed with periods of uncertainty leading to new course commitments. These varying choice processes support findings from the earlier work of Anna Cleaves, who used longitudinal interview data, and challenge the often stated assumption that commitment to science courses tends to be cemented in the early years of schooling (Cleaves 2005). This perspective also challenges the common metaphor of interest and participation in STEM subjects as a 'leaky pipeline' (Blickenstaff 2005). A more nuanced metaphor is one of 'shifting pathways' with bidirectional flows; out of, but also in some cases into, the STEM 'pipeline'. For example, these bidirectional flows are shown clearly in the 'Sankey diagram' generated by Sadler et al. (2012) to represent changes in students' career interest from the beginning of high school to the end.

Perhaps a less obvious feature of the choice process is its extension beyond key decision points. This point has been highlighted in interviews conducted with students concerning their choice of university course. For example, an analysis of interviews with 20 first year students following STEM programmes in Denmark identified the 'expectancy-experience gap' as a key feature of post-decision choice processes (Chap. 15). In choosing to enrol on a specific university programme, students show their expectations of the course. This analysis suggests, for many students, that their experiences of the programme are very different from their expectations. This disjuncture can lead to further choices and decision points (hence a choice process) for example to change course, or leave Higher Education altogether. For example, the Danish student, Emily, enrolled on an engineering degree. She experienced a large expectancy-experience gap that involved an ongoing process of reflection on her decision to follow this course, and ultimately resulted in a decision to leave the programme. Similar findings result from the analysis of Norwegian university students' written reflections on their choice of university course, as exemplified by the computer science student, Tina, who wrote 'I am still very uncertain. Am I really the right girl for this?' (Chap. 17).

A striking feature of the choice process, and one with significant methodological implications (as discussed later in this chapter), is the developing nature of students' accounts of the process of choice. This development is shown particularly in Chap. 3, which explores the significance of narrative approaches to studies of the choice process. The chapter provides an example, Christine, who originally stated that she did not want to follow a course leading to teaching, but then (following the decision to enrol on such a course) reconstructed her narrative to state that she had always wanted to become a teacher. This process is interpreted through the lens of narrative psychology as an attempt by the student to maintain a sense of stability in her understanding of herself, as her experiences develop. This chapter suggests a metaphor for reflections on choice in terms of the view from a car winding through the countryside with changing views of the countryside through both the front and rear car windows. The choice narrative may include perspectives on the future (looking through the front window) and retrospective accounts of how experiences in the past led up to the present situation (looking through the rear window). An important point here is that when the perspective through the front window changes (that is, the decision of which path to follow and therefore which choice to make) it also changes what is seen in the rear window. In other words, the interpretation and narrative concerning what happened in the past changes as the forward perspective changes. Therefore, an individual's conception and interpretation of both these views – her narrative of the choice – is constantly tried out and negotiated in the students' ongoing social relations.

The Role of Identity and Social Structure

Another starting point for the IRIS project was the key role that identity plays in students' choice of STEM higher education. The importance of the identity component has been corroborated by several of the studies within IRIS. Identity perspectives have clear links to narrative theory and late-modernity perspectives referred to above, for example Giddens (1991) who described identity as the process of keeping "a particular narrative going" (p. 54). Shanahan (2009) has highlighted the significance of interactions involving individual personality and broader social structures in the development of identities. This point relates to the classic discussion of structure and agency, concerning 'the degree to which the behaviour of individuals and groups can be attributed to social, political and economic forces or wilful, purposeful intentionality' (Shanahan 2009, p. 45). Identity is developed and acted within a social structure that provides opportunities and limitations for the identities that individuals can develop. Thus, when studies report that students' choices are based on 'personal interests' we need to recognise that these interests are embedded in, and interact with, a social context. Ryan (2012, p. 170) contests the suggestion that choices or preferences are freely made, asserting that "social structural constraints operate and further, that 'traces' of constraint can be discerned in accounts of 'choices'". The present volume provides examples of the importance of the structural level in understanding the way identity is constructed and how it affects recruitment and retention in STEM higher education. In this section, we explore how the studies in this book contribute to our understanding of these interactions.

Institutional Structures

One general structural level that affects students' choices is the educational system and the provision of programmes within it. Clearly, students' choices are limited by the courses offered by universities. Thus changes in the patterns of students' choices not only reflect changes in the attitudes and preferences of young people, but also changes in the provision of programmes. As explored in an Australian context, when new and different programmes are offered, students may move to different subject areas (Chap. 10). Hence, the structural level of the educational system is important in developing an understanding of the importance of identity in two ways. Firstly, there is a danger of interpreting shifts in student choices as representing changes in student identities without recognising that such shifts may have more to do with changes in the availability (or entry requirements) of specific courses. Secondly, the structure of the programme, and the choices it makes possible also present the students with particular elements that they can integrate into their identity construction. Changes in the provision of programmes can, therefore allow, or even call for, particular shifts in identities developed by STEM students.

Frequently, STEM programmes can be related to a single discipline (for instance a 'physics' degree programme is clearly identified with the discipline of 'physics'). In such cases students need to be able to recognise desirable identities related to this specific discipline. For some fields of study this is problematic because the identities that can be developed within the particular discipline are limited and perhaps unattractive to individuals (cf. Holmegaard et al. 2014). However, experiences in the UK suggest that programmes that include disciplinary components from both STEM and other fields (for instance, forensics) can attract more students, not least women. Similar tendencies can be found in Danish programmes combining engineering with other disciplines such as medicine, biotechnology, design, or architecture. However, it should be noted that these 'mixed' programmes frequently include aspects from biology and medicine, fields that otherwise also attract more women. Nevertheless, it is likely that programmes combining elements from different fields of STEM may allow for the construction of a broader variety of identities and therefore attract a more diverse group of students, including more women.

Another important aspect of the structural level is found in the results from the Slovenian study of students' choice to embark on a PhD (Chap. 11). The study shows that national funding programmes targeted at particular challenges facing female PhD students had an important impact in making a PhD within STEM more attractive to these students. This finding indicates how structural and economic factors affect student choice patterns.

The Significance of Curriculum

Another structural element affecting students' identity work is the curriculum. Here 'curriculum' refers to the content of teaching and the kinds of teaching and learning activities within educational programmes. The design of the curriculum offers particular opportunities for students in terms of what identities they are able to construct when they are on a particular programme, the way they can act as students, and the kind of participation that is possible. Therefore, a conflict may arise between the possible identities and practices made possible in STEM programmes and the identities that are recognisable and attractive to the students. In an analysis of the Danish upper-secondary school physics curriculum, Krogh (2006) used the concept of cultural border crossing (Aikenhead 1996) to compare the identity-related values of late-modern students and values in physics teaching, what he called 'the ethos of science teaching' (Krogh 2006). Krogh found a fundamental clash between values of young people and the ethos of physics teaching that impede students' identifying with doing science. The two students, Claire and Anya (Chap. 7), provide examples of how the same curricular element (socio-scientific issues) was valued differently by individual students. Whereas Claire preferred science content linked to 'facts' that were new to her, Anya appreciated more the possibility of linking science content to her everyday life. Thus, including socio-scientific issues in the curriculum may cater for the interest of some students whilst being at odds with the self-image of other students.

For students who have already entered a higher education STEM programme, the process of academic and social integration involves balancing, on the one hand, the possible interests and practices necessary to become involved in the curriculum and the social life within the programme, and, on the other, their personal interests and self-images. For the biochemistry student, Frida (Chap. 15), this balancing act involved a focus on the parts of the programme curriculum that allowed her to get a sense of 'turning into a professional' (e.g., putting on a lab coat) and prioritising social integration.

In summary, the construction of identity, and a viable narrative about who the students are and who they are to become, is closely related to the structural level of curriculum design. Programme content and teaching/learning activities need to provide room for the development of student identity. The different reactions of the students described above (e.g. Anya and Claire, Chap. 7) suggest that any degree curriculum should seek to open up multiple ways for students to engage with the subject.

Available Discourses

A pivotal element in the students' construction of identities is the set of available discourses through which social structure acts. Several chapters in this volume show that students' choices are related to existing discourses. For example Chap. 6 highlights a prevalent discourse that science disciplines are for 'the clever' and that learning science is particularly demanding and difficult. This finding suggests that students need to see themselves as 'clever' in order to pursue science disciplines, an identity that many students (including high attaining students) find difficult. Furthermore, engagement with science subjects also means that students may be considered "brainiacs" who "don't want to do anything else in their life", as the student, Celina, remarks in Chap. 6. A discourse that science is for clever and 'good' students means that the students who wish to engage in science also need to adopt a 'good student' identity. Thus, 'science as difficult and for the clever' is a discourse that restricts participation in science for students of both genders, and for students from particular social and ethnic backgrounds (Chap. 6).

Chapter 9 describes how personal interest is central in the discourses that young people engage in concerning educational choice. When prompted to describe their educational choice, young people choose to present a narrative focused on individual interest, downplaying other influence factors (parental influence, career prospects) and priorities that are arguably at play. A discourse and a disciplinary culture demanding strong personal interest and dedication (as also seen in Chap. 18) may turn some students away, notably those that have broader educational and career priorities than pursuing passionate interest alone (Bøe and Henriksen 2013; Hazari et al. 2010).

As noted in Chap. 15, many study programmes and disciplines can be said to have an 'implied student' (Ulriksen 2009). That is to say that the curriculum and the culture within the programme hold particular expectations and presuppositions concerning the interests, attitudes, and practices of the students involved. Even though there may be more than one implied student associated with a particular programme, it nevertheless means that the students need to relate to, and to some extent adapt to, the student implied by the programme. For some students, there is a conflict with their notions of who they believe they are or who they wish to become. At the same time, it is difficult for the students (or the teachers) to challenge these presuppositions because they are implicit rather than explicit.

As the statement by Celina quoted above suggests, students' choices are also affected by discourses outside their degree programme. Students balance the discourse of, for instance, what a 'proper' scientist or engineer is, with discourses from their social life within, and outside of, their educational institution. Consequently, when students make their choice of study, they are positioning themselves in relation to different discourses many of which originate outside their educational context. Hence, the issue of identity cannot be limited to a relation between the individual student and the subject; it reaches beyond the subject and beyond the educational world.

As argued earlier, choice is a continuous process over time. Consequently, when young people are deciding which programme to enter, they become entangled in a web of different discourses concerning higher education and other realms of life rooted in the personal history of the student. Clearly students' narratives about themselves can change, alongside the discourses that are culturally and socially available to them. However, there is an inevitable inertia associated with these changes. Thus, it is reasonable to assume that efforts to change the choice patterns of young people have to involve more than an isolated event. If the experiences of young people in different recruitment and outreach initiatives are to impact on the choice of young people, in spite of dominant and persistent contrary discourses, it is more likely to happen if the students are involved in the activity over a sustained period of time (Chap. 12). Furthermore, as the choice process continues after the students have entered their higher education programmes, these different discourses continue to act as the student is coping with the university experience and deciding whether to persist or not (Chaps. 13 and 15).

In summary, research in IRIS has demonstrated that the importance of identity and discourses for educational choice cannot be limited to the relation between the discipline and the individual, but involves discourses related to a number of fields. Moreover, identity and discourses continue to affect the experiences and decisions of students even after they have entered the science path. IRIS research demonstrates the importance of addressing issues of identity on several levels: institutional, social and cultural.

The Role of Gender Within Students' Experiences of Choice

The studies in this book corroborate previous research in showing clear differences, at the group level, between male and female student participation in many STEM subjects in higher education. In some subjects (e.g., physics and engineering) women are typically under-represented, whilst in other STEM subjects, gender participation disparities are much less pronounced. Indeed in some STEM programmes women may be over-represented (e.g., the molecular biomedicine programme examined in Chap. 19). One purpose of the IRIS study was to explore the experiences of students that underpin these different gendered participation patterns. In doing so we have emphasised the importance of not treating female (or male) experiences of choice in an undifferentiated manner. The monolithic (or 'essential') character of the concepts of male and female has to be abandoned in favour of an understanding that is sensitive to the differences within each group. Firstly, male or female is not something someone is. Rather, it is something that is performed. Secondly, each gender can be performed in numerous ways. Within-gender cases may differ more from each other than many between-gender cases (Butler 1990; Søndergaard 1996; Sinnes and Løken 2012). Thus, we have not sought to identify a 'female approach to choice' that can then be used to account for the gendered participation patterns. To do so would run the significant risk (as elaborated in Chap. 17) of reinforcing gender stereotypes, and thereby sustaining existing gendered participation patterns. Rather, our research has explored the different ways in which, for example, women experience the processes of subject choice. However, at the same time, we have avoided a ‘gender neutral’ perspective on subject choice. By this we mean an approach that, in an attempt to avoid stereotypes, refrains from addressing issues in terms of gender. Such an approach would contribute little to developing ways of reducing the male/female disparities in participation patterns within specific subjects.

In Chap. 4, it is argued that even if the overall discourse of science presents ‘science’ and ‘doing science’ as neutral and objective, the subject is often associated with being male due to a long-standing historical association of male gender with rationality, versus the ‘emotional’ female gender. This gendering is amplified by an ongoing discourse of science as being ‘a boy’s thing’. In Chap. 18 it is demonstrated how Italian male STEM students tend to rely on pre-established roles (which in the case of science and technology are easily available in the culture) when making their educational choice. The ASPIRES study (Chap. 6) provides an example of one girl who stated that she had stopped attending a school science club because it was mainly boys who attended. Girls who turn towards science, therefore, need to develop strategies that can balance interest in and intentions to enter science with this popular discourse.

Among female science students, the discourse of science as masculine leads to an ambiguous sense of being different and standing out because they belong to a minority, while simultaneously opposing being labelled in a particular way due to their gender (Chap. 17). These students therefore both identify themselves as ‘like any other girl’, as the female student, Stella, expressed it, and as someone who is always visible because of being a minority, as expressed by Maria in the same chapter. Many female students, therefore, have both to cope with a discourse questioning whether doing science is compatible with being a girl, and with the situation of being visible and standing out. Similarly, Danielsson (2009) claimed that taking on a physics identity for a female student requires distancing oneself from what is “traditionally” female. However, an important point here is that the pressure of becoming a particular kind of student that requires women to neutralise their gender expression may also be a challenge to particular ways of being a male student, a point also made by Walker (2001) in a study involving engineering students.

In some STEM programmes male students have the experience of being a minority group, e.g., male students in the female-dominated molecular biomedicine programme (Chap. 19). Even so, while men and women share some experiences when being a minority, Chap. 19 also reports that the conditions for coping with these experiences differ for men and women. Many of the female students adopted a strategy of ‘being as’; becoming like one of the boys, thereby ‘neutralising’ their gender. However, the strategy of many of the male students was that of ‘fitting in’; adjusting to the dominant culture and ways of behaviour, but doing so while remaining different, retaining their male gender. Furthermore, it was also found that the conditions for women being a minority varied across the two study programmes. Women in computer science programmes appeared to have more room for 'being a girl' in different ways, as compared to the culture within the physics and nanotechnology programme. This finding illustrates the nuanced outcomes of an analysis that looks for differences between science subjects, and which goes beyond an essentialist or monolithic perspective on women's experiences of choice, and yet is not gender neutral.

The results of the IRIS project corroborate previous studies that suggest that there are indeed group-level differences in male and female STEM participation as well as in the conditions and opportunities of men and women. Some of these differences are related to social discourses about men and women and to social structures that are detrimental to women choosing a STEM path. The studies in this volume also call for attention to be paid to how these gender differences are addressed in both future research and policy. The challenge for future research is to continue the exploration of social structures, discourses, curricular components, etc. that impede the participation of women in fields of science where until now they have only had a small representation. At the same time, this should be done in a way that does not imply an understanding of gender as having a monolithic or 'essential' character, and that is sensitive to the individual variations within gender groups.

Methodological Insights

We have conceptualised educational choice as a process over time, punctuated by multiple decision points. Furthermore, an individual is continually constructing (and re-constructing) accounts, or narratives, of these processes and decisions. This perspective has significant methodological implications. For example, 'snapshot' accounts of educational choice (collected at a single point in time), whilst providing important insights, are limited in capturing the 'process' character of educational choice. Rather, longitudinal studies, of the kind reported in Chap. 3 and by Cleaves (2005), are more suited to investigations of the processes of educational choice. Relatedly, retrospective accounts of choice may not reflect the narratives that students constructed which were influential at the time of an educational decision. Many of the research studies reported in this volume have used both retrospective and snapshot accounts. Such studies are also dominant in the wider research literature. For example, Sadler et al. (2012) is a recent example of a study into changes in career interest using a retrospective cohort study. While such studies do provide useful insights into educational choice processes (e.g. enabling the collection of data from larger groups of students to probe group-level differences over time) they need to be supplemented with longitudinal studies that are open to the potential for re-constructed narratives of student choice.

The IRIS research collaboration involved a cross-country questionnaire (with both closed and open response questions) and in-country case studies typically using more extended, qualitative methodologies. This use of multiple methods to explore educational choice has several advantages. We have been able to identify diverse influences on educational choice: the detail of extended personal accounts; indications of broader socio-cultural influences; and, more systemic influences resulting from institutional, regional and national educational policies. Furthermore, we have provided an overview of student choices, e.g., at a national level, whilst also probing more deeply into the experiences of individual students. However, we recognise that more could be made of the potential for mixed methods, i.e., studies that utilise insights from one approach to inform the design and analysis of another approach. One recent example of such a study used quantitative analysis of a large-scale national dataset to construct a sampling frame for the selection of case study schools (Bennett et al. 2011; Hampden-Thompson et al. 2011).

Several of the contributions in this book demonstrate the use of national datasets to identify trends in student participation over time. Chapter 14 reports on changing patterns in student choices in Denmark. This official, annual and ongoing dataset links students' educational choices to characteristics such as gender, academic attainment and parental educational background. Analysis results in fine-grained identification of educational trends linked to socio-economic and other factors, that extend beyond the more usual blanket identification of 'a shortage of science students'. England has a similar national pupil dataset that has been used in educational research studies (Homer et al. 2013). Gill and Bell (2013), for example, use multilevel modelling techniques to identify the effects of school type (e.g., mixed or single sex schools) on student participation in post-compulsory physics, whilst controlling for the effect of other variables (e.g., science attainment, socio-economic status). Given the significance of educational outcomes for individuals, societies, governments and economies, it is surprising that such datasets, and their use to inform educational policy, are not more widespread.

The Australian study reported in Chap. 10 demonstrates a different approach to identifying national trends. Here the authors have repeated the use of a carefully designed and trialled questionnaire, first used in the 1970s, to challenge a common assumption of declining student enjoyment of science. This analysis led the researchers to consider alternative influences on educational choice, resulting in the identification of the likely significance of systemic policy structural changes beyond science education in the Australian national context. Again, such use of well-designed instruments, repeated over time, could be more widespread.

Few studies, either in this volume or in the research literature more broadly, involve the research-informed design and evaluation of interventions that aim to change educational participation. Such intervention studies provide the opportunity to test, and refine, hypotheses on how educational participation can be changed. The ongoing ASPIRES project (Chap. 6) promises to be an exception. Again, such studies would be of great value to both researchers and policymakers. This volume does include examples of research into the impact of pre-existing educational interventions (e.g., the inclusion of teaching about socio-scientific issues in Chap. 7; ENT3R and 'the girls' day' in Chap. 12). Whilst these studies do provide insights, the strength of findings would be much greater were researchers able to design the intervention from the outset to test specific hypotheses.

Future Research Directions

A central theme of the studies reported in this book, and one that surfaces in several places in the overview above, is that of choice as a process that develops over time. We have drawn a distinction between this ongoing choice process, and specific decision points that punctuate students' lives (e.g., completing and submitting a university course application form by the required deadline, deciding not to return to enrol on the second year of a university chemistry degree programme). Our use of theoretical perspectives from narrative psychology has also emphasised the shifting nature of students' retrospective accounts of subject choice, up to and beyond specific decision points. These research insights highlight some of the limitations of 'snapshot' accounts of educational choice (collected at a specific point in time) and analyses of retrospective accounts of choice experiences provided by students. To further develop our understanding of choice processes, and the impact of specific intervention strategies, would require the use of longitudinal research designs that examine student experiences up to, and beyond, key decision points.

Whilst recognising their limitations, our research does recognise the important contribution of 'snapshot' accounts of educational choice. For example, there is the feasibility of collecting data from large (and perhaps representative) groups of students. An important part of the background for IRIS is the under-representation of females in many STEM disciplines – in itself a group-level phenomenon. To look for explanations and remedies to this situation, looking at group-level differences in for instance expectancies and values does give insight into which factors (for instance in the school-science curriculum) may on average attract more girls to STEM – and thus over time contribute to changing the persistent group-level difference in STEM participation. Furthermore, it is difficult for large-scale recruitment interventions to be tailored to individuals. It is important then to be aware of group-level differences, for instance between genders or other subpopulation groups, in order to appeal to the majority of their target group. "Snapshot" accounts of the interests, expectancies and priorities of large groups therefore have value in such a context.

Longitudinal, qualitative research studies need to recognise the range of factors influencing student choice. We have emphasised the role of student identity construction and the interaction of this process with social structures (e.g., the responses of peers and parents) and institutional structures (e.g., the availability of specific programmes and gender balance within subjects) and associated discourses. An important way forward is to not only address this variety of factors, but to scrutinise the way they intersect and interact. Taking a holistic account of this range of factors is likely to require qualitative approaches. That said, the IRIS study has pointed to the potential value of mixed research methods, for example using large-scale national/regional quantitative datasets (e.g., recording student characteristics, the outcomes of specific decision points) to identify targeted cases for longitudinal, qualitative data methodologies.

The nature of choice is also likely to be changing. In terms of institutional and disciplinary structures, many of the boundaries of STEM subjects within higher education are shifting. 'New' subjects such as forensic science, biophysics and nanotechnology are becoming prominent. Such subjects provide the potential for new subject discourses, e.g., around 'cleverness', difficulty and gender. Furthermore, youth itself is a developing theme, as portrayed by developments in sociological theories of late-modernity. Research studies are needed that identify, and explore, choice experiences around these 'new' subject disciplines, examining how identity and gender are being played out in distinctive ways.

We have also emphasised the need to break down the monolith of gender, to consider within-gender differences, whilst at the same time avoiding a gender neutral perspective on choice processes. Within IRIS, this approach has been most successful within case studies using qualitative methodologies. By contrast, despite their value in providing important group level insights, large-scale survey analyses run the danger of reproducing gender monolith accounts of choice. Again, we would emphasise the potential value of the use of large-scale quantitative data analysis mixed with more nuanced qualitative data analysis.

Several chapters in the present volume have highlighted the influence of out-of school experiences, media and popular culture, and popular science in students' educational choice. The use of electronic and social media has accelerated even during the short time since the studies reported on here were designed, and future research could explore how social media as well as web sites such as for instance YouTube (which several higher education institutions now use to advertise their programmes) enter into students' choice processes. Another direction which might be further pursued in future studies is how the structural level – higher education policy, funding mechanisms, application and acceptance procedures – impact on educational choice processes.

Given the wealth of research studies conducted to date, it could be argued that we already know all we need to know about how young people make educational choices. From this perspective, the main challenge now is to develop research-informed educational interventions and associated practices that impact on how young people see STEM in relation to their educational and career aspirations and on gender equity in terms of opportunity and participation across STEM subjects. Our perspective is that research activity is still needed, but that more effort needs to be placed on the design and long-term evaluation of educational interventions aiming to impact on subject choice. Chapter 22 presents some insights, based on theoretical perspectives and empirical findings from IRIS, that we believe designers of such interventions need to consider.

References

Aikenhead, G. S. (1996). Science education: Border crossing into the subculture of science. *Studies in Science Education, 27*(1), 1–52. doi:10.1080/03057269608560077.

Archer, L., DeWitt, J., Osborne, J., Dillon, J., Willis, B., & Wong, B. (2012). Science aspirations, capital and family habitus: How families shape children's engagement and identification with science. *American Educational Research Journal, 49*(5), 881–908.

Atkinson, W. (2008). Not all that was solid has melted into air (or liquid): A critique of Bauman on individualization and class in liquid modernity. *The Sociological Review, 56*(1), 1–17.

Bennett, J., Lubben, F., & Hampden-Thompson, G. (2011). Schools that make a difference to post-compulsory uptake of physical science subjects: Some comparative case studies in England. *International Journal of Science Education, 35*(4), 663–689.

Blickenstaff, J. C. (2005). Women and science careers: Leaky pipeline or gender filter? *Gender and Education, 17*(4), 369–386.

Bøe, M. V. (2012). Science choices in Norwegian upper secondary school: What matters? *Science Education, 96*(1), 1–20.

Bøe, M. V., & Henriksen, E. K. (2013). Love it or leave it. Norwegian students' motivations and expectations for post-compulsory physics. *Science Education, 97*(4), 550–573.

Butler, J. (1990). *Gender trouble: Feminism and the subversion of identity*. New York/London: Routledge.

Cleaves, A. (2005). The formation of science choices in secondary school. *International Journal of Science Education, 27*(4), 471–486.

Danielsson, A. T. (2009). *Doing physics – Doing gender*. Uppsala: Uppsala Universitet.

Eccles, J. (2007). Where are all the women? Gender differences in participation in physical science and engineering. In S. J. Ceci & W. M. Williams (Eds.), *Why aren't more women in science? – Top researchers debate the evidence* (pp. 199–210). Washington, DC: American Psychological Association.

Eccles, J., Vida, M. N., & Barber, B. (2004). The relation of early adolescents' college plans and both academic ability and task-value beliefs to subsequent college enrollment. *Journal of Early Adolescence, 24*(1), 63–77.

Fredricks, J. A., & Eccles, J. S. (2002). Children's competence and value beliefs from childhood through adolescence: Growth trajectories in two male-sex-typed domains. *Developmental Psychology, 38*(4), 519–533.

Furlong, A. (2009). Revisiting transitional metaphors: Reproducing social inequalities under the conditions of late modernity. *Journal of Education and Work, 22*(5), 343–353.

Giddens, A. (1991). *Modernity and self-identity. Self and society in the late modern age*. Cambridge: Polity Press.

Gill, T., & Bell, J. F. (2013). What factors determine the uptake of A-level physics? *International Journal of Science Education, 35*(5), 753–772.

Hampden-Thompson, G., Lubben, F., & Bennett, J. (2011). Post-16 physics and chemistry uptake: Combining large-scale secondary analysis with in-depth qualitative methods. *International Journal of Research and Method in Education, 34*(3), 289–307.

Hazari, Z., Sonnert, G., Sadler, P. M., & Shanahan, M. (2010). Connecting high school physics experiences, outcome expectations, physics identity, and physics career choice: A gender study. *Journal of Research in Science Teaching, 47*(8), 978–1008.

Holmegaard, H. T., Madsen, L. M., & Ulriksen, L. (2014). To choose or not to choose science: Constructions of desirable identities among young people considering a STEM higher education programme. *International Journal of Science Education, 36*(2), 186–215. doi:10.1080/09500693.2012.749362

Homer, M., Ryder, J., & Donnelly, J. (2013). Sources of differential participation rates in school science: The impact of curriculum reform. *British Educational Research Journal, 39*(2), 248–265.

Jacobs, J., Davis-Kean, P., Bleeker, M., Eccles, J., & Malanchuk, O. (2005). "I can, but I don't want to". The impact of parents, interests, and activities on gender differences in maths. In A. Gallagher & J. Kaufman (Eds.), *Gender differences in mathematics. An integrative psychological approach* (pp. 246–263). New York: Cambridge University Press.

Krogh, L. B. (2006). *'Cultural border crossings' i fysikundervisningen – unges forhold til fysik i et kulturelt perspektiv*. Aarhus: Aarhus University.

Ryan, L. M. (2012). "You must be very intelligent...?": Gender and science subject uptake. *International Journal of Gender, Science and Technology, 4*(2), 167–190.

Sadler, P. M., Sonnert, G., Hazari, Z., & Tai, R. (2012). Stability and volatility of STEM career interest in high school: A gender study. *Science Education, 96*(3), 411–427.

Schreiner, C. (2006). *Exploring a ROSE-garden. Norwegian youth's orientation towards science – Seen as signs of late modern identities.* Doctoral thesis, University of Oslo, Oslo.

Shanahan, M.-C. (2009). Identity in science learning: Exploring the attention given to agency and structure in studies of identity. *Studies in Science Education, 45*(1), 43–64. doi:10.1080/03057260802681847.

Sinnes, A., & Løken, M. (2012). Gendered education in a gendered world: Looking beyond cosmetic solutions to the gender gap in science. *Cultural studies of science education*, 1–22. doi:10.1007/s11422-012-9433-z.

Søndergaard, D. M. (1996). *Tegnet på kroppen. Køn: Koder og konstruktioner blandt unge voksne i akademia*. København: Museum Tusculanums Forlag, Københavns Universitet.

Ulriksen, L. (2009). The implied student. *Studies in Higher Education, 34*(5), 517–532. doi:10.1080/03075070802597135.

Walker, M. (2001). Engineering identities. *British Journal of Sociology of Education, 22*(1), 75–89. doi:10.1080/01425690124860.

Chapter 22
Improving Participation in Science and Technology Higher Education: Ways Forward

Ellen Karoline Henriksen, Justin Dillon, and Giuseppe Pellegrini

Improving STEM Participation

In this chapter we present and discuss important insights for those wanting to improve participation in science, technology, engineering and mathematics (STEM). As was discussed in Chap. 1, by improved STEM participation we mean a situation where:

(a) A larger and more diverse group of young people take into account reliable information and realistic impressions of their STEM education and career opportunities and, as a result, consider STEM a viable possibility when making their educational choice; and,
(b) A greater proportion of students continue in their chosen STEM higher education until graduation (that is, improved course retention).

The key principle involved in improving participation is free and well-informed choices. We believe that nothing will be gained by 'recruiting' young people to STEM based on false premises, but we do believe that certain obstacles are currently restricting young people's free and informed choices. These obstacles relate to a lack of information about the variety of contexts and professions where STEM is used, as well as cultural stereotypes, gender roles, experiences of school

E.K. Henriksen (✉)
Department of Physics, University of Oslo, P.O. Box 1048, Blindern, N-0316 Oslo, Norway
e-mail: e.k.henriksen@fys.uio.no

J. Dillon
Department of Education and Professional Studies, King's College London, Waterloo Bridge Wing, Franklin-Wilkins Building, Waterloo Road, London SE1 9NH, UK
e-mail: justin.dillon@kcl.ac.uk

G. Pellegrini
Observa Science in Society, Viale A. Fusinieri, 65, 36100 Vicenza, Italy
e-mail: pellegrini@observanet.it

E.K. Henriksen et al. (eds.), *Understanding Student Participation and Choice in Science and Technology Education*, DOI 10.1007/978-94-007-7793-4_22

science and mathematics as unengaging, and (for some students) unrealistically low expectation of success. In some cases young people who lack an informed view of STEM, may consider it as 'not for me'.

The statements introducing each of the sections below express insights that have been achieved through the IRIS project – both from empirical results generated within the project and from a review of relevant research literature – which we believe may be useful for those working to improve STEM participation. The chapter is aimed at stakeholders such as policy-makers at the European and national levels; faculty and administrative staff at universities; companies and professional societies wanting to improve STEM participation and extend the future workforce; teachers and administrators at schools; textbook writers; curriculum designers; career advisers; the media, and the informal science sector (museums, science centres, science fairs, etc.). Rather than simply asserting 'what works' or giving detailed recommendations about the design of interventions, we have chosen to describe insights that we think are important for all stakeholder groups.

Stimulating Interest in STEM Is Necessary, but on Its Own It Will Not Improve Participation

Interest is the strongest factor emerging from students' own account of their educational choice, and stimulating interest in STEM and its applications is an obvious way to improve participation. However, the stories students tell about their educational choice, often describing an interest that 'has always been there', are told retrospectively. The strong focus on personal interest is in line with late modern ideals for how educational choices should be made, but it may not always provide an accurate description of the actual influences on the choice process. Stimulating interest is, therefore, not sufficient in itself when designing interventions aimed at improving STEM participation. Particularly, we believe that in order to be effective in promoting STEM educational choice, efforts appealing to students' interest need to be coupled with a focus on STEM career opportunities and how STEM can contribute to a person's identity development.

For students already 'on the STEM track', providing interesting experiences may reinforce their dedication to STEM education and careers. However, the target group for interventions aimed at improving participation, namely students for whom STEM is *not* the obvious choice, may be motivated by other strategies compared with the typical STEM choosers. The notion that STEM is only for the most interested and dedicated, those aiming to be research scientists, may prevent some young people from considering a STEM career. To reach them, it may be a good idea to downplay the level of interest and passion required in order to choose a STEM education or career and, instead, to promote choices based on broader motivations. For example, some students may find the applications of STEM more interesting than the 'pure science' – and this perspective can be communicated as being an acceptable and welcome reason for choosing STEM.

Having made the important point that stimulating interest is not *sufficient*, we may add that appealing to young people's interests (in a broad sense) is probably *necessary* in order for them to make a STEM choice. The question then is, how can interest in STEM and its applications be supported through different measures and for different age groups? Several chapters in this book have pointed to the influence of factors such as school science experiences, leisure-time activities and experiences, family background and popular science and media in young people's educational choice processes. Initiatives using channels of influence such as parents, teachers, media and arenas for informal science communication to strengthen interest and enjoyment are likely to promote STEM choice. This strategy may be particularly important for young people from families who lack 'science capital'.[1] Out-of school experiences may be particularly important for building up interest during primary and lower secondary school age, whereas for older students it becomes important to sustain interest by providing school science contents and contexts that are perceived as interesting and relevant. In late adolescence, popular science and, in particular, targeted information and out-reach activities might work as important influences in connections with educational decision points and might therefore profit from giving direct examples of a range of interesting STEM careers and of a variety of 'STEM identities' that the target audience may identify with.

Even if recruitment material describing higher education programmes needs to portray STEM education and career as *interesting*, institutions must also take care not to promote programmes on false premises. If students are to stay in their chosen programme until graduation, they have to experience their expectations for interest and relevance being met. One way of achieving this outcome may be to let students engage with subject content that matches their interests early in their first year of STEM higher education. This approach may involve reconsidering the sequencing of first-year courses (for example moving some of the compulsory mathematics away from the first semester), and/or introducing short modules or guest lectures presenting contemporary research or applications.

Choosing an Education Is Related Not Only to What You Want to Do, but to Who You Want to Be

The dominant cultural values of self-realisation, free choice and identity development are manifested through young people's educational choice processes, and identity is thus a central issue for understanding and influencing those choices. This issue concerns both the initial choice of course or programme, and the choice to stay in a STEM higher education programme until graduation. Initiatives to improve

[1] 'Science capital' is defined by Archer and DeWitt in Chap. 6 as 'the material and cultural science-related resources that a family may be able to draw on, such as science-related qualifications, knowledge, scientific literacy and social contacts'.

participation are more likely to succeed if they allow students *to see themselves* as a STEM student or professional.

Traditionally, and in popular culture, the scientist/technologist is associated with a range of stereotypical identity traits (caricatured as a white, male, middle-aged person wearing glasses, socially incompetent, extremely dedicated, and unaware of his outward appearance as well as of the societal impacts of his research activities). Such stereotypes still appear to impact on young people's choices. Interventions aimed at making more young people see STEM as 'something for me' therefore need to display the diversity of possible identities people can take on and still be a STEM professional. By providing opportunities for *personal* meetings between young people and tertiary STEM students or professionals, such initiatives can contribute to increasing the perceived fit between the participant's identity and a STEM identity.

A partnership with the media, or with institutions such as museums and science centres, could be a fruitful way to disseminate the idea that science is more inclusive and promotes more diverse identities. In recent years, fiction has presented young people with a range of new images of scientists; for instance, the television series 'CSI' has been important in inspiring choices of biological laboratory science and related disciplines and illustrates the power of the popular media.

Students construct and negotiate the story of their educational choice through interplay with parents, teachers, peers and others. Thus, these groups may need to be made aware of the important role they can play in young people's identity work and educational choice process. Such people can help young people to explore their personal interests, capabilities and values as well as the various options for education and careers, and they are a potential target group for efforts to improve STEM participation.

The dominant associations of science with masculinity make it particularly challenging for many girls, especially of working-class background, to see science as something 'for me'. Boys, on the other hand, may easily fall into pre-established roles which, in the case of science and technology, are easily available and provide reassurance. Initiatives, therefore, must visualize a range of roles available for both genders within STEM. However, care should be taken not to use gender as a defining term when communicating with young people in the target group since communicating broad generalisations based on gender might reinforce gender stereotypes and counteract diversity.

The construction of identity, and a viable narrative about who the students are and who they may become, is closely related to the structural level of curriculum design. For students who have already entered STEM education, the disciplinary culture they meet provides possible positions for the students to take and, as a result, makes some identities more legitimate and recognisable than others. Particularly, some female students experience a challenge to reconcile their gender identity with an identity as a STEM student in subjects such as physics and computer science. Therefore, the institutional culture in science and engineering departments needs to be addressed. Higher education institutions working to increase retention in their STEM programmes need to develop their institutional cultures in the direction of greater diversity and openness, displaying a variety of professional identities among their staff and supporting students from different backgrounds and with different interests in developing 'STEM identities'. Through programme content, teaching/learning activities and support for student social activities, STEM departments need to seek to open up multiple ways for students to engage with the subject.

Engaging with Socio-scientific Issues Within the School Science Curriculum Encourages Students to Consider Further STEM Education

School science curriculum content is one influence amongst many within students' educational choice processes. Many young people, and particularly those who do not initially have STEM on top of their list of possible educational paths, have experienced a school science course that did not offer content or contexts that was interesting and relevant. A focus on socio-scientific issues and the nature of science appears to encourage many students to consider pursuing STEM further. It appears that including debates about socio-scientific issues in school science may improve the view that many students have about science, and does not appear to turn students who are already interested in STEM careers away. In order to improve STEM participation, it would be useful to include a wider variety of socio-scientific topics, and to include them as part of the curriculum at an earlier stage.

Many socio-scientific issues are related to questions of great importance to the well-being of humans, such as health and environmental issues. We know that many young people, and, it has been argued, females in particular, seek an education and career where they can contribute to helping other people, doing something important for society, and promoting sustainable lifestyles and global justice. Such applications of STEM, therefore, need to be clearly visible in school science and mathematics as well as in higher education programmes. A closer cooperation with the industrial and organisational sector, demonstrating how the work of different STEM professionals relates to societal and environmental challenges, would be one way forward.

Students Are Concerned About the Workload and Difficulty of STEM Education

Students perceive the STEM subjects to be particularly difficult and work-intensive, and a STEM choice is often associated with 'cleverness'. For many girls in particular, the effect of this association is reinforced by unrealistically low self-efficacy in science and mathematics. Thus, supporting the self-efficacy of students and downplaying the associations of STEM as 'only for the brightest' are likely to contribute to improving participation.

Initiatives may benefit from presenting STEM education in terms of a positive challenge rather than a difficulty to be dealt with – in much the same way as training for a challenging sports event is perceived as a positive challenge by many. Initiatives promoting STEM education and careers can focus on the fulfilment that can be experienced when succeeding with a challenging task one has set for oneself. Such an approach is also in line with many young people's ideal of seeking to develop themselves through challenges.

Initiatives may aim at reducing young people's anxiety about the difficulty and workload involved in STEM by promoting mastery experiences in a safe learning environment where trying and failing is accepted and welcome. This approach applies to school science and mathematics education as well as to informal science activities and higher education. Information and recruitment material from educational institutions should provide a realistic account of the workload involved and the pre-qualifications expected of students, but should also reassure students that (given sufficient entrance qualifications and reasonable effort) it is possible for them to succeed and that the institution provides support for student learning. In higher education, in particular, effective and easily accessed student-staff interactions are important in supporting undergraduate students and reducing the anxiety related to difficulty, workload and achievement expectations, thus increasing the retention of STEM students.

A Variety of STEM-Related Experiences Meets the Needs of Diverse Student Groups

Young people are different as individuals and any initiative or intervention aiming to reach a broad group of youth should address their target group through offering several kinds of information and experiences. For some individuals, examples of STEM careers and reassurance that STEM education leads to a safe job is what they need to make up their mind; for others, supporting their expectation of success and reassurance that the study situation will be manageable in terms of workload and difficulty are needed; others need to see how a STEM identity can be combined with identity traits that they wish to take on, and so forth. Thus, we recommend *variety* in the strategies employed in any intervention aimed at improving STEM recruitment.

Young Women and Men Refuse to Be Defined Through Their Gender; Yet There Are Group-Level Differences Between the Priorities of Men and Women Making Educational Choices

Interventions to improve recruitment should be designed with sensitivity to gender differences but without making essentialist claims about gender or reproducing self-fulfilling prophesies about gender and STEM. Such essentialist claims ('women are like this, men are like that') do not acknowledge individual variations and the different ways of enacting a gender identity, and are not likely to resonate well with young people's sense of identity. An exaggerated focus on gender differences in attitudes and interests related to STEM runs the risk of overlooking individual differences as well as other important group differences, for instance in terms of class/socio-economic status or ethnicity.

Having given these important caveats, we proceed to point out that there are still social discourses that position males and females differently in relation to STEM and that there are also average differences in how females and males at the group level rate some priorities and values related to STEM as an educational option. One of these differences is females' relatively greater focus on applications of STEM in idealistic contexts within health, climate, environmental protection, etc. This trend is related to a tendency of young men to be relatively more concerned with the subject itself, whereas young women are more interested in learning STEM in relation to a broader world view and view of human well-being. Showing how STEM contributes in these respects is therefore likely to be particularly appealing to many females.

Another consistently observed difference is in the expectation of success that young people express in relation to STEM studies: males express greater confidence in their abilities to succeed and lower perceived cost (in terms of difficulty and workload). Therefore, interventions providing mastery experiences, a welcoming learning environment and reassurance of support from teaching staff may be particularly well suited for increasing female STEM participation.

Insight into group differences such as these may be useful for targeting large-scale interventions specifically to increase the participation of women. However, assumptions about gendered preferences should not be displayed in the actual information given to the target group. For example, including examples of medical technology in information material about an engineering study may be wise if the aim is to recruit more women to the programme; however, the student or professional exemplifying a medical technology career may well be male.

Upon entering higher education STEM departments, female students may experience a tension between enacting an identity as 'female' and enacting an identity as 'STEM student'. In order to improve the retention of female STEM students, therefore, attention needs to be given to the culture of the institution and to the identities that are 'available' for students to take on.

Students Have Limited Knowledge About STEM-Related Applications and Professions

In order to evaluate and prioritise between various options, young people need an overview of the courses and occupations available. However, the range of occupations that they may be aware of and would consider pursuing is often limited and includes mainly well known professions such as engineer or doctor. Together with gender role expectations and cultural stereotypes connected with different professions, this restricted level of awareness is likely to limit the number of possibilities that young people actually consider when choosing an education. The range of professions where STEM knowledge and skills are used in one way or the other is therefore among the most important messages that can be conveyed through interventions. Most young people seek an education that will lead to a safe career in terms of employability and income. Thus, the level of current and predicted demand for STEM professionals in the labour market should be communicated to young people.

Interest and self-realisation are top of the list of values young people seek in a future career, and examples should, therefore, display how interest, challenges and meaning are met in STEM-related professions. What is regarded as a 'self-realising' pursuit may vary between individuals so that some see a self-realisation potential in pursuing pure, academic science whereas others seek challenges in more applied disciplines, often with an idealistic purpose. Females, more often than males, tend to aim at medically related and other idealistically oriented professions. Pointing out how STEM professionals (also from the engineering and physical science disciplines) contribute to medical diagnostics and treatment, climate research, renewable energy development, providing clean water and cheap energy in less economically-developed countries, etc., might be effective in improving female participation in STEM. Also, as discussed above, young people must be able to see how a STEM career fits into their identity development and examples of STEM-related professions should, therefore, present a range of different 'STEM personalities' and should be as diverse as possible with respect to gender, ethnicity and other identity traits. Various forms of work-placement and/or cooperation between schools and professional organisations/industry are also worth exploring.

Information about STEM-related careers could be conveyed through a number of different channels: on educational institutions' web pages and printed information material; in school science as integrated parts of STEM curriculum or teaching approaches; through out-of-school activities and informal science channels; through the media (documentaries as well as fiction/drama featuring STEM practitioners), etc. Notably, material promoting the STEM professions may be targeted at groups of 'significant persons', such as parents and teachers, who often play important roles in young people's educational choice processes. Developing curriculum-relevant material that displays how topics in the school science or mathematics curriculum are actually employed in STEM professions would make it easier for teachers to choose to spend classroom time on such material.

The Media and the Informal Science Sector Influence Young People's Images of Science and Technology

Media coverage of science and technology transmits social roles and behaviours. Given this influence, the media can, for example, challenge some traditional stereotypes regarding gender roles. Thus it could be useful to support a closer cooperation between educational authorities, research communities and the media.

For example, journalists and scientists might be involved in common training courses with the aim of developing news selection capabilities and narrative skills. Popular science, as well as fiction or drama in which STEM plays a role, contribute to shaping young people's images of the disciplines and of the practitioners. Such channels may be used in targeted efforts to influence attitudes to STEM.

Social media and Internet channels such as YouTube and science blogs are in rapid development and are likely to shape young people's attitudes to STEM. The potential of such new technologies for interacting and collaborating for science education and STEM participation purposes is relevant since it offers new opportunities for learning and interaction. Different formats and offer different opportunities. For example, a scientist's blog exerts a different influence from a cognitive and emotional point of view than a portal offering experiments and educational tools; a discussion forum or wiki still others.

Museums, science centres and other informal science arenas also have great potential for communicating information as well as for influencing values and priorities. They are at the interface between science, technology and society and have an institutional mandate to communicate science and technology developments and their applications. Such institutions may provide opportunities for dialogue between young people and relevant STEM stakeholder groups and allow students to encounter diverse professional identities. As mentioned above, these channels may be of particular importance for students from low science capital backgrounds.

The Nature of the STEM Participation Challenge May Necessitate Large-Scale, National Interventions Over a Sustained Period

Educational choice is related to the structure of the educational system, to youth culture and to values and attitudes prevalent in society at large. Imparting a lasting influence on young people's educational choice patterns is therefore a demanding task that requires long-term work. Developing effective policy to increase STEM enrolment requires attention to the interrelationships between curriculum, societal, school and student factors associated with educational choice. Improving STEM participation may require a systemic approach, for instance concerning entrance requirements, grading practices, the number and nature of courses and educational programmes on the market, and so on.

STEM recruitment interventions are most likely to have an impact if they are developed, evaluated and improved through several cycles over an extended period of time. Also, a series of initiatives involving individual students over months or years are more likely to influence choice compared to single events (although the latter may also have some effect, particularly when focused on educational decision points). An example of an initiative which combines many of the above concerns is the ENT3R project, a Norwegian mentor programme in which students aged 14–17 come to their local university for weekly mathematics trainings led by mentors who are tertiary mathematics and science students.

Finally, for the purpose of keeping undergraduate STEM students on their chosen educational track, large-scale, national interventions may be necessary in addition to institutional measures as described below. The *'Young researchers programme'* in Slovenia has been successful in this regard, particularly for encouraging women.

Students Whose Experiences in Undergraduate STEM Programmes Match Their Expectations Are More Likely to Complete Their Course of Study

Students' educational choice processes continue even after they have entered the programme they have decided to study. Their choice narratives – how they explain and defend their educational choice to themselves and to others – develop as they try to balance their expectations with what they experience. Most students experience a gap between what they had expected and what they experience in the STEM higher education programme. Consequently, institutions need to be aware that students have to adjust their expectations in relation to becoming a higher education student and that this adjustment process may also require some effort and support from the institution. Higher education programmes should be sufficiently flexible to allow students to re-orient their course and switch to a related programme if they develop different orientations during the first years of STEM studies.

Students' first-year experiences in STEM highlight the combined importance of the social and academic integration that constitute a core feature of Tinto's model (see Chap. 13). STEM higher education programmes need to focus on students' experience of meaning and coherence in their studies. The students should be able to see how different courses contribute to the overall goal, and the teaching and learning activities they are involved in need to relate to the subject they opted for when choosing their course. Both the sequencing of the courses and the methods of teaching have an impact in this respect.

The students' sense of a match between themselves and the programme appears to be important in promoting persistence. Higher education institutions should, therefore, provide frames for the students' construction of identity as STEM students and future STEM practitioners. These frames relate both to the social environment and the academic practices such as teaching methods and content, the degree of support from teaching staff and the available arenas for students to interact with each other and with academic staff.

Well-prepared teachers able to guide, motivate and support their students through varied and student-active approaches are decisive. Informal peer groups, laboratory work-groups and teachers who give feedback and who transmit their passion for their subject are key factors in motivating students to continue.

Finally, it is essential that institutions and their teachers attempt to specify clearly what they expect of students, not only in terms of skills but also in terms of interests, motivation and attitudes. Such explication of expectations may serve two purposes: first, the surfacing of institutions' implicit expectations may lead to revision or modification of those expectations to bring them more in line with students' educational aims, their background from previous education, their preferred work forms, etc. Secondly, expectations should be made explicit to show students exactly what they are facing. These expectations should also be communicated in course material provided to students, both prior to the choice of study programme and in connection with embarking on the programme.

Conclusion: A View to an Equitable Society with Sufficient STEM Expertise in the Workforce

As discussed in the Introduction (Chap. 1) to this book, a sufficiently large and diverse workforce with education in STEM is considered vital for solving global challenges and promoting economic growth and equitable societies. In an equitable society, everyone has a real, not just a formal, free choice of education. This entitlement requires that young people have access to sufficient and reliable information about the various educational and occupational options available to them. Similarly, there should be no formal or informal obstacles to a free choice. In this chapter, summing up experiences from the research described throughout this volume, we have identified insights that we consider important for stakeholders aiming to improve young people's participation in STEM. It is our hope that these insights may ultimately help more young people to see how a STEM education may contribute to their personal development and career – that is, to be able to say that STEM is 'something for me'.

Appendix: The IRIS Questionnaire – Instrument Development, Data Collection and Respondents

The IRIS Questionnaire (IRIS Q)

The IRIS questionnaire (IRIS Q) was developed as a common data collection instrument for all six IRIS consortium partners. All partners contributed to instrument development and collected data in their respective countries. In addition, international associated IRIS partners[1] outside the IRIS consortium have been invited to collect data in their own countries by using IRIS Q and following the data collection procedures described here and detailed in "Guidelines for IRIS Q translation, sampling, data collection and coding".[2] As of January 2014, data has been collected using the IRIS Q in a total of at least 15 countries in addition to the five countries in the IRIS consortium. Some results from the IRIS Q are found in the present book (Chaps. 9, 16, 18 and 20). Results have also been published in conferences, articles and reports,[3] and more IRIS Q-based publications are likely to appear in the future.

In this Appendix, methodological information about the IRIS Q and data collection procedures are briefly outlined; details are found in the publication "Guidelines for IRIS Q translation, sampling, data collection and coding". The final section of the Appendix concerns the IRIS PhD questionnaire, which was developed to investigate the choice process of PhD students within STEM disciplines in Slovenia.

---

[1] The international associated IRIS partners are coordinated by Dr. Anders Jidesjö at Linköping University, Sweden.

[2] Available (as of August 2014) at http://iri.uni-lj.si/data/Projekti/IRIS/irisarhiv/iris-documents/index.html

[3] For instance: Cerinsek, G., Hribar, T., Glodez, N., & Dolinsek, S. (2012). Which are my Future Career Priorities and What Influenced my Choice of Studying Science, Technology, Engineering or Mathematics? Some Insights on Educational Choice – Case of Slovenia. *International Journal of Science Education*, 1–27, doi:10.1080/09500693.2012.681813.

Lyons, T., Quinn, F., Rizk, N., Anderson, N., Hubber, P., Kenny, J., et al. (2012). *Starting out in STEM. A study of young men nd women in first year science, technology, engineering and mathematics courses*. University of New England: SiMERR National Research Centre.

E.K. Henriksen et al. (eds.), *Understanding Student Participation and Choice in Science and Technology Education*, DOI 10.1007/978-94-007-7793-4

Instrument Development and Format

IRIS Q was based on the theoretical perspectives described in Part I in this book and on previous projects, notably the ROSE study and the Lily study.[4] IRIS Q was designed to tap into important aspects of educational choice and equity considerations in STEM and to cover areas of research interest for all IRIS partners and all the project work packages. Questionnaire items were suggested and discussed in several rounds by IRIS research consortium members, and a preliminary version was piloted in all IRIS consortium countries in early 2010. Piloting was done in various ways; in some countries through an electronic solution; in others using pencil and paper, some places in combination with a focus group discussion of the questionnaire. In most consortium countries, members of the project's reference group among stakeholders in education, government and industry were consulted during instrument development and testing. As a result of this development and testing process, the English-language master version of IRIS Q (see section "The IRIS Q Master Version") was finalised in March 2010.

The questionnaire comprised a total of 65 items covering school science experiences, sources of inspiration for choice of education, expectations for future job, first year experiences as a STEM student, and attitudes to gender equity in STEM. A few questions in IRIS Q were open-ended, offering the respondents a field where they could answer in their own words. The majority of IRIS Q, however, consisted of questions with fixed pre-structured responses. The response categories were mainly five-point Likert scales, ranging from "Not important" to "Very important" or from "Strongly disagree" to "Strongly agree", with text headings only above the extreme categories and all categories numbered from 1 to 5. A few items had three-point scales, labelled "worse than expected", "as expected" and "better than expected" (see section "The IRIS Q Master Version").

When developing IRIS Q, an important aim was to keep it relatively short. However, some versions of IRIS Q have been extended with one or more additional items in the end of the questionnaire. Notably, the last item in the IRIS Q master did not appear in the Norwegian and the Danish versions, since this question (regarding the economic cost of study) is not applicable in the Norwegian and Danish contexts. Also, when IRIS Q was implemented in Italy, a number of questions were added in order for the Italian IRIS team to collect additional data for their research.

The IRIS Q master version was translated to the language of STEM instruction in each of the IRIS consortium countries following a standardised procedure to ensure the quality of the translation (see "Guidelines for IRIS Q translation, sampling, data

---

[4] Schreiner, C. (2006). *Exploring a ROSE-garden: Norwegian youth's orientations towards science – seen as signs of late modern identities. Based on ROSE (The Relevance of Science Education), a comparative study of 15 year old students' perceptions of science and science education.* (Doctoral thesis), University of Oslo, Oslo. Retrieved from www.ils.uio.no/forskning/pdh-drgrad/doktoravhandlinger/docs/schreiner_thesis.pdf Schreiner, C., Henriksen, E. K., Sjaastad, J., Jensen, F., & Løken, M. (2010). Vilje-con-valg: valg og bortvalg av realfag i høyere utdanning [Lily: Choosing – or not choosing – science, mathematics and technology in higher education]. *KIMEN, 2010* (2), http://www.naturfagsenteret.no/kimen

collection and coding"). IRIS Q was administered electronically in most IRIS consortium countries, but on paper in some (see details in sections "IRIS Q Denmark", "IRIS Q Italy", "IRIS Q Norway", "IRIS Q Slovenia" and "IRIS Q England").

Target Population

The criteria used for defining the target population for the IRIS Q survey were:

- Central or typical STEM education/subject
- Documented recruitment challenge
- Gender imbalance
- Easy to identify corresponding education/programme across countries
- Well-defined education/programme (not cross-discipline programmes such as "technology, organization and learning")
- Preferably large numbers of students in a small number of institutions/programmes (to facilitate sampling and administration).

Following discussions in the IRIS consortium, the target population was defined as *all first-year students who enrolled in Fall 2009 in all 3-year bachelor programmes and all 5-year integrated master programmes with the following International Standard Classification of Education (ISCED) codes*[5]:

- 421 Biology and biotechnology
- 441 Physics
- 442 Chemistry
- 461 Mathematics
- 481 Computer science
- 521 Mechanics and metalwork
- 523 Electronics and automation
- 524 Chemical and process

Data Collection

IRIS data collection was undertaken during the period March–June 2010, with one exception (see section "IRIS Q Slovenia"). At this time, students were well into their second semester as first-year STEM students. In the IRIS countries with small

[5] The International Standard Classification of Education (ISCED) was developed by UNESCO to facilitate comparisons of education statistics and indicators across countries on the basis of uniform and internationally agreed definitions, see http://www.uis.unesco.org/Education/ (accessed December 2012). We have used EUROSTAT's "Fields of education and training manual", based on the ISCED system (accessed December 2012 from http://circa.europa.eu/Public/irc/dsis/edtcs/library?l=/public/measuring_lifelong/classifications/isced97_fields/education_training/_EN_1.0_&a=d).

populations (Slovenia, Denmark and Norway), the entire target population was invited to respond to the questionnaire (a so-called census). In Italy and the UK, samples were drawn (see sections "IRIS Q Italy" and "IRIS Q England").

The IRIS research team in each country approached the administrative staff at higher education institutions in order to establish co-operation about IRIS Q data collection. In most countries, the students were invited by e-mail to take part in the survey by the study administration staff at each educational institution. The QuestBack online survey application (http://questback.co.uk/) was used for harvesting data in three of the five IRIS countries and has also been used by many IRIS associated partners. Reminders were sent out electronically to the invited students. In Italy and Slovenia, questionnaires were mainly administered on paper.

Section "Overview of Target Populations, Data Collection and Response Rates" in this Appendix gives an overview of target populations, data collection procedures and response rates in each of the IRIS consortium countries.

In order to maximise response rates, students inn all five countries were offered an incentive to take part in the study: The final page in the online questionnaire invited students to participate in a prize draw and redirected them to a separate questionnaire asking for contact information (e-mail address). Rewards differed somewhat between countries, but in most cases consisted of general (or widely applicable) gift vouchers or cheques or similar rewards. The total value of the incentives provided (to be distributed between four winners in each country) was weighted with the gross domestic product (GDP) per capita in each country, and varied between 2,000 Euro for Norway and 1,000 Euro for Slovenia. Prize draws were performed in each country after data collection had closed.

National IRIS Q data files were "cleaned" by the IRIS research team in each country. "Cleaning" entails checking for (and possibly deleting) responses that appear invalid, for instance because the respondent is outside the target population. An ISCED code was assigned manually to each respondent. This procedure sometimes involved some degree of judgement, since the question where respondents identified their chosen programme was open ended and individual respondents used different terms and abbreviations to describe the same programmes. This procedure may thus have introduced some small degree of error in the categorisation.

The IRIS Q Master Version

The IRIS Q master version is shown on the following pages. Here it is shown as it appeared on the screen for UK respondents, using the QuestBack electronic solution for data gathering.

IRIS
Interests & Recruitment in Science

IRIS questionnaire

This questionnaire is about you and your choice of course. Your answers are important to the study! The information you provide to the IRIS research project may help us improve teaching and develop more targeted information for future students. All replies are confidential, and no information will be traceable to you as an individual. Thank you very much!

Are you... *

○ Female? ○ Male?

In what year were you born? *

- Select answer -

At which university/college are you studying? *

- Select answer -

Degree programme/Course *

Next >>

9 % completed

IRIS
Interests & Recruitment in Science

IRIS questionnaire

Have you previously been enrolled in a higher education programme?

○ No ○ Yes

If your answer is yes above, in which programme?

How important was each of the following school experiences in choosing your course?

	Not important 1	2	3	4	Very important 5
Your interest in related subjects	○	○	○	○	○
Your previous attainment in related subjects	○	○	○	○	○
Experiments/laboratory work	○	○	○	○	○
Field work or excursions	○	○	○	○	○
Lessons showing the relevance of the subjects to society	○	○	○	○	○
Lessons showing practical applications of the subjects	○	○	○	○	○
Using mathematics in lessons	○	○	○	○	○
Clear feedback on whether you got the right answer	○	○	○	○	○

Next >>

IRIS
Interests & Recruitment in Science

IRIS questionnaire

How important were the following persons in choosing your course?

	Not important 1	2	3	4	Very important 5
Mother or step-mother	○	○	○	○	○
Father or step-father	○	○	○	○	○
Good teachers	○	○	○	○	○
Friends (including boyfriend/girlfriend)	○	○	○	○	○
Siblings or other relatives	○	○	○	○	○
Careers advisors in school	○	○	○	○	○

Other people (who?)

[ ]

Next >>

27 % completed

IRIS
Interests & Recruitment in Science

IRIS questionnaire

How important were each of the following in choosing your course?

	Not important 1	2	3	4	Very important 5
Popular science books and magazines	○	○	○	○	○
Science fiction or fantasy books/films	○	○	○	○	○
Computer games	○	○	○	○	○
Museum/science centre	○	○	○	○	○
Popular science television channels/programmes (Discovery channel, Horizon)	○	○	○	○	○
Films or drama on television (CSI, Numbers, Grey's Anatomy and so on)	○	○	○	○	○
Science, Technology or Maths competitions	○	○	○	○	○
Other kinds of outreach activities (e.g. science festivals, science summer schools)	○	○	○	○	○

Other/comment

Next >>

36 % completed

IRIS
Interests & Recruitment in Science

IRIS questionnaire

Please describe how you came to choose this course

Next >>

45 % completed

IRIS
Interests & Recruitment in Science

IRIS questionnaire

To what extent do you agree with the following statements about your experiences as a student so far?

	Strongly disagree 1	2	3	4	Strongly agree 5
I enjoy the company of the other students on my course	○	○	○	○	○
I feel I fit in socially	○	○	○	○	○
I feel I can keep up with the pace of the teaching	○	○	○	○	○
I get personal feed-back from lecturers and teachers when I need it	○	○	○	○	○
I feel my teachers care about whether students learn or not	○	○	○	○	○
The university/college offers good working conditions (equipment, library, common areas, cafés, technical support)	○	○	○	○	○
I can see the relevance of what I learn	○	○	○	○	○
I feel that my course suits the kind of person that I am	○	○	○	○	○
I have become more interested in the subject since I started	○	○	○	○	○

Next >>

IRIS

Interests & Recruitment in Science

IRIS questionnaire

Have the following aspects of your everyday life as a student been as expected, better than expected or worse than expected?

	Worse than expected	As expected	Better than expected
The overall experience of being a student on this course	○	○	○
Your social relationship with your fellow students	○	○	○
The overall quality of the teaching	○	○	○
How interesting you find the content of the course	○	○	○
The effort you have to spend on studying ("worse" means "greater effort")	○	○	○

Next >>

64 % completed

IRIS
Interests & Recruitment in Science

IRIS questionnaire

To what extent do you agree with the following statements?

	Strongly disagree 1	2	3	4	Strongly agree 5
I will do better than average in this course	○	○	○	○	○
I easily learn the subject matter in this course	○	○	○	○	○
I am confident that I am good enough at the subjects in this course	○	○	○	○	○
I am very motivated to study this course	○	○	○	○	○
I will probably decide to leave this course before I finish	○	○	○	○	○

If someone you know was thinking about enrolling on your course and asked you about it, what would you say to her or him?

Next >>

IRIS

Interests & Recruitment in Science

IRIS questionnaire

Regarding your priorities for the future; how important are the following factors to you?

	Not important 1	2	3	4	Very important 5
Getting a secure job	○	○	○	○	○
Opportunities to earn a high income	○	○	○	○	○
Starting to make money as soon as possible	○	○	○	○	○
Working with something that is important for society	○	○	○	○	○
Helping other people	○	○	○	○	○
Contributing to sustainable development and protection of the environment	○	○	○	○	○
Doing something I am interested in	○	○	○	○	○
Using my talents and abilities	○	○	○	○	○
Developing myself	○	○	○	○	○

Next >>

82 % completed

IRIS

Interests & Recruitment in Science

IRIS questionnaire

Do you attend a course where one gender is over-represented?

○ Yes ○ No

If yes: Why do you think this is the case?

Do you see any reason why the situation described above should be changed – and if so, what do you think could be done to change it?

Next >>

91 % completed

IRIS
Interests & Recruitment in Science

IRIS questionnaire

Cost

	Not important 1	2	3	4	Very important 5
How important was the cost of study in your choice of course?	○	○	○	○	○

Please explain your answer

Send

100 % completed

Overview of Target Populations, Data Collection and Response Rates

IRIS Q Denmark

Statistics Denmark (the Danish central government agency for statistics) provided a list of current degree programmes which in the Danish system corresponded to the chosen ISCED codes. This resulted in a list of 15 educational institutions and a total of 46 degree programmes. It must however be considered that the institutions are not obligated to report to Statistics Denmark which degree programs they offer. Also, there was some judgement involved concerning programmes belonging to a previous degree structure and whether or not these belonged in the target group. The list therefore has some margin of error. The research group found it prudent to deselect one institution from the list due to difficulties of categorisation because of the organizational structure (in this institution, students do not chose their 'final' degree programme until after 2 years). The data extract also included information about the number of students in each degree programme; however, Statistics Denmark did not provide students numbers distributed on genders, and it has therefore been impossible to determine response rates for female and male students separately (Table A.1).

To secure institutional support of the project, contact was made with the Study Director at each institution. The IRIS project was introduced and the relevant degree programmes at the institution in question were listed. The vast majority of directors were positive towards participating in the survey and most frequently the research group was referred to a study secretary, who would handle the practicalities of distribution.

The data was collected electronically via QuestBack. A total of 3,034 students received an invitation, and 954 responses were received, giving an initial response rate of 31 %. In the process of data file "cleaning", 69 respondents were taken out because they appeared not to belong to the target population. Moreover, the number obtained from Statistics Denmark of STEM students in the IRIS target population was not exactly the same as the number of invited respondents as obtained from the institutions. In Table A.1, the population data from Statistics Denmark are used, and the numbers of respondents correspond to the valid respondents – those in the "cleaned" data file.

IRIS Q Italy

As mentioned in section "Instrument Development and Format", the Italian IRIS questionnaire had an add-on section containing 16 items to be used in related research in which the Italian IRIS partner (Observa) was involved.

Table A.1 Target population (based on figures from Statistics Denmark), valid IRIS respondents and response rates for the Danish IRIS Q survey

Denmark		Total target pop (national)	Number of valid IRIS respondents	Response rate
421 Biology and biotechnology	Males		85	
	Females		182	
	Total	706	267	38 %
441 Physics	Males		55	
	Females		45	
	Total	296	100	34 %
442 Chemistry	Males		17	
	Females		16	
	Total	148	33	22 %
461 Maths	Males		45	
	Females		65	
	Total	289	110	38 %
481 Comp. science	Males		223	
	Females		31	
	Total	1,596	254	16 %
521 Mechanics and metal work	Males		0	
	Females		0	
	Total		0	0 %
523 Electronics and automation	Males		22	
	Females		2	
	Total	134	24	18 %
524 Chemical and process	Males		37	
	Females		49	
	Total	247	86	35 %
Sum	**Males**		**484**	
	Females		**390**	
	Total	**3,416**	**874**	**26 %**

To establish the overall number of students in different parts of the IRIS Q target population, the Observa team consulted the Anagrafe Nazionale Studenti (National Registry Office of Students) of MIUR – the Italian Ministry of University and Research. Thanks to the MIUR National Registry data, it was possible to obtain the target population's distribution on ISCED cohorts and at the same time by geographical areas. Data presented here were updated as of June 1st, 2010.

Starting from the overview of the target population of first-year students starting in the academic year 2009–2010, the research team designed a stratified sampling plan, looking at the students' distribution on 68 different universities and clustering the twenty Italian regions into five geographical areas: North West, North East, Centre, South and Islands. This is a widely used classification, capturing the main geographical differences existing within Italy.

Universities eligible for the IRIS Q survey were selected on the basis of two criteria: University size and geographical area. We defined three categories of Universities: small (up to 10,000 students); medium (10,000–30,000 students) and large (more than 30,000 students). Considering university size, a sample of universities within each geographical area was randomly selected. The research team designed a quota sample, considering both ISCED cohorts and geographical areas.

A letter requesting collaboration was sent to chancellors of the sampled institutions, in order to facilitate the task of the data collectors. The request was to be allowed to administer the questionnaire to students during lectures (authorized by the professors). Only two universities declined to participate in the survey. To complete the sample, in one case a university was replaced with another one of the same territorial area, and in the other case questionnaires were administrated outside lectures.

Data was collected through paper and pencil administration of questionnaires within the sampled Universities (a small subsample of the students responded via an electronic web solution). 25 data collectors were trained by Observa in February-March 2010 and assigned the task of administering paper questionnaires in the sampled universities. Data collection was carried out in 34 universities from the end of March until the last week of June 2010.

At the end of the field work, all IRIS Q data (paper and online) was gathered in an electronic data file, which was "cleaned" and ISCED-coded as described in section "Data Collection". It was also possible to redefine the number of valid cases through a careful analysis, identifying mistakes and incoherencies. The exclusion of 228 cases was due to: lack of information about STEM discipline (for ISCED coding); missing gender information; lack of geographical information; uncertainty about the year of enrolment. Finally, 2,667 cases were defined as valid (2,584 paper questionnaires filled out during lectures and 83 online).

Comparing the distribution of the target population of students and that of valid cases, some combinations of ISCED category, gender, university size and geographical area are over-represented among Italian respondents (indicated by a coverage rate of more than 100 % in Table A.2) whereas other such combinations are under-represented. This was an unavoidable consequence of the sampling design, since entire institutions were sampled rather than individuals or classes of students. All in all, however, the procedure of administering the questionnaire in lectures on paper resulted in the Italian data set having the highest overall response rates within the IRIS consortium. Females are slightly over-represented among the respondents (see Table A.2).

IRIS Q Norway

Permission to collect and store data was obtained from the Norwegian Social Science Data Services. Statistics Norway was contacted and provided a list of all educational programmes under the eight selected ISCED codes in all public higher education institutions. Twenty-four public higher education institutions in Norway

Table A.2 Target population (based on numbers obtained from the National Registry Office of Students), total quota sample, valid IRIS respondents and coverage rates (col. III: col. II) for the Italian IRIS Q survey. Population, sample and respondent numbers are distributed on genders and educational categories (ISCED codes)

Italy		I: total target pop (national)	II: total quota sample	III: number of valid IRIS respondents	IV: coverage rate
421 Biology and biotechnology	Males	4,706	184	189	103 %
	Females	10,513	411	385	94 %
	Total	15,219	595	574	96 %
441 Physics	Males	1,779	176	181	103 %
	Females	902	89	102	114 %
	Total	2,681	265	283	107 %
442 Chemistry	Males	1,461	137	111	81 %
	Females	1,308	123	106	86 %
	Total	2,769	260	217	83 %
461 Maths	Males	1,615	177	159	90 %
	Females	1,980	218	202	93 %
	Total	3,595	395	361	91 %
481 Comp. science	Males	7,931	304	269	88 %
	Females	1,328	51	70	137 %
	Total	9,259	355	339	95 %
521 Mechanics and metal work	Males	7,433	347	407	117 %
	Females	609	28	69	243 %
	Total	8,042	375	476	127 %
523 Electronics and automation	Males	2,864	347	215	62 %
	Females	352	43	55	129 %
	Total	3,216	390	270	69 %
524 Chemical and process	Males	841	143	84	59 %
	Females	420	72	63	88 %
	Total	1,261	215	147	68 %
Sum	**Males**	**28,630**	**1,772**	**1,615**	**91 %**
	Females	**17,412**	**1,078**	**1,052**	**98 %**
	Total	**46,042**	**2,850**	**2,667**	**94 %**

offered these courses or programmes. Invitations were sent out in February 2010 through letters to members of the administration at the eligible institutions. Follow-up was done through e-mail and telephone. Of the 24 institutions that were eligible and received an invitation, 19 consented to participate.

When institutions had agreed to participate, they received (in March 2010) a new letter giving detailed instructions about how to invite students (see "Guidelines[6]..."). The electronic questionnaire (using "QuestBack") was open during the period March 16th – June 15th, 2010. Each institution was instructed to invite their students to

[6] http://iri.uni-lj.si/data/Projekti/IRIS/irisarhiv/iris-documents/index.html

participate (via e-mail) at a time during this period that was convenient in relation to local schedules, but that allowed students a period of 3 weeks to respond. Reminders were sent to students as suggested in the IRIS Q Guidelines.

Of the 19 institutions that consented to participate, one appeared (from inspection of the data file) not to have invited all target groups and possibly to have invited wrong target groups; thus, 18 institutions are represented in the final data file. The data file was cleaned and ISCED-coded as described in section "Data Collection".

According to data from Statistics Norway, 5,663 students were in the IRIS Q target population in the academic year 2009–2010. In the letter to administrators, each institution was asked to provide (*after* data collection at their institution) information about how many students, and from which programmes/courses, they had actually invited. This information indicated that a total of 4,270 students actually received an e-mail invitation from their institution to participate in the online IRIS questionnaire. The difference is probably mainly due to students having left their study during the first months of study. After "cleaning", the number of valid respondents in the Norwegian IRIS Q data file was 1,314 (Table A.3). Thus, the IRIS respondents in Norway constitute slightly less than one fourth of the target population and slightly less than one third of those who actually received an invitation to participate. Females are slightly over-represented among the respondents. Also, the "academic sciences" mostly have higher response rates than the more applied sciences and engineering studies.

IRIS Q Slovenia

It was decided to collect data from the three largest public universities in Slovenia, hosting around 99 % of the total student population within STEM subjects. In the beginning of 2010, communication with target faculties was initiated in order to obtain administrative support for inviting students to reply to the electronic version of IRIS Q. In March and April 2010, eligible study programmes within the 8 target ISCED categories were identified in dialogue with institutions, and the invitation for students to respond to the online IRIS Q (using QuestBack) was sent in April and May to the chosen 12 faculties. By the end of May and in the beginning of June, a second invitation was sent.

After data collection was closed, the Slovenian data file comprised 417 responses, giving a response rate of 19.5 %, which was considered less than desirable. It is a known problem in Slovenia that many students are so-called "ghost students" who only register in order to obtain economic benefits that students are entitled to; they are not actually following any course. There is reason to believe that the population of students actually following STEM education was considerably lower than 2000, which would make the real response rate higher. Nevertheless, it was decided to perform a new data collection 1 year later, in spring 2011, this time using a pen-and-paper solution, in order to obtain a higher response rate. Despite the fact that there might be differences between student cohorts that might impact on comparisons with the other national IRIS data sets collected in 2010, a good response rate was considered more important.

Table A.3 Target population (based on numbers obtained from Statistics Norway), valid IRIS respondents and response rates for the Norwegian IRIS Q survey. Population and respondent numbers are distributed on genders and educational categories (ISCED codes)

Norway		Total target pop (national)	Number of valid IRIS respondents	Response rate
421 Biology and biotechnology	Males	224	64	29 %
	Females	480	179	37 %
	Total	704	243	35 %
441 Physics	Males	267	97	36 %
	Females	103	42	41 %
	Total	370	139	38 %
442 Chemistry	Males	54	17	31 %
	Females	42	18	43 %
	Total	96	35	36 %
461 Maths	Males	126	23	18 %
	Females	76	16	21 %
	Total	202	39	19 %
481 Comp. science	Males	1,516	287	19 %
	Females	212	59	28 %
	Total	1,728	346	20 %
521 Mechanics and metal work	Males	581	131	23 %
	Females	121	34	28 %
	Total	702	165	24 %
523 Electronics and automation	Males	1,032	183	18 %
	Females	113	27	24 %
	Total	1,145	210	18 %
524 Chemical and process	Males	465	86	18 %
	Females	251	51	20 %
	Total	716	137	19 %
Sum	**Males**	**4,265**	**888**	**21 %**
	Females	**1,398**	**426**	**30 %**
	Total	**5,663**	**1,314**	**23 %**

During the first months of 2011, data was collected in the same three institutions, which again provided the research team with current student numbers within each of the eight chosen ISCED categories. According to these register data (obtained directly from the Enrolment offices of the three universities), the Slovenian target population in spring 2011 was 2107 students. Questionnaires were administered by members of the IRIS research team during ordinary classes within the target STEM study programmes. Students were informed of the purposes of the project and that their responses were anonymous and to be used for research purposes only. The students were willing to participate in the survey and they completed the IRIS Q individually without interacting with each other.

Eleven different STEM faculties participated in the study. The sample consisted of 861 male and 420 female undergraduates. As Table A.4 shows, response rates

Table A.4 Target population (based on numbers obtained from the Enrolment offices of the three participating universities), number of valid IRIS respondents and response rates for the Slovenian IRIS Q survey. Population, sample and respondent numbers are distributed on genders and educational categories (ISCED codes). Note that the Slovenian data was collected in spring 2011, 1 year after the other IRIS Q national data sets (see above)

Slovenia		Total target pop (national)	Number of valid IRIS respondents	Response rate
421 Biology and biotechnology	Males	90	50	56 %
	Females	177	141	80 %
	Total	267	191	72 %
441 Physics	Males	107	75	70 %
	Females	26	20	77 %
	Total	133	95	71 %
442 Chemistry	Males	72	40	56 %
	Females	98	60	61 %
	Total	170	100	59 %
461 Maths	Males	91	39	43 %
	Females	120	58	48 %
	Total	211	97	46 %
481 Comp. science	Males	387	234	60 %
	Females	62	56	90 %
	Total	449	290	65 %
521 Mechanics and metal work	Males	351	170	48 %
	Females	28	13	46 %
	Total	379	183	48 %
523 Electronics and automation	Males	337	215	64 %
	Females	24	16	67 %
	Total	361	231	64 %
524 Chemical and process	Males	66	38	58 %
	Females	71	55	77 %
	Total	137	93	68 %
SUM	**Males**	**1,501**	**861**	**57 %**
	Females	**606**	**419**	**69 %**
	Total	**2,107**	**1,280**	**61 %**

were indeed considerably higher (with an overall response rate of 61 %) in this second data collection (2011) where the questionnaire was administered on paper. The response rate was higher for females than for males.

IRIS Q England

The UK IRIS team developed a rationale for sampling across higher education institutions (HEIs) in England, recruited HEIs and communicated with them in identifying the relevant student populations (ISCED codes are not used directly within HEIs in England). Institution contacts were asked to arrange for emails to be sent out to the target population with a personalized message from the host department (to encourage a higher response rate).

Data was collected using the QuestBack electronic submission system during the period April–June 2010. Population data, sample sizes and respondent numbers are given in Table A.5.

Table A.5 Target population (based on numbers obtained from institution contacts), sample, valid IRIS respondents and response rates for the UK version of the IRIS Q survey. Population, sample and respondent numbers are distributed across genders and educational categories (ISCED codes)

UK		I: total target pop (national)	II: total sample	III. Number of valid IRIS respondents	IV: response rate (III:II)
421 Biology and biotechnology	Males	4,455	475	103	22 %
	Females	5,775	575	222	39 %
	Total	10,230	1,050	325	31 %
441 Physics	Males	2,840	235	59	25 %
	Females	725	70	22	31 %
	Total	3,565	305	81	27 %
442 Chemistry	Males	2,325	285	30	11 %
	Females	1,645	200	48	24 %
	Total	3,970	485	78	16 %
461 Maths	Males	4,185	450	73	16 %
	Females	2,905	355	74	21 %
	Total	7,090	805	147	18 %
481 Comp. science	Males	14,935	580	59	10 %
	Females	2,785	105	17	16 %
	Total	17,720	685	76	11 %
521 Mechanics and metal work	Males	5,690	469	34	7 %
	Females	480	35	7	20 %
	Total	6,170	504	41	8 %
523 Electronics and automation	Males	4,420	130	21	16 %
	Females	555	30	3	10 %
	Total	4,975	160	24	15 %
524 Chemical and process	Males	1,335	80	17	21 %
	Females	490	15	6	40 %
	Total	1,825	95	23	24 %
Sum	**Males**	**40,185**	**2,704**	**396**	**15 %**
	Females	**15,360**	**1,385**	**399**	**29 %**
	Total	**55,545**	**4,089**	**795**	**19 %**

IRIS Total

Table A.6 presents an overview of IRIS Q respondents across the IRIS consortium countries. The overall response rate for IRIS Q in the five IRIS consortium countries is 38 %; however, response rates vary greatly between countries, as sections “IRIS Q Denmark”, “IRIS Q Italy”, “IRIS Q Norway”, “IRIS Q Slovenia”

Table A.6 Overview of the total target population, sample, valid IRIS respondents and response rates for the IRIS Q survey carried out in the five consortium countries. The table is assembled based on the information given in Tables A.1, A.2, A.3, A.4 and A.5. Population, sample (where applicable) and respondent numbers are distributed on genders and educational categories (ISCED codes). Response rates could not be calculated here for each gender separately due to lacking gender distribution data from Denmark (see section “IRIS Q Denmark”)

IRIS all		I: total target pop	II: total sample (for No, Dk, and Slo, sample = target pop)	III: number of valid IRIS respondents	Response rate, III:II
421 Biology and biotechnology	Males			491	
	Females			1,109	
	Total	27,126	3,322	1,600	48 %
441 Physics	Males			467	
	Females			231	
	Total	7,045	1,369	698	51 %
442 Chemistry	Males			215	
	Females			248	
	Total	7,153	1,159	463	40 %
461 Maths	Males			339	
	Females			415	
	Total	11,387	1,902	754	40 %
481 Comp. science	Males			1,072	
	Females			233	
	Total	30,752	4,813	1,305	27 %
521 Mechanics and metal work	Males			742	
	Females			123	
	Total	15,293	1,960	865	44 %
523 Electronics and automation	Males			656	
	Females			103	
	Total	9,831	2,190	759	35 %
524 Chemical and process	Males			262	
	Females			224	
	Total	4,186	1,410	486	34 %
Sum	**Males**			**4,244**	
	Females			**2,686**	
	Total	**112,773**	**18,125**	**6,930**	**38 %**

and "IRIS Q England" have shown. In many countries, response rates are lower than what is desirable; however, they are comparable to response rates in similar projects, such as the Futuretrack survey.[7] Moreover, the population that response rate calculations are based on may be too high in some cases due to drop-out during the first months of the academic year (see sections "IRIS Q Norway" and "IRIS Q Slovenia"), possibly resulting in artificially low response rates. In most countries, female students are slightly over-represented, which should be kept in mind when interpreting results. A merged file with data from all IRIS consortium countries was produced to enable future comparative analyses.

The IRIS PhD Questionnaire

As part of the IRIS project work package assigned to the Slovenian project team, an investigation of PhD students' educational choice process was carried out. For this purpose, an IRIS PhD questionnaire (IRIS PhD Q) was designed as outlined in Chap. 11. Below is the questionnaire administered to PhD students in Slovenia in 2011.

[7] See Purcell, K., Elias, P., Ellison, R., Atfield, G., Adam, D., & Livanos, I. (2008). Applying for Higher Education – the diversity of career choices, plans and expectations. Findings from the First Futuretrack Survey of the 'Class of 2006' applicants for Higher Education. Coventry: IER, University of Warwick (http://www2.warwick.ac.uk/fac/soc/ier/news/wfreport0408.pdf)

IRIS PHD Q

IRIS Interests & Recruitment in Science

Survey about PhD educational choices http://iris.fp-7.org/

This questionnaire is about you and your PhD choice of course. Your answers are important to the study! The information you provide to the IRIS research project may help us improve teaching and develop more targeted information for future students. All replies are confidential, and no information will be traceable to you as an individual. Thank you very much!

1. **Gender:** ☐ female ☐ male

2. **Year of Birth:** 19........

3. **Where do you study?** (University and Faculty): ....

4. **Your field of doctoral study:** ....

5. **Where did you finish your Bachelor degree (University, Faculty and study field)?**

....

6. **Where did you finish your Master's degree (University, Faculty and study field)?**

....

7. **Do you have a Young Researcher status?** ☐ yes ☐ no

8. **When did you first get interested in the field of your doctoral study? How important was a specific time period?**

	Not important 1	2	3	4	Very important 5
a. Early childhood (6-11 years old)........................	☐	☐	☐	☐	☐
b. Puberty (12-18 years old)........................	☐	☐	☐	☐	☐
c. Early adulthood (19-27 years old)........................	☐	☐	☐	☐	☐
d. Later (27 and more)........................	☐	☐	☐	☐	☐

IRIS is an EU-supported research project. In Slovenia it is conducted by the IRI UL. Participation in this questionnaire survey is voluntary. All information is treated confidentially The project has been reported to the Privacy Ombudsman for Research, Norwegian Social Science Data Services.

IRIS PHD Q

9. How important were the following persons in choosing your doctoral study program?

	Not important 1	2	3	4	Very important 5
a. Colleagues, fellow students	☐	☐	☐	☐	☐
b. Mentors	☐	☐	☐	☐	☐
c. Friends and partners	☐	☐	☐	☐	☐
d. Family members	☐	☐	☐	☐	☐
e. Career advisors	☐	☐	☐	☐	☐
f. Professors at undergraduate study	☐	☐	☐	☐	☐
g. High school teachers	☐	☐	☐	☐	☐

10. How important were the following wishes in choosing your study?

	Not important 1	2	3	4	Very important 5
a. A wish to become a member of the scientific community	☐	☐	☐	☐	☐
b. A wish to contribute to scientific community	☐	☐	☐	☐	☐
c. A wish to transfer the scientific knowledge on younger generations	☐	☐	☐	☐	☐
d. A wish to contribute to sustainable development	☐	☐	☐	☐	☐
e. A wish to contribute to development of society in general	☐	☐	☐	☐	☐

11. How important are the following characteristics of the career development you?

	Not important 1	2	3	4	Very important 5
a. Improvement of professional competencies	☐	☐	☐	☐	☐
b. Pursuing ideals and values	☐	☐	☐	☐	☐
c. Getting the promotion	☐	☐	☐	☐	☐
d. Getting a secure job	☐	☐	☐	☐	☐
e. Helping others	☐	☐	☐	☐	☐
f. Realising my own potential	☐	☐	☐	☐	☐

12. To what extent does your PhD work weaken the following relationships and areas of life?

	Not at all 1	2	3	4	Very much 5
a. Social life (friends)	☐	☐	☐	☐	☐
b. Relationships (family, partners)	☐	☐	☐	☐	☐
c. Business life and career	☐	☐	☐	☐	☐
d. Hobbies	☐	☐	☐	☐	☐

IRIS PHD Q

13. When choosing the field of your doctoral study, how important was the fact that you are a woman/man by your opinion?

Not important				Very important
1	2	3	4	5
☐	☐	☐	☐	☐

Please explain:

....

....

14. How important or how much did your relationship with your mentor influence your choice of the doctoral study field?

Not important				Very important
1	2	3	4	5
☐	☐	☐	☐	☐

15. Do you think some fields of doctoral studies are more attractive for women and others for men? Please explain your answer.

....

....

16. Do you think the general efforts striving for both genders to be evenly represented across all scientific fields are worth working for?

Not important				Very important
1	2	3	4	5
☐	☐	☐	☐	☐

17. Do you attend a doctoral program where one gender outnumbers the other? What do you think is the reason, please explain?

....

....

....

18. Tell us how and why you came to choose your PhD.

....

....

....

Thank you for your help!

About the Editors

Ellen Karoline Henriksen is associate professor of physics education at the Department of Physics, University of Oslo. Her research falls mainly within two strands: (1) young people's educational choices related to science, technology, engineering and mathematics (STEM); and (2) student understanding and learning in physics. Ellen was the co-ordinator of the research project Interests and Recruitment in Science (IRIS), upon which this book is based. She is a member of the Executive Board of the European Science Education Research Association (ESERA).

Justin Dillon is professor of science and environmental education and head of the Science and Technology Education Group at King's College London. He was president of the European Science Education Research Association (ESERA) from 2007 to 2011 and is a co-editor of the *International Journal of Science Education*. His research interests include educational choice as well as science teaching and learning in schools, museums, science centres and in the outdoor classroom.

Jim Ryder is professor of science education at the School of Education, and Director of the Centre for Studies in Science and Mathematics Education (CSSME), at the University of Leeds, UK. He is a member of the Executive Board of the European Science Education Research Association (ESERA) and Editor of the international review journal *Studies in Science Education*. His research examines attempts to develop a school science education that supports people's engagement with science outside of formal schooling. He is currently examining the varying ways in which schools, teachers and students respond to major systemic curriculum reform initiatives.

E.K. Henriksen et al. (eds.), *Understanding Student Participation and Choice in Science and Technology Education*, DOI 10.1007/978-94-007-7793-4

About the Contributors

Alessandra Allegrini is a researcher and consultant in the field of gender/feminist studies, with a specific interest in feminist approaches to gender, science and society issues. Her work includes empirical and theoretical research, project planning, educational training and communication. She has been working with universities as well as private and public organisations. Her latest research concerns the historical-theoretical reconstruction of the Italian feminist events and elaborations about gender, science and technology from the late 1970s to the early 2000s. She is a member of the national Women and Science Association's directing board.

Jaume Ametller is a lecturer in Science Education at the University of Girona. He has previously worked at the Autonomous University of Barcelona, where he studies Physics and completed a PhD in Science Education, and at the University of Leeds, where he did the work presented in this book and where he remains a visiting research fellow. His main area of research is teaching and learning science in the classroom with an emphasis on classroom discourse, dialogic education, the use of ICT, and the design of research informed teaching sequences.

Louise Archer is Professor of Sociology of Education at King's College, London. Her research focuses on educational identities and inequalities of gender, ethnicity and social class. She leads the ASPIRES project, is lead coordinator of the TISME Science and Mathematics Education research programme (tisme-scienceandmaths.org) and directs the research strand of Enterprising Science (kcl.ac.uk/enterprisingscience).

Maria Vetleseter Bøe completed her PhD in science education at the University of Oslo (Norway) in 2012. She previously completed a Master of Science in Physics at the Norwegian University of Science and Technology, where she also received her qualification as a teacher in science and mathematics. She is currently employed as a researcher at the Norwegian Centre for Science Education, working mainly with issues related to physics education and participation in science and technology.

E.K. Henriksen et al. (eds.), *Understanding Student Participation and Choice in Science and Technology Education*, DOI 10.1007/978-94-007-7793-4

Jennifer DeWitt is a research fellow at King's College London, focusing on the 'Science Aspiration and Career Choice: Age 10–14 (ASPIRES)' project. Prior to that and building on work in children's museums and educational publishing, she completed a PhD at King's, where her thesis focused on school trips. Her research interests include identity, learning and engagement in families and school groups in informal science environments.

Slavko Dolinšek obtained his PhD in the field of automation and production cybernetics and went on to postdoctoral studies and an international MBA degree at the University of California, Berkeley. Since 1998 he is a professor of at the University of Ljubljana (Faculty of Mechanical Engineering) and University of Primorska (Faculty of Management), and is currently the chief executive for IRI UL (Institute for Innovation and Development of University of Ljubljana). For several years he has been involved in various basic and applied research projects in the fields of production engineering and management.

Henriette Tolstrup Holmegaard is an assistant professor at the Department of Science Education, University of Copenhagen, Denmark. She holds a Ph.D. in Science Education. Her Dissertation is a longitudinal study of upper secondary school students' transition-process into higher education STEM study programmes. Together with the Danish co-authors of this book she has published on students' choices of and transition into higher education in general and in particular on why students' do not choose a future within STEM. In the near future she plans to carry out research on STEM-students' transition into first employment.

Tina Hribar is advisor and researcher at Institute for Innovation and Development of University of Ljubljana (IRI UL), where she works on national and European projects such as IRIS, CERIM, CRP, and EnergyViLLab. She graduated from the Department of Sociology at the University of Ljubljana, Slovenia, with a diploma thesis on how an alternative educational system, "educare", can benefit the public school system in Slovenia.

Fredrik Jensen has a master's degree in science education from the University of Oslo and is qualified as a science teacher. From 2008 to 2012, he was employed by the Norwegian Centre for Science Education, where worked towards a PhD with research related to the IRIS project. His research interests are STEM choices in general, and particularly how out-of-school experiences and recruitment interventions influence STEM motivations.

Terry Lyons is from 2014 associate professor in science education at Queensland University of Technology, Australia. While working with the contribution for this book, he was Associate Director of the National Centre of Science, ICT and Mathematics Education for Rural and Regional Australia (SiMERR Australia). He taught high school science for 15 years in Australia, Botswana and Kiribati and is currently a senior lecturer in science education at the University of New England. From 2006 to 2010 he was Chair of the International Organization for Science and Technology Education (IOSTE), and recently co-authored two significant Australian reports on science education: *Choosing Science* and *Starting Out in STEM*.

Marianne Løken is a PhD student in science education at the University of Oslo. Her research interest is on gender issues in science education with a particular focus on narratives of females choosing science subjects where females are underrepresented. Løken is trained in social anthropology, special needs education and communication studies, and has years of experience in science communication, both as an editor and a writer. Currently she is working with tasks related to educational research in The Norwegian Directorate for Education and Training.

Lene Møller Madsen is an Associate Professor in science education at the Department of Science Education, University of Copenhagen. She holds a Ph.D. in geography and has published on the development of spatial thinking skills, students' meeting with teaching in geography and how institutional cultures interact with researchers' perceptions of fieldwork. Together with the Danish co-authors in this book she is involved in a number of research projects concerning student's academic integration within higher education, non-completion and gender issues in science, as well as educational choices.

Giuseppe Pellegrini (Ph.D. Sociology) is lecturer in Methodology and Social Research at the University of Padova. His current research focuses on public participation with specific regard to techno-scientific issues. He is the coordinator of the research area "Science and Citizens" at Observa – Science in Society, member of the EASST (European Association for the Study of Science and Technology) and 4S (Society for the Social Studies of Science).

Frances Quinn has worked as a scientist and science educator for close to 25 years, in secondary and higher education sectors. After teaching secondary school science, and biological science and ecology at the University of New England, Australia, she proceeded to a research and teaching focus in Science Education. Her research interests span students' perceptions of learning science, socio-scientific issues in science education, Education for Sustainability and teaching and learning in science.

Elaine Regan is a Postdoctoral Research Associate in the Department of Education and Professional Studies, King's College London. Her research interests include student participation in science, inquiry-based science, effective teaching, professional development for teachers and informal science educators, and learning outside the classroom. She is currently working in partnership with Professor Louise Archer and the Science Museum, London on the Enterprising Science project. Recent publications include the co-authored book Effective Classroom Practice (2014), Open University Press.

Chiara Segafredo obtained a Master's degree in International Relations in 2013 from the University of Padova. She has been junior research assistant at Observa from 2009 to 2012 for science in society research, working on the European research project ACCENT on climate change and on the IRIS project on young people's participation in STEM studies. Her current research interests concern the right to health, equity and women's empowerment.

Lars Ulriksen is a Professor in higher education teaching and learning at the Department of Science Education, University of Copenhagen, Denmark. He has been leading the Danish part of the IRIS project on retention and gender issues. His research areas include the transition of students into higher-education and the meeting of the students' prior knowledge and experiences with the culture and the curriculum of the programmes. He has developed the notion of *the implied student* and together with the Danish co-authors of this book published on student retention, transition and educational choices.

The manufacturer's authorised representative in the EU is Springer
Nature Customer Service Centre GmbH, Europaplatz 3, 69115 Heidelberg,
Germany. If you have any concerns regarding our products, please
contact ProductSafety@springernature.com

Printed and bound by CPI Group (UK) Ltd, Croydon, CR0 4YY
30/09/2026
02225447-0015